Energy, Environment, and Sustainability

Series Editor

AIMS AND SCOPE

This books series publishes cutting edge monographs and professional books focused on all aspects of energy and environmental sustainability, especially as it relates to energy concerns. The series is published in partnership with the International Society for Energy, Environment, and Sustainability. The books in these series are edited or authored by top researchers and professional across the globe. The series aims at publishing state-of-the-art research and development in areas including, but not limited to:

- Renewable Energy
- Alternative Fuels
- Engines and Locomotives
- Combustion and Propulsion
- Fossil Fuels
- Carbon Capture
- Control and Automation for Energy
- Environmental Pollution
- Waste Management
- Transportation Sustainability

Review Process

The proposal for each volume is reviewed by the main editor and/or the advisory board. The chapters in each volume are individually reviewed single blind by expert reviewers (at least four reviews per chapter) and the main editor.

Ethics Statement for this series can be found in the Springer standard guidelines here https://www.springer.com/us/authors-editors/journal-author/journal-author-helpdesk/before-you-start/before-you-start/1330#c14214

More information about this series at https://link.springer.com/bookseries/15901

Avinash Kumar Agarwal · Dhananjay Kumar ·
Nikhil Sharma · Utkarsha Sonawane
Editors

Engine Modeling and Simulation

Springer

Editors
Avinash Kumar Agarwal
Department of Mechanical Engineering
Indian Institute of Technology Kanpur
Kanpur, India

Dhananjay Kumar
Department of Mechanical Engineering
Indian Institute of Technology Kanpur
Kanpur, India

Nikhil Sharma
Department of Mechanical Engineering
Malaviya National Institute of Technology
Jaipur (MNIT)
Jaipur, India

Utkarsha Sonawane
Department of Mechanical Engineering
Indian Institute of Technology Kanpur
Kanpur, India

ISSN 2522-8366 ISSN 2522-8374 (electronic)
Energy, Environment, and Sustainability
ISBN 978-981-16-8617-7 ISBN 978-981-16-8618-4 (eBook)
https://doi.org/10.1007/978-981-16-8618-4

This Springer imprint is published by the registered company Springer Nature Singapore Pte Ltd.
The registered company address is: 152 Beach Road, #21-01/04 Gateway East, Singapore 189721, Singapore

Preface

Energy requirement has been rising drastically due to the increasing population and globalization. The global economy and society significantly depend on energy availability because it touches every aspect of human life and activities. The living standards of human beings have improved due to the availability of power sources and transportation facilities. This is possible due to advancements in the automotive and energy sector. Industry and academia are working together to achieve energy security for the future.

The International Society for Energy, Environment and Sustainability (ISEES) was founded at the Indian Institute of Technology Kanpur (IIT Kanpur), India, in January 2014 to spread knowledge/awareness and catalyze research activities in the fields of energy, environment, sustainability, and combustion. Society's goal is to contribute to the development of clean, affordable, and secure energy resources and a sustainable environment for society and spread knowledge in the areas mentioned above, and create awareness about the environmental challenges the world is facing today. The unique way adopted by ISEES was to break the conventional silos of specializations (engineering, science, environment, agriculture, biotechnology, materials, fuels, etc.) to tackle the problems related to energy, environment, and sustainability in a holistic manner. This is quite evident by the participation of experts from all fields to resolve these issues. ISEES is involved in various activities such as conducting workshops, seminars, and conferences in the domains of its interests. Society also recognizes the outstanding works of young scientists, professionals, and engineers for their contributions in these fields by conferring them awards under various categories.

The Fifth International Conference on 'Sustainable Energy and Environmental Challenges' (V-SEEC) was organized under the auspices of ISEES from December 19 to 21, 2020, in virtual mode due to restrictions on travel because of the ongoing COVID-19 pandemic situation. This conference provided a platform for discussions between eminent scientists and engineers from various countries, including

India, Spain, Austria, Bangladesh, Mexico, USA, Malaysia, China, UK, Netherlands, Germany, Israel, and Saudi Arabia. At this conference, eminent international speakers presented their views on energy, combustion, emissions, and alternative energy resources for sustainable development and a cleaner environment. The conference presented two high-voltage plenary talks by Dr. V. K. Saraswat, Honorable Member, NITI Ayog, on 'Technologies for Energy Security and Sustainability' and Prof. Sandeep Verma, Secretary, SERB, on 'New and Equitable R&D Funding Opportunities at SERB.'

The conference included nine technical sessions on topics related to energy and environmental sustainability. Each session had 6–7 eminent scientists from all over the world who shared their opinions and discussed future trends. The technical sessions in the conference included emerging contaminants: monitoring and degradation challenges; advanced engine technologies and alternative transportation fuels; future fuels for sustainable transport; sustainable bioprocessing for biofuel/non-biofuel production by carbon emission reduction; future of solar energy; desalination and wastewater treatment by membrane technology; biotechnology in sustainable development; emerging solutions for environmental applications and challenges and opportunities for electric vehicle adoption. Five hundred plus participants and speakers from all over the world attended these three days conference.

The conference concluded with a high-voltage panel discussion on 'Challenges and Opportunities for Electric Vehicle Adoption,' where the panelists were Prof. Gautam Kalghatgi (University of Oxford), Prof. Ashok Jhunjhunwala (IIT Madras), Dr. Kelly Senecal (Convergent Science), Dr. Amir Abdul Manan (Saudi Aramco), and Dr. Sayan Biswas (University of Minnesota, USA). Prof. Avinash Kumar Agarwal, ISEES, moderated the panel discussion. This conference laid out the roadmap for technology development, opportunities, and challenges in energy, environment, and sustainability domain. All these topics are very relevant for the country and the world in the present context. We acknowledge the support received from various agencies and organizations for the successful conduct of the Fifth ISEES conference V-SEEC, where these books germinated. We want to acknowledge SERB (special thanks to Dr. Sandeep Verma, Secretary) and our publishing partner Springer (special thanks to Ms. Swati Meherishi).

The editors would like to express their sincere gratitude to a large number of authors from all over the world for submitting their high-quality work on time and revising it appropriately at short notice. We would like to express our special gratitude to our prolific set of reviewers Dr. Sameera Wijeyakulasuriya (Convergent Science), Dr. Rajesh Kumar Prasad (ICFAI University), Dr. Krishnamoorthi M. (IIT Kanpur), Dr. Kelley Senecal (Convergent Science), Dr. Rohit Kumar Singla (TIET Patiala), Dr. Balaji Mohan (Saudi Aramco), Mr. Harshimran Singh (IIT Kanpur), Dr. Anoop Kumar Shukla (Amity University), Dr. Nikhil A. Baraiya (SVNIT Surat), Mr. Ashutosh Jena (IIT Kanpur), Dr. Ali Turkcan (Kocaeli University), Dr. Thokchom Subhaschandra Singh (NIT Manipur), Dr. Xiao Fu (Progress Rail Service), Dr. Jinlong Liu (West Virginia University), Dr. Carlos Rodrigues Pereira Belchior (Federal University of Rio de Janeiro), Dr. Bipin Bihari (Argon National Lab), Dr. Ben Akih-Kumgeh (Syracuse University), Dr. Jinlong Liu (West Virginia University),

Dr. Erdiwansyah (Universitas Serambi Mekah Aceh Indonesia), Prof. S. Srinivasa Rao (NIT Warangal), and Dr. Javad Hosseinpour (Michigan State University), who reviewed various chapters of this monograph and provided their valuable suggestions to improve the manuscripts.

This book is based on the simulation and modeling of internal combustion engines. This book includes various aspects of diesel and gasoline engine modeling and simulation such as spray, combustion, ignition, in-cylinder phenomena, emissions, and exhaust heat recovery. This book also explored engine models and analysis of cylinder bore piston stresses and temperature effects. This book includes recent literature and focuses on current modeling and simulation trends for internal combustion engines. In this book, readers will get an idea about engine process simulation and modeling, which will be helpful for the development of efficient and emission-free engines. Few chapters of this book are based on the review of state-of-the-art models for spray, combustion, and emissions, focusing on the theory, models, and their applications from an engine point of view. We hope that the book would greatly interest the professionals and postgraduate students involved in alternative fuels, IC engines, engine modeling and simulation, and environmental research.

Kanpur, India — Avinash Kumar Agarwal
Kanpur, India — Dhananjay Kumar
Jaipur, India — Nikhil Sharma
Kanpur, India — Utkarsha Sonawane

Contents

Editors and Contributors

About the Editors

Avinash Kumar Agarwal joined IIT Kanpur in 2001. He worked at the Engine Research Center, UW@Madison, the USA as a Post-Doctoral Fellow (1999–2001). His interests are IC engines, combustion, alternate and conventional fuels, lubricating oil tribology, optical diagnostics, laser ignition, HCCI, emissions, and particulate control, 1D and 3D Simulations of engine processes, and large-bore engines. Prof. Agarwal has published 435+ peer-reviewed international journal and conference papers, 70 edited books, 92 books chapters, and 12200+ Scopus and 19000+ Google Scholar citations. He is the associate principal editor of FUEL. He has edited *Handbook of Combustion* (5 Volumes; 3168 pages), published by Wiley VCH, Germany. Prof. Agarwal is a Fellow of SAE (2012), Fellow of ASME (2013), Fellow of ISEES (2015), Fellow of INAE (2015), Fellow of NASI (2018), Fellow of Royal Society of Chemistry (2018), and a Fellow of American Association of Advancement in Science (2020). He is the recipient of several prestigious awards such as Clarivate Analytics India Citation Award-2017 in Engineering and Technology, NASI-Reliance Industries Platinum Jubilee Award-2012; INAE Silver Jubilee Young Engineer Award-2012; Dr. C. V. Raman Young Teachers Award: 2011; SAE Ralph R. Teetor Educational Award -2008; INSA Young Scientist Award-2007; UICT Young Scientist Award-2007; INAE Young Engineer Award-2005. Prof. Agarwal received Prestigious

CSIR Shanti Swarup Bhatnagar Award-2016 in Engineering Sciences. Prof. Agarwal is conferred upon Sir J. C. Bose National Fellowship (2019) by SERB for his outstanding contributions. Prof. Agarwal was a highly cited researcher (2018) and was in the top ten HCR from India among 4000 HCR researchers globally in 22 fields of inquiry.

Dhananjay Kumar is a research scholar at the Engine Research Laboratory, Indian Institute of Technology (IIT) Kanpur, India. He received his B.Tech. in Mechanical Engineering from National Institute of Technology (NIT) Mizoram, India, in 2017. His research area includes laser ignition technology, alternative fuels, combustion simulations, particulate characterization, etc. He has published international journal papers and authored book chapters in the areas of laser ignition, combustion simulations, and alternative fuel like methanol, H2, CNG, DME, etc.

Nikhil Sharma is Assistant Professor in the Department of Mechanical Engineering, Malaviya National Institute of Technology Jaipur, India. He completed his postdoctoral at the division of Combustion and Propulsion Systems, Chalmers University of Technology, Sweden. He also worked as CSIR Pool Scientist at the Engine Research Laboratory, Indian Institute of Technology (IIT) Kanpur, India. He was involved in gasoline spray and particulate investigations from direct injection engines. His areas of research include spray characteristics, exhaust gas aftertreatment system, particulate filters, and renewable fuels. He has published more than 23 technical articles in international journals.

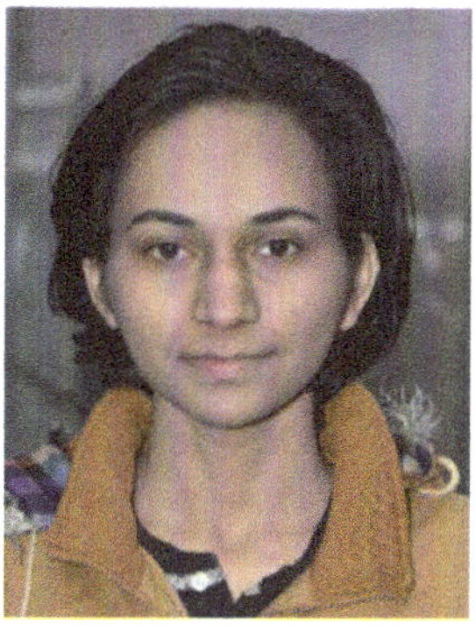

Utkarsha Sonawane is a research scholar at the Engine Research Laboratory, Indian Institute of Technology (IIT) Kanpur, India. She has received her M.Tech. in Automotive Technology from the College of Engineering Pune (COEP), India, in 2018. Her research areas include internal combustion engines, combustion, alternate and conventional fuels, spray and combustion characterization, and computation fluid dynamics (CFD). She has published international journal papers and authored book chapters.

Contributors

Avinash Kumar Agarwal Engine Research Laboratory, Department of Mechanical Engineering, Indian Institute of Technology Kanpur, Kanpur, India

Sami M. M. E. Ayad Department of Mechanical Engineering, Federal University of Rio de Janeiro, Rio de Janeiro, Brazil

Carlos R. P. Belchior Department of Mechanical Engineering, Federal University of Rio de Janeiro, Rio de Janeiro, Brazil

Pravesh Chandra Shukla Department of Mechanical Engineering, Indian Institute of Technology Bhilai, Raipur, India

Andrea Coraddu Department of Naval Architecture, Ocean and Marine Engineering, University of Strathclyde, Glasgow, UK;
Department of Maritime and Transport Technology, Delft University of Technology, Delft, The Netherlands

P. W. Deshmukh Department of Mechanical Engineering, College of Engineering Pune (CoEP), Pune, Maharashtra, India

Sudipta De Jadavpur University, Kolkata, India

Atul Dhar School of Engineering, Indian Institute of Technology Mandi, Mandi, India

Rinze Geertsma Department of Maritime and Transport Technology, Delft University of Technology, Delft, The Netherlands

Arpan Gupta School of Engineering, Indian Institute of Technology Mandi, Mandi, India

Miltiadis Kalikatzarakis Department of Naval Architecture, Ocean and Marine Engineering, University of Strathclyde, Glasgow, UK

Alli Anil Kumar Department of Mechanical Engineering, NIT Warangal, Warangal, Telangana, India

Dhananjay Kumar Engine Research Laboratory, Department of Mechanical Engineering, Indian Institute of Technology Kanpur, Kanpur, India

Parmod Kumar School of Engineering, Indian Institute of Technology Mandi, Mandi, India

Vikas Kumar Sahu Department of Mechanical Engineering, Indian Institute of Technology Bhilai, Raipur, India

Subhash Lahane Department of Mechanical Engineering, College of Engineering Pune (CoEP), Pune, Maharashtra, India

Pradeep Menon Indian Institute of Technology Madras, Chennai, India

Mayank Mittal Indian Institute of Technology Madras, Chennai, India

Subha Mondal Aliah University, Kolkata, India

Kotha Madhu Murthy Department of Mechanical Engineering, NIT Warangal, Warangal, Telangana, India

Sarthak Nag School of Engineering, Indian Institute of Technology Mandi, Mandi, India

M. R. Nandgaonkar Department of Mechanical Engineering, College of Engineering Pune (CoEP), Pune, Maharashtra, India

Luca Oneto DIBRIS, University of Genova, Genoa, Italy

Rajesh Kumar Prasad Department of Mechanical Engineering, ICFAI University Jharkhand, Ranchi, India

Inderpal Singh School of Engineering, Indian Institute of Technology Mandi, Mandi, India

Rahul Kumar Singh Engine Research Laboratory, Department of Mechanical Engineering, Indian Institute of Technology Kanpur, Kanpur, India

José R. Sodré Department of Mechanical Engineering and Design, Aston University, Birmingham, UK

Utkarsha Sonawane Engine Research Laboratory, Department of Mechanical Engineering, Indian Institute of Technology Kanpur, Kanpur, India

Nikhil Sharma Malaviya National Institute of Technology Jaipur, JLN Marg, Jaipur, Rajasthan, India

Gerasimos Theotokatos Department of Naval Architecture, Ocean and Marine Engineering, University of Strathclyde, Glasgow, UK

Part I
Introduction

Chapter 1
Introduction to Engine Modeling and Simulation

Avinash Kumar Agarwal, Dhananjay Kumar, Nikhil Sharma, and Utkarsha Sonawane

Abstract Stricter emission norms and increasing pollution have motivated researchers to find an optimal way to explore and deploy advanced engine technology and alternative fuels. Engine combustion is a complex process that is yet to understand significantly. Computational modelling has enabled the researcher to understand complex underlying processes quickly and economically. This book covers major aspects of internal combustion (IC) engine modelling. The first two sections of this book focus on various engine models, mathematical modelling of injection processes, and spray breakup modelling. The third section of this book is dedicated to the modelling of diesel engine combustion. Emphasis has been made on soot and NOx modelling. The fourth section of this book is dedicated to model the spark ignition engine processes. Various 1D and 3D tools that are currently used by the automotive community are discussed in detail. The last section of this book covers the prediction of exhaust heat recovery from IC engines. Overall, this book emphasises mathematical modelling of the different processes involved in IC engines.

Keywords Simulation · Modelling · IC engine · Spray · Emissions · Exhaust heat recovery

The first part of the book introduces modelling and simulation of IC engines. One of the chapters discusses the development of engine models and analyses of stresses and temperature. This chapter covers the simulations and analyses of the structural and thermal consequences of specified bore piston and cylinder block models. This section effectively analyses stress and temperature using finite element methods using various tools such as ANSYS workbench, ADSL, and Deform 3D.

A. K. Agarwal (✉) · D. Kumar · U. Sonawane
Engine Research Laboratory, Department of Mechanical Engineering, Indian Institute of Technology Kanpur, Kanpur 208016, India
e-mail: akag@iitk.ac.in

D. Kumar
e-mail: dhanu@iitk.ac.in

N. Sharma
Malaviya National Institute of Technology Jaipur, JLN Marg, Jaipur, Rajasthan 302017, India

A. K. Agarwal et al. (eds.), *Engine Modeling and Simulation*, Energy, Environment, and Sustainability, https://doi.org/10.1007/978-981-16-8618-4_1

The second part of the book covers the fundamentals of spray modelling. One of the chapters discusses the mathematical modelling of fuel injection processes and spray characteristics of diesel engines. In a diesel engine, the injection phenomenon is vital because it impacts fuel spray characteristics and mixture formation processes. The current research aims to review computational/ mathematical models of spray characteristics and empirical models of the engine's fuel injection and spray processes. The challenges for modelling the spray characteristics are also highlighted in this section. Numerous research works have been analysed for the effect of fuel injection parameters. Another chapter in this section covers spray breakup modelling for internal combustion engines. This chapter attempts to capture widely used spray breakup models and research studies involving IC engines. Fundamentals of spray breakup and atomisation have been discussed at the beginning of the section. In the end, the section has comprehensively examined the key contributions in spray modelling to provide an overall idea about the spray models and their application for IC engines.

The third part of this book has four chapters covering different simulation aspects of diesel engines. This section comprehensively covers complete modelling and simulation processes, starting from the basic input data required, typical assumptions, fundamental governing equations and discretisation, model formulation, initial conditions, boundary conditions, model validation, various sub-models used, and the analysis of results. The first chapter of this section is about the modelling and simulation of diesel engines using CFD. The second chapter of this section is focused on a large-bore marine diesel engine. This chapter discusses physical and data-driven models' hybridisation for modelling the dynamic state of a four-stroke marine diesel engine. Due to strict emission norms, low complex emissions prediction models are needed, actively controlling the combustion phase and after-treatment systems. The last two chapters in this section are about emissions from a diesel engine. This section focuses on the simulation and experimental investigations of a simplified 0-D model for the prediction of NOx emissions with unburned zones, burned zones, adiabatic flame temperatures, and species concentrations, especially atomic oxygen [O], Nitrogen [N_2], and nitric oxide [NO]. To understand the development of a model for soot formation, the Hiroyasu-NSC model, Waseda model, Gokul model, and Dalian model with their capabilities are discussed. For NOx formation, the two most fundamental models based on the Zeldovich mechanism and prompt NOx (Fenimore mechanism), which are semi-empirical, are discussed.

The fourth part of the book has four chapters covering simulation aspects of the spark-ignition (SI) engine. The first chapter covers the fundamentals, evolution, and modelling of ignition systems. This chapter summarises modelling aspects of spark ignition engines using 1-D and 3-D simulation tools. The working of these advanced ignition systems has been discussed in detail, along with their challenges. The second chapter in this section focuses on the calibration of combustion and heat transfer models and exergy analysis in the SI engines. This chapter reports on a workflow that aims at obtaining more profound insights into SI engines using thermodynamic modelling. This workflow is divided into two steps: (i) Auto-calibration of combustion and heat transfer models using AVL Boost and AVL Design Explorer;

and (ii) In-cylinder exergy analysis using Wolfram Mathematica. This analysis can be instrumental in determining the best fuel mixture, operating conditions, and areas for improvement. The last chapter discusses the effect of injection timings on combustion, performance, and emissions of methanol fuelled direct-injected SI engines through 1-D simulations.

The fifth part of the book covers exhaust heat recovery from IC engines. This part of the book discusses generalised principles for the simulation of IC engine waste heat recovery. Automotive exhaust thermoelectric generator unit is integrated into the exhaust noise muffler, heat recovery, and noise attenuation simulations. The first chapter discusses harvesting mufflers that utilise thermoelectric generators (TEG) to convert waste heat from the engine exhaust into electricity and simultaneously reduce the engine noise. The recovered electricity can power the auxiliary units in automobiles and can effectively improve the system's overall efficiency. In this section, an automotive exhaust thermoelectric generator (AETEG) unit is fabricated to extract the waste exhaust heat from the engine. To further enhance the performance of the AETEG unit, changes in the internal geometry of the unit are proposed, and a systematic computational study is carried. The last chapter is about heat recovery from diesel engines. The present chapter explores the simulation results of possible schemes and effect of waste heat recovery from diesel engines.

This book presents both fundamental and applied emerging technologies for emission reduction using a simulation approach. Several mathematical models to predict the engine spray, combustion, and emissions are discussed. The topics covered in this book include:

- Introduction to engine modelling and simulation,
- Development of engine models and analysis of cylinder bore piston stresses and temperature effects in IC engines,
- Mathematical modelling of injection and spray characteristics of a diesel engine: a review,
- Spray breakup modelling for internal combustion engines,
- Modelling and simulation of diesel engines using CFD and its applications in optimising various in-cylinder techniques,
- Physical and data-driven models hybridisation for modelling the dynamic state of a four-stroke marine diesel engine,
- Soot particles and NOx modelling for diesel engine,
- NOx emission prediction model for DI diesel engine,
- Fundamentals, evolution, and modelling of ignition systems for spark-ignition engines,
- Methods in SI engine modelling: auto-calibration of combustion and heat transfer models and exergy analysis,
- Investigation on the effect of injection timings on combustion, performance, and emissions of pure methanol fuelled DISI engine through 1-D simulations,
- Automotive exhaust thermoelectric generator unit integrated to exhaust noise muffler: heat recovery and noise attenuation simulations,

- Diesel engine waste heat recovery schemes for improved fuel economy and reduced emissions: simulation results.

These mentioned topics are categorised into five sections: (i) Introduction, (ii) Spray Modelling, (iii) CI Engine Modelling, (iv) SI Engine Modeling, and (v) Exhaust Heat Recovery.

Chapter 2
Development of Engine Models and Analysis of Cylinder Bore Piston Stresses and Temperature Effects in Internal Combustion Engine

Alli Anil Kumar and Kotha Madhu Murthy

Abstract The engine block is a crucial component of internal combustion engines since it provides the source of power for the vehicle. The engine block is a solid cast element that contains the cylinders and their related components inside a cooled and lubricated crankcase. The majority of engine blocks are cast iron, a brittle rich material that can sustain greater variations and weights while also being vibration resistant. Energy is changed from potential chemical energy to mechanical energy in this step, which is heat produced from burning fuel in the form of mechanical energy. A large amount of energy may be lost during the firing of an engine. This must be assessed using CFD prediction techniques. For starters, significant thermal and structural stresses may be produced over the cylinder blocks and pistons. Prior to the fabrication of the specific Internal Combustion engine, the structural and thermal analysis aids in the failure analysis. The solid modeller Parametric-Creo is used to create the current and new engine block models. This technical chapter will cover the simulation and analysis of the structural and thermal consequences of each specified bore piston and cylinder block model. The effective stress and temperature analysis using the finite element method will be performed using current tools such as ANSYS workbench, ADSL, and Deform 3D, with the processes and analytical functions described utilising supplementary literature theories, which can be found here.

Keywords Internal combustion engine · Solid modelling · ANSYS · Simulation · von Mises stress · Displacement

---

A. A. Kumar (✉) · K. M. Murthy
Department of Mechanical Engineering, NIT Warangal, Warangal, Telangana 506004, India
e-mail: anilkumar927@student.nitw.ac.in

K. M. Murthy
e-mail: madhu@nitw.ac.in

A. K. Agarwal et al. (eds.), *Engine Modeling and Simulation*, Energy, Environment, and Sustainability, https://doi.org/10.1007/978-981-16-8618-4_2

2.1 Introduction

2.1.1 Introduction of the Engine Models

The practice of evaluating the performance and efficient operation of an internal combustion engine is known as engine testing. In order to evaluate Internal Combustion engines, a variety of parameters can be monitored. Engine testing is required to comprehend the engine's and its components' effective working. The engine block, which serves as the vehicle's powerhouse, is the foundation of internal combustion vehicles. The term "block" refers to a solid cast automobile part that houses the cylinders and their accessories inside a cooled and oiled crankcase. The new engine's performance is determined by the surface finish, tolerance, and lubrication and cooling system. The performance of the engine may be increased by adjusting different parameters and properly configuring all of the engine's components. This item is extremely strong and durable because its failure causes the car to stop working until the engine block is replaced or rebuilt. The majority of engine blocks are made of cast iron, but in the late 1990s, some varieties of experimental materials were used in prototype cars with the goal of producing more lightweight, efficient automobiles. Cylinder blocks are subjected to high thermal and structural loads. The project's primary goal is to investigate thermal and structural load measurement in engine blocks. Solid works simulation is used to design the engine block, and thermal and mechanical analysis is also done in solid works simulation. Internal-combustion engine, any of a group of systems in which the combustion reactants (oxidizer and fuel) and combustion products serve as the engine's operating fluids. Such an engine is powered by the heat generated during the combustion of the non-reacted working fluids, the oxidizer-fuel mixture. This process occurs within the engine and is part of the device's thermodynamic cycle.

Engines are classified based on the number of strokes each cycle, the number of cylinders, the kind of fuel utilised, and other factors. The cylinder and valve design, the ignition method, the kind of cooling, and the valve trains.

Classification based on the following categories:
Based on design:

- Reciprocating Engine (Piston engine)
- Rotary Engine.

Based on the number of strokes:

- Four-Stroke Engines
- Two-Stroke engines.

Based on the fuel usage:

- Petrol Engine
- Diesel Engine.

Based on ignition method:

- Compression-Ignition (CI) method
- Spark Ignition (SI) method.

Based on the number of cylinders:

- Single-Cylinder engine
- Multi-Cylinder Engine.

The term "inline" refers to the arrangement of the cylinders in the engine block. An inline engine's cylinders are placed in a straight line and maybe built as inline 2, 3, 4, 5, 6, or 8 cylinder engines, as shown in Fig. 2.1a. Four-cylinder engines are referred to as "inline-four engines" by manufacturers, whereas engines with six or eight cylinders are referred to as "straight engines." Inline engines are used in a variety of applications, including aviation, and are less expensive to manufacture than the more sophisticated but ever-popular V-engines or smaller horizontally opposed engines. A V engine is a type of internal combustion engine distinguished by the "V" shape formed by the cylinders when viewed from the front of the engine, perpendicular to the crankshaft axis, as illustrated in Fig. 2.1b. The cylinders are often placed at a 60 or 90° angle.

The W engine refers to those engines that are part of a hybrid of two VR engines based on the modular design principle. The W engines are formed by placing the cylinders of one bank at a 15° angle with respect to each other, while the other two VR banks are at a 72° V-angle. The designs of the V and VR engines were combined

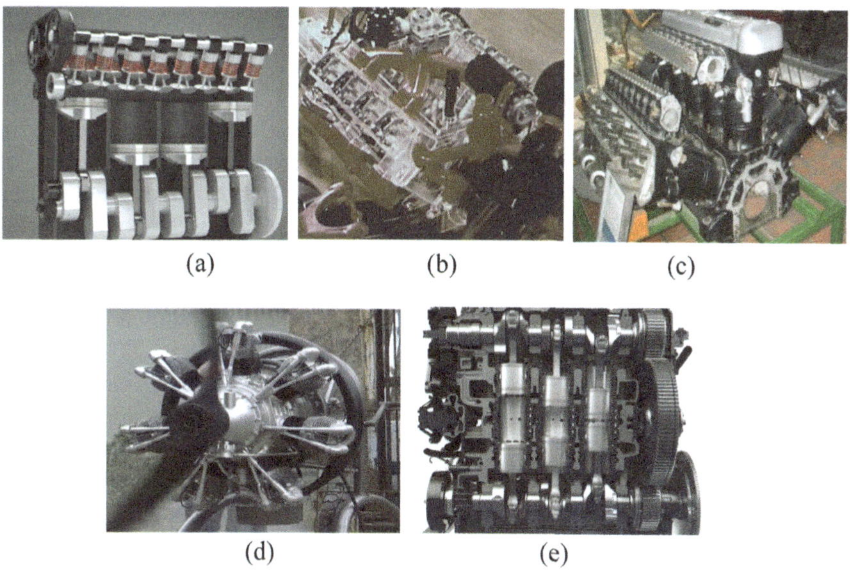

Fig. 2.1 **a** In-line engine, **b** V type engine, **c** W type engine, **d** radial engine, and **e** opposed cylinder engine

to create a highly efficient engine known as the W engine. The engine was designed with the goal of producing a more compact engine with a large number of cylinders. A V engine's cylinders are divided into two banks. Cylinder banks in W8 and W12 engines are positioned at a V-angle of 72° with respect to one another. Similarly, the cylinders inside the VR engine's bank form a 15° V-angle. When viewed from the front, the cylinder arrangement of the W engine appears to be a double-V, as shown in Fig. 2.1c. The two Vs of the right and left cylinder banks are combined to form a W engine. This is where the W engine earned its name.

Figure 2.1d depicts a radial engine, which is a type of reciprocating internal combustion engine in which the cylinders "radiate" outward from a central crankcase like wheel spokes. This type of radial engine is a type of internal combustion engine that is mostly used in small aeroplanes, with cylinders ranging from five to as many as 28, depending on engine size, placed in a circle around the crankshaft in banks of two or more.

Figure 2.1e depicts an opposed-piston engine, which is a piston engine with pistons on both ends of each cylinder but no cylinder head. Petrol and diesel opposed-piston engines have been widely used in large-scale applications such as ships, military tanks, and factories. In contrast to conventional four-stroke engines, OPEs have two pistons per cylinder that move in opposing, reciprocating motion. This removes the cylinder head and valve train, two of the most complex and costly components in conventional engines, as well as the primary sources of heat and friction losses.

2.1.2 Description of an Internal Combustion Engine

In this chapter, the classification of engines can be discussed based on their models and types. The engine is otherwise known as a motor. This machine is used to convert one form of energy into other forms of energy. For various domains, different engines are used to utilize the transfer of energy. Particularly, the converting of Mechanical energy and Electrical energy. Internal combustion engines are classified as either compression ignition engines or spark-ignition engines. Suction, compression, power or working stroke, and exhaustion stroke are the four cycles that are typically employed. Based on fuel usage, petrol and diesel engines are frequently used by human resources. Nowadays, electric vehicles are very familiar throughout the world without spending money on fuel consumption.

A heat engine is a device that converts a fuel's chemical energy into thermal energy and then uses that energy to perform mechanical work. It is divided into two types:

(a) Externally combustible engine

The results of combustion of air and fuel in this engine transfer heat to a second fluid, which is the cycle's working fluid.

(b) Internally combustible engine

The combustion of air and fuels takes place inside the cylinder in this engine and is employed as the direct motive power.

Main components of an Internal Combustion engine:

Cylinder: It is the major component of the engine within which the piston reciprocates back and forth. It is advised and enough to endure high pressures of more than 50 bar and temperatures of more than 2000 °C. The design of maximum allowable pressure is 180 bar for conventional diesel engines. However, currently, the pressure increased due to heavier load requirements. Heavy-duty engines are composed of steel alloys or aluminium alloys, while standard engines are made of cast iron. The cylinders in a multi-cylinder engine are cast in a single block known as the cylinder block.

Cylinder head: The cylinder head covers the top end of the cylinder and houses the inlet and exhaust valves, spark plugs, and injectors. To create an airtight joint, a copper or asbestos gasket is installed between the engine cylinder and cylinder head.

Piston: The force exerted by the burning of charge is transmitted to the connecting rod by the piston. Typically constructed of an aluminium alloy, which has good heat conductivity and strength at high temperatures.

Piston rings: These are composed of steel alloys that retain their elastic qualities even at high temperatures and are housed in circumferential grooves on the outside surface of the piston. There are two kinds of rings: compression rings and oil rings. The compression ring is the piston's upper ring that provides an airtight seal to prevent burnt gases from leaking into the bottom part. The oil ring is the lowest ring that offers a good seal to prevent oil leaks into the engine cylinder.

Connecting rod: In the working stroke, the connecting rod transfers the reciprocating action of the piston into the circular motion of the crankshaft. The gudgeon pin connects the connecting rod's smaller end to the piston, and the crankpin connects the connecting rod's larger end to the crank. For the production of connecting rods, specific steel alloys or aluminium alloys are employed.

Crankshaft: With the help of the connecting rod, it turns the reciprocating action of the piston into rotary motion. Special steel alloys are used in the crankshaft's construction. It is made up of an eccentric part known as the crank.

2.1.3 Literature Review of the Stresses and Temperature of Internal Combustion Engines

Ansari et al. (2020) used the Solid Works simulation platform to create and analyze an engine block model that clearly displays structural and thermal load findings. During the static study, the magnesium alloy demonstrated the least stress, while the aluminium metal matrix demonstrated the least deformation and strain value. Grey cast iron, on the other hand, has a higher temperature distribution, indicating that it has a lower minimum temperature and a larger maximum heat flux than aluminium metal matrix under the same thermal load situation. Because it can withstand high temperatures and stresses under high load and thermal conditions, the aluminium metal matrix is preferred for both thermal and structural load distribution. Although aluminium metal matrix can withstand higher pressures, grey cast iron can withstand higher temperatures. Aluminium is a Nonferrous material that delivers lesser weight and higher tensile strength, its tensile strength is relatively more, even the ductility also has been increased at a certain level is better utilized for the making of piston components of an IC engine. Gray Cast Iron is a ferrous material that contains more percentage of iron, having more weight used to resist vibrations which applies to crankshaft, engine block etc. Zhang et al. to Narayana et al. (Zhang et al. 2013; Ahmed and Basim 2009; Szurgott and Niezgoda 2011; Kurbet and Krishnakumar 2004; Hamzwhei and Manochehr 2006; Rahman et al. 2019; Narayana et al. 2014; Jayale et al. 2017) have discussed that the engine component design and analysis has gotten increasingly difficult. The engine piston is one of these components. A diesel engine's piston is frequently subjected to fluctuating thermal and mechanical loads on a regular basis. Piston simulations and strength analysis has been a popular topic of study, attracting a lot of attention. Sanket et al. have studied the use of Finite Element Method to determine the stress distribution and maximum stress of aluminium alloy and cast iron pistons, as well as evaluate the superior performance among them.

Chigrinova et al. (2004) investigated the heat stresses of combustion-chamber parts, as well as the development of a method for estimating the appropriate thickness of the heat-protective coating on the heat-absorbing surface of the piston and cylinder head in an internal-combustion engine. Under the same conditions, a piston with a heat-protective and heat-resistant AMDO coating made of Tzoref et al. (1978) investigated the effect of power outputs and their causes on the structure of the piston due to excessive thermal aspects. Different temperature field points and radial stresses were investigated at various points along with the piston structure. Kumar et al. (2017) finished the piston modelling in CATIA software before moving on to the Finite Element Analysis in ANSYS workstation. The von Mises stresses criterion is used in finite element analysis. Using the thermal boundary conditions, the temperature distribution and thermal stress in the piston were computed, and the maximum and minimum stress locations in the piston were discovered. Ghagare and Sandeep (2015) have investigated by the review article, the different stresses that operate to enhance the quality of a piston's ability to endure high thermal and structural stresses while also reducing stress concentration at the piston's upper end. The FEA for a conventional four stroke engine piston is proposed, and the results of the study are compared for maximum stress. Krishna and Aravind (2014) investigated the thermal and stress distribution of an Internal Combustion engine during combustion. Pro E Wild fire 5.0 was used to model the piston head part, and then ANSYS (APDL11.0) was used for analysis. Likewise, these simulations of these literature reviews, the Creo made piston model has been analyzed by using ANSYS ADSL 15 as shown in Fig. 2.2. Here, the piston should be considered in the form of hollow structural component.

The thermal study of the modelled piston with the related boundary conditions was examined by Sharma et al. (2015).The displacements and stresses, in this case, have

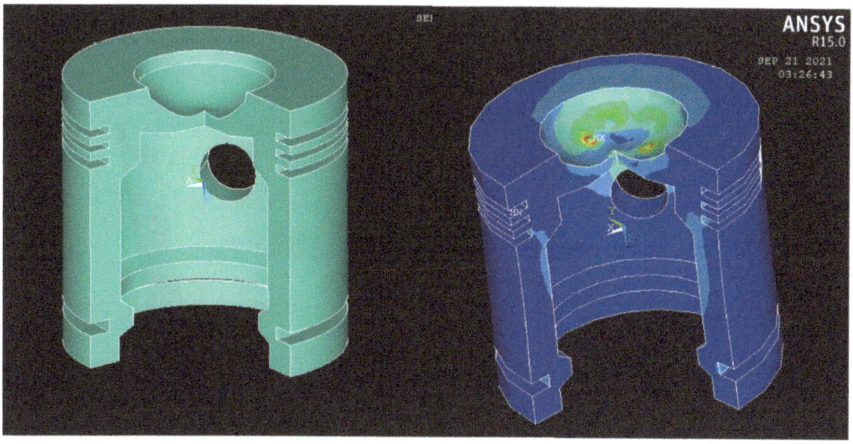

(a) (b)

Fig. 2.2 FE analysis of **a** Piston and solid modelling, **b** ANSYS simulation

been calculated in relation to the load conditions as well as the temperature fields. The radial and vertical displacements of the pistons were calculated at various points on the pistons. The greatest radial and vertical displacements were determined to be due to full load conditions. The three-quarter load conditions caused the irregular displacement fluctuations. For the half load circumstances, steady and uniformly distributed displacements were produced. The importance of relative or convergence mesh generation has also been discussed here in this research article. In this regard, Fig. 2.3a depicts that the solid modeling of piston structure has generated by the revolve option with the reference of geometric section of Fig. 2.3d. The 270° cross sectional view of this revolved section can be seen for the detailed view of the inner meshed areas of each elements as mentioned in Fig. 2.3b, c.

Bhagat et al. (2012) discussed the various thermal stresses acting on the piston heads, as well as variations in mesh attributes and total deformations. The shear stress and thermal flux were also measured using the ANSYS ADSL, and a mathematical model of optimization was obtained to obtain reduced values of stress concentration and better FEA analysis results. Sathishkumar et al. (2016) determined that cast iron is the appropriate material for medium and heavy-duty applications using Finite Element Analysis, as shown in Fig. 2.3. A new type of aluminium composite material has been developed that can withstand temperatures as high as 1573 K and pressures as high as 6195 kPa. The piston's structure was modelled using CATIA V5R20 software. For finite element modelling and analysis, ANSYS 14 was used. The maximum withstanding temperature and the maximum pressure of the piston were determined, and their stress distributions were examined. Furthermore, Pandey et al. (2020) discovered that an Internal Combustion engine Piston of various materials has multiple static and thermal stresses.

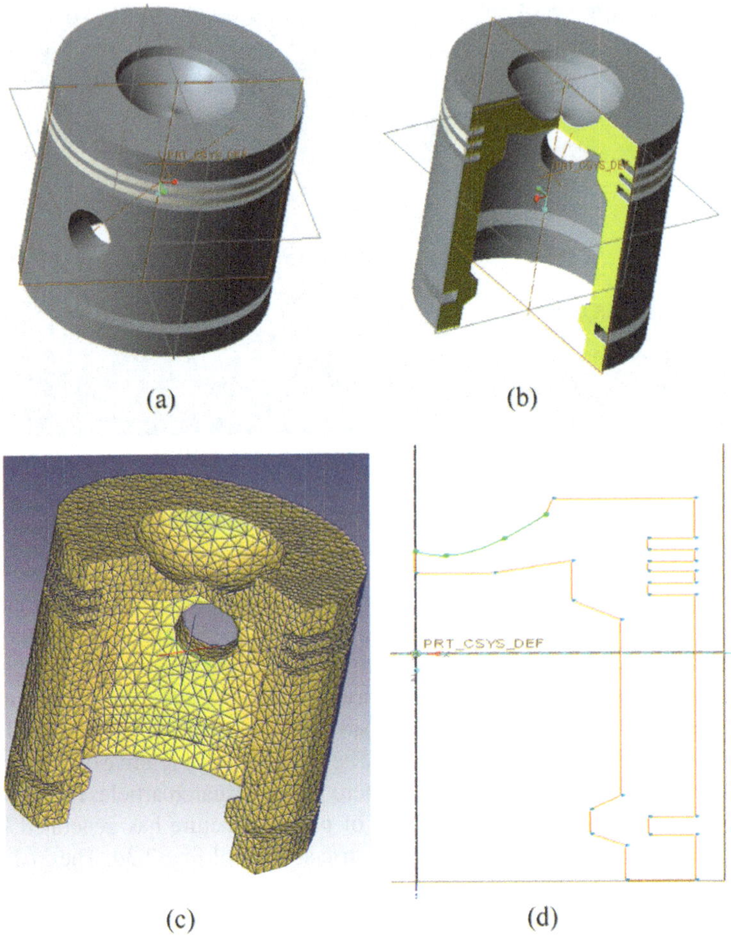

Fig. 2.3 **a** Solid modelling of Piston, **b** sectional view of the piston, **c** mesh generation, **d** geometric sections for revolving function

Singh et al. (2015) discovered that a maximum stress value smaller than the maximum tensile stress and yield strength of the material, as well as a minimal factor of safety, clearly shows that the design is safe. Balamurugan et al. (2017) has determined that the flow of heat at various notches is available and the analysis is done by using ANSYS—CFD Fluent software (Iruthayaraj et al. 2021; Sathishkumar and Ugesh 2016). Sathishkumar et al. (2016) and Iruthayaraj et al. (2021), Finite Element Analysis of a Shaft Subjected To a Load the author tells about the objective is to build a model and assemble the part files and to analyze the various stress and deformation. The part files and assembly are done by using CREO software and the analysis is done by using ANSYS software. Challen and Baranescu (1999) have been discussed that the temperature distribution at Maximum levels in a gasoline and oil cooled diesel piston has been discussed through the simulation.

2.2 Modelling and Drafting Techniques of Engine Models

2.2.1 Modelling and Simulation Tools

In general, both businesses and institutions place a premium on the usage of contemporary tools. In this sense, modelling refers to the process of creating a part from a description of any component or part for exact dimensions. Before creating the prototype, change or construct the part model using graphical manipulation. A "virtual prototype" is one in which each component may be modelled separately before being combined to create the final shape of the machine or engine. This kind of modelling is known as part modelling. We may use the software application to simulate or animate this component or assembled model.

2.2.2 Part Modelling of Engine Block and Accessories

Using various control commends, the Internal Combustion engine blocks are built with complicated shaped geometries. It is composed of cast iron components and can resist high vibration during an engine's operating cycle while running. Any solid modeller (Parametric Creo, Autodesk Inventor, Solid Works, etc.) may be used to create the intricately formed cylinder block. Complex or intricately formed things in Parametric Creo may be created as two-dimensional diagrams at first. The extrusion will then modify it, revolute commends.

The engine block's two-dimensional sketch outline has been created here utilising the line and spline controls. Following the extruded length of the block for the required length, the holes for the sequence intervals for the reciprocating piston with a diameter of 40 mm are produced. Figure 2.4 shows how the sketch plane may be chosen for each component model and extruded for the desired length.

The line and curve choices were used to create the section of one side of the piston. Lines are used to linking all of the boundaries in relation to the central axis. The thickness may be specified based on the piston's standard size. The top fin-like provisions may be modelled for conduction analysis. Figure 2.5 shows how the whole half portion drawing may be rotated with regard to its axis (a). Then, at about 360°, a full rotation of the half cross sectional drawn sketch will produce the whole piston body. With the through all option, the hole tool may be used to construct the diametrical piston pin provision. The 2D drawing may be started or located using a single sketching plane, which can be front, rear, side, or top. The necessary edges for the well-finished solid models may be chamfered or shaped using the chamfer,

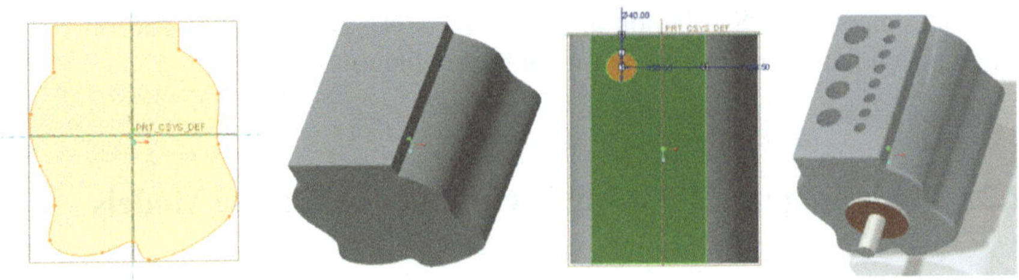

(a) (b) (c) (d)

Fig. 2.4 **a** 2D drawing of complex-shaped engine block cross-section, **b** solid part modelling of rigid engine block, **c** creation of piston boreholes, **d** solid modelling of the engine block with piston bores

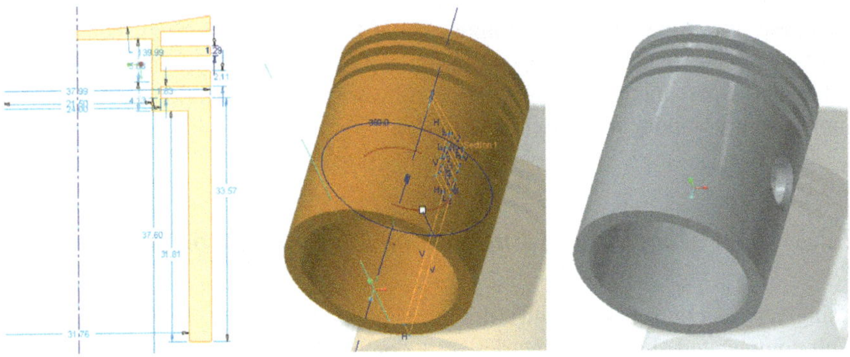

(a) (b) (c)

Fig. 2.5 **a** 2D sketch of piston cross-section, **b** revolute piston model, **c** piston model with piston pinholes

round edge choices accessible in the toolbar. This may be used for static analysis if the component file is in.IGES or .stl format.

2.2.3 Assembling of Engine Components

The assembling or dismantled sections also have been obtained here in this modelling and simulation platforms for the well-known dimensional accuracies. The parts to be mixed together may be opened in separate windows in the assembling section. We can simply assemble the component to its necessary location by using the part model's mate, move, and rotate arbitrary movements. The disassembled view of the engine and piston assembly is shown in Fig. 2.6.

Fig. 2.6 Disassembled view of the engine and piston assembly

2.2.4 Meshing Attributes

Discretization of structure is one of the FEA processes used. Basic data generation, Node and element numbering, when it comes to defining the degree of freedom, Determine the displacement function, the stiffness matrix, and the nodal stresses and displacements. A mesh is the whole set or assembly of pieces. The act of expressing a component as an assembly of finite elements, known as discretization, is the first of several critical steps in comprehending FEM analysis. To begin running an FEA simulation, a mesh is created, which contains millions of microscopic elements that make up the overall shape. This is a method of converting a three-dimensional object into a collection of mathematical points that may subsequently be studied. The density of this mesh can be adjusted depending on how complex or simple the simulation is required. Calculations are performed on each individual element or point of the mesh, and the results are then merged to get the overall final result for the construction. Because the calculations are performed on a mesh rather than the full real object, some interpolation between the points is required. These assumptions are usually within the parameters of what is required. Nodal points are spots on the mesh where the data is known mathematically, and they are typically grouped around boundaries or other areas of change in the design of an object.

The simulation portion includes the modelled cylinder and piston blocks. The meshing characteristics are the first stage in each simulation. The closely spaced combined rectangles are connected one by one across the solid model's surface. Based on the subdivision of each element, this discretization technique may be split into a limited number of elements. The aspect ratio is the proportion between the shortest and longest lengths of an element. A good mesh, on the other hand, will have almost all of its components with a small aspect ratio or a number less than three. Meshing is classified into two types: convergent mesh and relative mesh. Mesh convergence specifies the number of elements required in a model to guarantee that changing the mesh size has no impact on the analysis results as shown in Fig. 2.7.

The system response (stress, deformation) will converge to a repeatable solution as element size decreases. Convergence is all about energy conservation and the difference between input and output energy. ANSYS use the Newton–Raphson

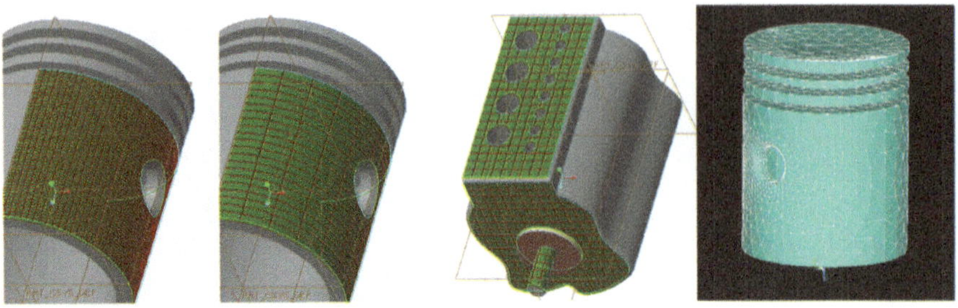

(a) (b) (c) (d)

Fig. 2.7 **a, b** Meshing of piston with higher amount of discretized elements, **c** meshing of the cylinder block with higher amount of discretized elements, **d** convergence mesh generation in ANSYS ADSL

method to "predict" the results at each iteration and whether or not the result has converged. We can simply determine if the mesh is convergent or not.

- Analyze the model by generating a mesh with the fewest and most acceptable number of components.
- Analyze the mesh again with a denser element distribution and compare the results to the previous mesh.
- Increase the mesh density and re-analyze the model until the results converge.

2.3 Procedure of Simulation

2.3.1 Introduction of Finite Element Analysis

The FEM's fundamental ideas are straightforward. Consider a body or an engineering component through which a field variable, such as displacement or stress, must be distributed. Components under load, temperatures subject to heat input, and so on are examples. The body, which is a one-, two-, or three-dimensional solid, is modelled as an assemblage of discrete bits known as elements—'finite elements.' The term "finite" refers to the limited, or finite, number of degrees of freedom utilised to simulate each element's behaviour. The elements are assumed to be linked, but only at interconnected joints known as nodes. It's worth noting that the elements are conceptually small regions rather than independent entities like bricks, and there are no cracks or surfaces between them. Finite element analysis (FEA) is a numerical technique that uses a variety of physics-based calculations to address engineering problems. Although the system of equations describing the distribution of structural stresses is known, it cannot be easily or directly solved for a complex structure like a bridge. As a result, the system must be broken down into small elements like the "finite elements" that can be easily calculated and solved. Finite elements can be one-dimensional, two-dimensional, or three-dimensional. The Finite Element Method (FEA) is commonly used in engineering to computing approximate solutions to boundary problems (field problems). As the main domain, "the field" usually represents a physical structure. The variables of interest that are controlled by a differential equation are referred to as "field variables." They must accomplish a boundary condition for particular variable values.

Finite Element Analysis, or FEA, is the simulation of a physical phenomenon using a numerical mathematic technique known as the Finite Element Method, or FEM. This method is essential in mechanical engineering and a variety of other disciplines. It is also one of the basic principles used in the development of simulation software. Although partial differential equations can be used to analyse the majority of these events, Finite Element Analysis is the preferred mathematical technique in complex scenarios requiring several highly variable equations. FEA techniques are embedded in simulation tools such as Autodesk Inventor Nastran or the ANSYS software suite. These tools are generally linked to computer-aided design (CAD) software, making the transition from design to complex structural analysis easier for engineers. Vijayarangan and Ganesan (1994), Rao and Muthuveerappan (1993), Zheng et al. (2008) the internal combustion engine is the most significant section of the engine, and its strength and dynamic qualities are the engine's assurance for normal operation. It is essential to guide the design and improvement of the engine block by researching the structural strength of the engine block and properly displaying the stress and deformation distribution.

The main applications of the Finite Element Method are stress analysis of automobiles, aviation, structures, and buildings, transient thermal analysis of heat flux, fluid flow, and thermal flux, etc. This FEM can handle stress analysis of pistons, connecting rods, gears, and other components. Vibration analysis is an extra component used to determine structure and column models.

Thermal analysis within a material or shape can also be done using FEA. If you know the temperature at one location in an object, you can use the FEA to approximate the temperature at these points using various degrees of accuracy. There are three types of approximations: square, polynomial, and discrete. Each of these strategies improves in terms of accuracy and complexity.

Pre-processing: You must define the domain problem, material properties, geometric properties, mesh model, boundary constraints, and element types at this step. Typically, the engineer will simplify the problem as much as feasible at this step in order to obtain quick yet accurate results. Prior to simulation, for example, a CAD model may include a number of fillets, holes, or other characteristics that aren't necessary for the simulation but consume computer resources and raise render times. Engineers usually eliminate these characteristics in order to accelerate simulations.

Solution: Computations for the unknown values of the field variables are performed here. These numbers could be used to calculate derived variables such as stresses and reaction forces.

Post Processing: The final set of activities is sorting, charting, and printing, in which your chosen results are applied as solutions to your finite element analysis.

2.3.2 *Structural and Thermal Simulation of Engine Components*

The engine models may be used to create the simulated results in the simulation section. Figure 2.8 shows how the meshing process can be used to refine the component boundary conditions. Preferences may be specified in the structural category, and the element type can be selected from the element type library. The Solid brick 8 node element type was chosen for the appropriate engine cylinder part model.

In the material model, the material characteristics of young's modulus and poison's ratio may be specified. The displacement can be assigned to the piston model's fixed or clamped end. The piston pinhole can be used to stop the entire displacement. As shown in Fig. 2.8, the pushing pressure during the power stroke has been considered for the acting pressure over the crown area of the piston. The characteristics of this isotropic kind of material have been provided for obtaining simulation results. The finite number of elements can be split over the solid model after the mesh was generated over the piston body.

The linear structural elastic type of analysis was used during the structural analysis of this piston body. A static conditional structural analysis was performed. Following the definition of the displacement values, pressure or force can be applied to the specified edges or surfaces. Figure 2.9 depicts colourful graphs used to identify outputs based on this solution stage following preprocessing. The nodal solution of

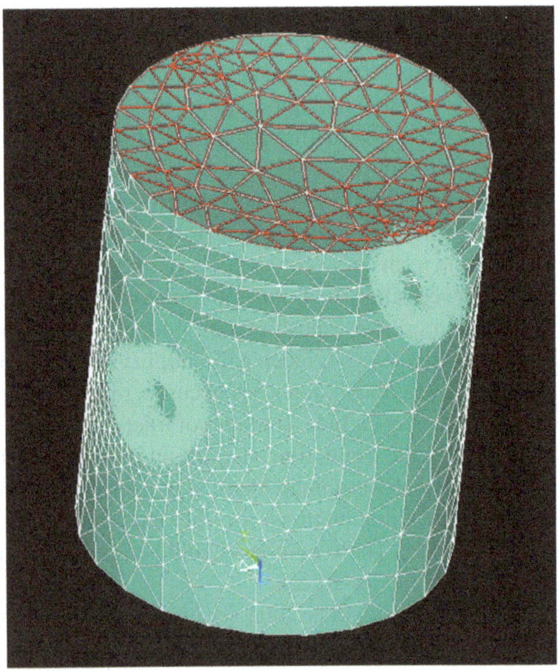

Fig. 2.8 Displacement and pressure distribution on the crown of the piston

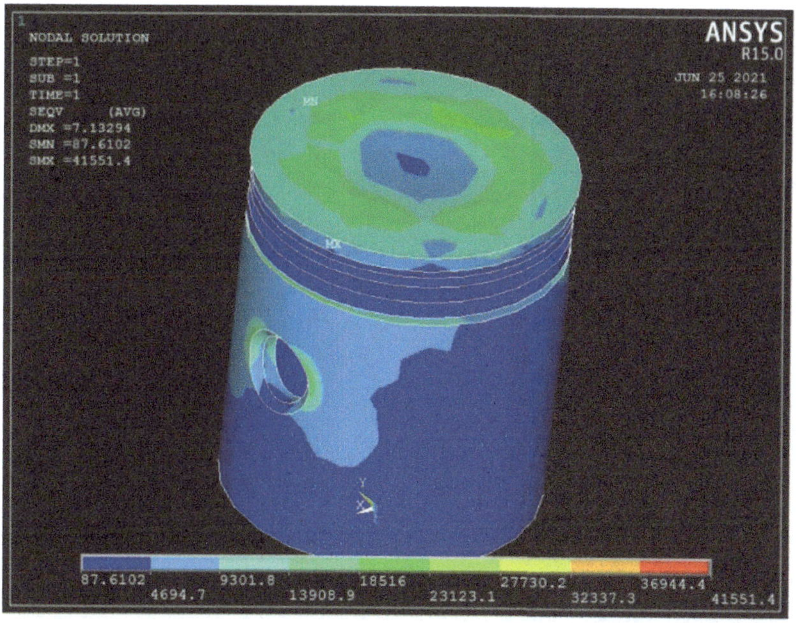

(a)

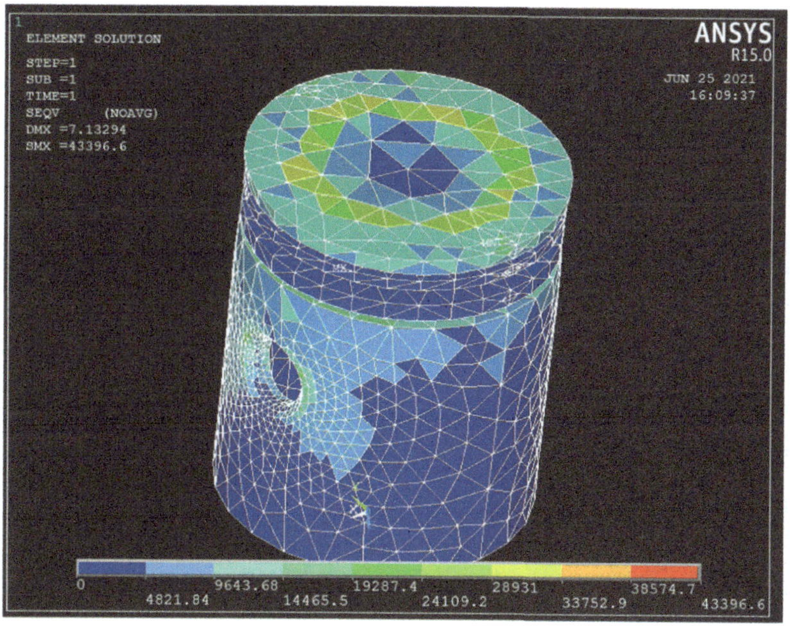

(b)

Fig. 2.9 Simulation outcomes (nodal and elemental solution) of the piston part

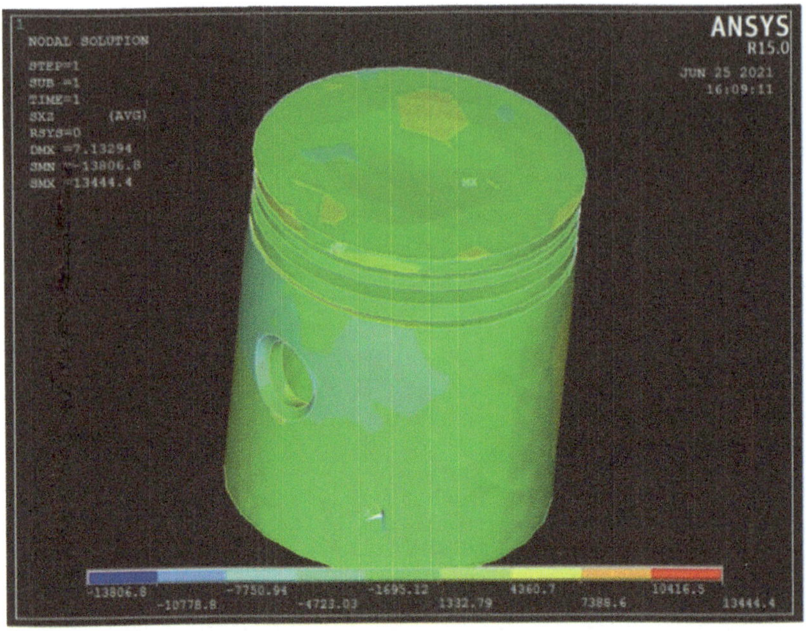

(c)

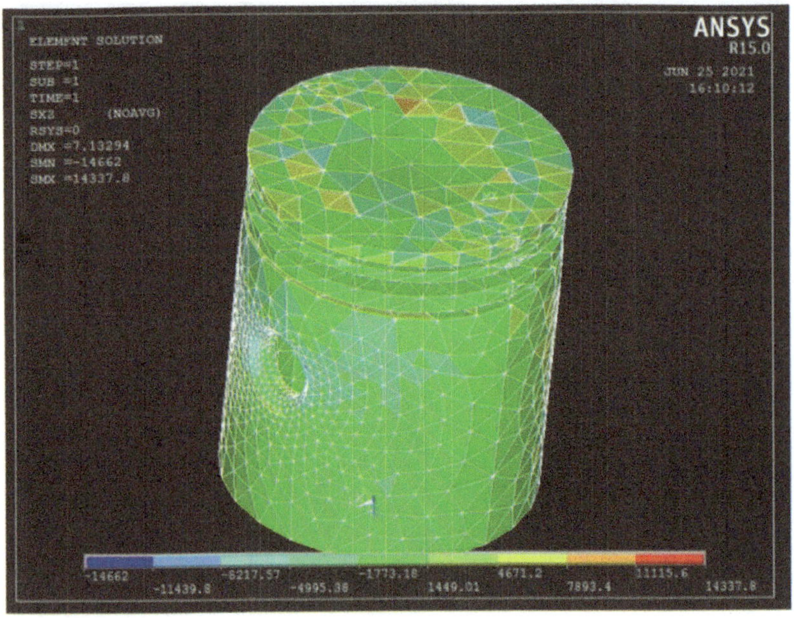

(d)

Fig. 2.9 (continued)

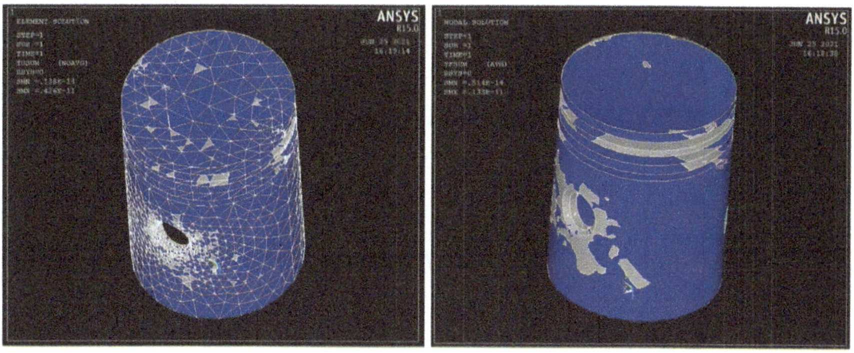

Fig. 2.10 Thermal analysis of piston in ANSYS

stresses and displacements, as well as the elemental solutions, were discovered in these simulated results. The von Mises stress and shear stress can be calculated to obtain precise prediction values for maximum and minimum stress components. The refined elemental solution of von Mises and shear stresses can be obtained by the finer mesh generations.

Similarly, the thermal forecast may be achieved using the same piston component part model as shown in Fig. 2.10. It depicts the estimated temperature gradient and heat flow from the anticipated ANSYS simulations. At the start of the thermal analysis, the preferences were thermal rather than structural. Thermal distribution and analysis have been conducted under steady-state conditions. During the solution stage, the temperature 350° acting over the crown of the piston was entered into the temperature field.

2.3.3 Comparison of Simulated Outcomes

Understanding the importance of these simulation results of both structural and thermal analysis can be compared with the graphical representation by using these outputs. Here Table 2.1 shows that the nodal values and elemental values of von Mises stress, shear stress (XZ) and displacement values. By the consideration of

Table 2.1 Nodal and elemental stresses and displacement values

S. No.	Particulars	Nodal solution	Elemental solution
1	von Mises stress	41,551.4 MPa	43,396.6 MPa
2	Thermal flux	0.133e−11 W/m^2	0.426e−11 W/m^2
3	Shear stress (XZ)	14,337.8 MPa	
4	Displacement	7.13294 mm	
5	Temperature	350°	

elemental solution, the stress has been attained as 43,396.6 MPa and the stresses were recorded in the nodal solution is 41551.4 MPa for the 7.13294 mm displacement. For the steady-state thermal analysis, the temperature has indicated over the crown of the piston is 350°. Thermal flux can be predicted as 0.133 and 0.426e−11 W/m^2 respectively.

2.3.4 Benefits and Outcome of the Simulated Results

FEA is one of the most powerful and widely utilised CAD design features. It is essential in CAD for analyzing an object using finite elements such as small rectangular and triangular structures. To study the entire object, each finite element in the entire object network is evaluated, and the results are merged to plot the overall behaviour of a complex object. Each element can be examined for specific property thresholds such as stress–strain, dynamics, and thermal characteristics. Engineers can use FEA to simulate a designed model rather than creating a physical model. This kind of simulation is used to make predictions prior to the development of a product.

- Using this platform, the maximum withstanding of product capabilities may be readily anticipated.
- Understanding the material models and meshing characteristics by applying different loads during the structural analysis and different temperature values during the thermal steady state study.
- This may be used to generate the unknown force components as well as displacements along the portion.
- The economic advantages of utilising these simulations may also be increased.

2.4 Conclusion

As a consequence, this simulation-based study chapter may be highly useful in learning and improving the analysis of one's own thoughts and designs of numerous component parts. A basic solid modelling approach was used to simulate a piece of an Internal Combustion engine. The better shear stress and von Mises stress which has been identified through the ultimate strength of each component and has been obtained with the safe design. Similarly, the simulations of each structural and thermal analysis of part models have been effectively created when compared to other traditional forms of analysis. In each nodal and elemental solution of stresses and displacements, the post-processing stage results of each simulation have been separated.

Acknowledgements First and foremost, I would like to thank the ISEES team for conducting the "Energy, Environment and Sustainability Series" with the book entitled "Engine Modelling and Simulations". My special thanks to professor K Madhu Murthy Garu (NITW), for giving me the best opportunity, He suggested and supported me to write a book chapter entitled "Development of Engine Models and Analysis of Cylinder Bore Piston Stresses and Temperature Effects in IC Engine". I am truly privileged to have had his assistance. I would like to thank my parents (Jagannatham-Sharada), Brothers, Life partner (Nagasudha), and my friends for their enduring support from the beginning.

References

Ahmed A, Basim (2009) Thermal effects on diesel engine piston and piston compression ring. Eng Technol J 27(8)

Ansari MB, Ul Haque MF, Joshi PS (2020) Design and analysis of engine block. Int Res J Eng Technol (IRJET) 07(05). e-ISSN: 2395-0056

Bhagat AR, Jibhakate YM, Chimote K (2012). Thermal analysis and optimization of I.C. engine piston using finite element method. Gas 2

Challen B, Baranescu R (1999) Diesel engine reference book, 2nd edn. In: Pistons and engine testing, Vieweg + Teubner Verlag, SAE, Warrendale, PA, Mahle, 2012. https://doi.org/10.1007/978-3-8348-8662-0

Chigrinova NM, Kuznechik OO, Chigrinov VV (2004) Analysis of heat stresses of the parts of the cylinder-piston group with heat-protective coatings in an internal-combustion engine. J Eng Phys Thermophys 77:578–589

Hamzwhei M, Manochehr R (2006) Determination of piston and cylinder head temperature distribution in a 4-cylinder gasoline engine at actual process. In: Proceedings of the 4th WSEAS international conference on heat transfer, thermal engineering and environment, Elounda, Greece, Aug 2006

Iruthayaraj R, Palani S, Ajay Vishal R, Living Rockson T, Anand K (2021) Design and analysis of rocker arm shaft in IC engine hino series for reduction of assembling dismantling time. In: Akinlabi E, Ramkumar P, Selvaraj M (eds) Trends in mechanical and biomedical design. Lecture notes in mechanical engineering. Springer, Singapore. https://doi.org/10.1007/978-981-15-4488-0_11

Jayale SR, Kadam GA, Pathan U (2017) Stress analysis of piston at different pressure load. Int Res J Eng Technol (IRJET) 04(05)

Kourav SK, Ghagare VB (2015) Design and analysis of piston by using finite element analysis. Int J Eng Res Technol (IJERT) 4(09):296–301. ISSN: 2278-018

Krishna S, Aravind A (2014) Thermal and stress distribution of different I.C. engine piston combustion chambers using 3-D finite element analysis method 3(2)

Kumar M (2017) A finite element thermo-mechanical stress analysis of internal combustion engine piston 04(06)

Kurbet SN, Krishnakumar R (2004) A finite element study of piston tilt effects on piston ring dynamics in IC engines. In: Proceedings of Institution of Mechanical Engineers (IMECHE) Part K, pp 107–117

Narayana KS, Phani Kumar S, Srinivasa Rao MSS (2014) Thermo-structural finite element analysis of I.C. engine pistons. In: International colloquium on materials manufacturing and metrology, ICMMM 2014, 8–9 Aug, IIT Madras, Chennai, India, pp 882–884

Pandey K, Gupta K, Papaiya V, Dwivedi M (2020) Analysis of static stresses and thermal stresses of an internal combustion engine piston of different materials

Rahman MM, Ariffin AK, Jamaludin N, Haron CHC (2019) Finite element based vibration fatigue analysis for a new free piston engine component. Arab J Sci Eng 3

Rao CRM, Muthuveerappan G (1993) Finite element modelling and stress analysis of helical gear teeth. Comput Struct 49(6):1095–1106

Sathishkumar K, Ugesh N (2016) Finite element analysis of a shaft subjected to a load in ARPN. J Eng Appl Sci 11(9). ISSN 1819-6608
Sathishkumar S, Kannan M, Raguraman V (2016) Finite element analysis of internal combustion engine piston using thermo mechanical approach 1(3):2394–9333
Sathishkumar K, Vignesh K, Ugesh N, Sanjeevaprasath PB, Balamurugan S (2017) Computational analysis of heat transfer through fins with different types of notches. Int J Adv Eng Res Sci 4(2):175–183. ISSN: 2349-6495, 2456-1908
Sharma SK et al (2015) Experimental thermal analysis of diesel engine piston and cylinder wall. J Eng 2015:1–10
Singh L, Rawat SS, Hasan T, Kumar U (2015) Finite element analysis of piston in ANSYS. Int J Mod Trends Eng Res (IJMTER) 02(04):619–626
Szurgott P, Niezgoda T (2011) Thermo mechanical FE analysis of the engine piston made of composite material with low hysteresis. J KONES Power Trains Transp 18(1)
Tzoref J, Stotter A, Zvirin Y (1978) Dynamic response of the temperature and stress fields in an I.C. engine piston. Int J Mechan Sci 20(9):581–592. ISSN 0020-7403
Vijayarangan S, Ganesan N (1994) Static contact stress analysis of a spur gear tooth using the finite element method, including frictional effects. Comput Struct 51(6):765–770
Zhang H, Lin Z, Xu D (2013) An analysis to thermal load and mechanical load coupling of a gasoline engine piston. J Theor Appl Inf Technol 48(2):911–917
Zheng ZC, Gao Y, Liu B (2008) Finite element analysis of zh195 diesel engine block. Internal Combustion Engine Power Plant 8:5–9

Part II
Spray Modeling

Chapter 3
Mathematical Modeling of Injection and Spray Characteristics of a Diesel Engine: A Review

Subhash Lahane, P. W. Deshmukh, and M. R. Nandgaonkar

Abstract In a diesel engine, the injection phenomenon is vital because it impacts fuel spray characteristics and mixture formation process. Accurate control of operating fuel injection parameters (injection timing, flow rate shaping and fuel in-line pressure) is necessary as it affects mixture formation method of air and fuel. If the energy content of fuel is less, to sustain the desired power output, the injection duration would increase. The increased injection duration would affect the important combustion parameter predominantly the combustion duration. So, the optimization of the injection system is needed for a diesel engine for better performance and emission characteristic. The fuel atomization characteristics, and subsequently spray development compete a crucial part in the progress of better combustion and engine performance, as it affects the mixture formation in the combustion chamber. There are studies and models available for diesel fuel spray with and without swirl. If the bulk modulus of fuel is higher, the fuel inline pressure increases significantly. The bulk modulus implies compressibility of a fuel. The longer fuel spray penetration observed with rise in the fuel inline pressure further it also increases the chances of wall impingement. It is one of the main durability issues of a diesel engine. On the other hand, it is important to note that the rise in pressure of injection is a useful approach to enhance fuel atomization characteristics and to enhance mixing of air and fuel. The higher bulk modulus of fuel results in advanced injection timing which further leads to high NOx emission. Furthermore, the increased spray penetration increases NOx emission which forms generally around periphery of spray. Therefore, it is necessary to evaluate the spray development and fuel atomization characteristics for reduction of exhaust emissions and subsequently, enhance the efficiency of the engine. The current research work is aimed at the extent review of computational/mathematical models of spray characteristics and empirical models of injection and spray process of engine. The challenges for modeling of spray characteristics are also highlighted. Numerous research works have been referred and analyzed for the effect of injection parameters and fuel properties on injection and spray characteristics of the engine.

S. Lahane (✉) · P. W. Deshmukh · M. R. Nandgaonkar
Department of Mechanical Engineering, College of Engineering Pune (CoEP), Pune, Maharashtra 411005, India

A. K. Agarwal et al. (eds.), *Engine Modeling and Simulation*, Energy, Environment, and Sustainability, https://doi.org/10.1007/978-981-16-8618-4_3

Keywords Injection · Spray · Penetration · Diesel engine · Empirical modeling

Nomenclature

C_{pg}	Specific heat of gas (J/kg K)
d_n	Sac diameter (m)
D_n	Hole diameter (m)
dP	Fuel pressure difference (Pa)
dV	Change in volume (m^3)
L	Latent heat (J/kg)
L_b	Break-up length (m)
L_n	Nozzle length (m)
L_θ	Length w. r. t. crank angle (m)
m_a	Air entrainment (kg)
m_f	Liquid mass (kg)
m_S	Air entrainment with swirl (kg)
Q	Fuel discharge (mm^3/st)
Re	Reynolds number
R_S	Swirl ratio
r_c	Compression ratio
S, X	Spray penetration (m)
t	Injection time (s)
t_b	Break-up time (s)
T_g	Air temperature (K)
T_s	Droplet temperature (K)
V_i	Injection velocity (m/s)
X_S	Penetration distance with swirl (m)
X_{32}	Sauter mean diameter (m)

Greek Letters

ρ_l	Fuel density (kg/m^3)
ρ_a	Air density (kg/m^3)
μ_l	Fuel viscosity (Pa s)
μ_a	Air viscosity (Pa s)
σ	Surface tension of liquid (N/m)
θ	Spray cone angle (°)
θ_S	Spray cone angle with swirl (°)
λ_g	Thermal conductivity of air (W/m K)

3.1 Introduction

As initiator for transportation, regionalized power production source, agricultural uses, construction of building and mining, a diesel engine have been proved as a primarily vital prime mover. There exists a complexity in the combustion process of diesel engine is turbulent, diffusion, heterogeneous, and unsteady. A fuel–air mixing in the engine cylinder plays important part in the combustion (Kozina et al. 2020).

During the suction stroke of the diesel engine, only air is drawn, and heated due to large compression ratio. In this pressurized heated air, the fuel is injected, and a combustible mixture of fuel and air need to be formed, which can immediately auto-ignite. The efficiency of the cycle increases with higher values of compression ratio, but the maximum pressure reached in the cylinder also increases. The air–fuel ratio used in the diesel engine lies between 16 and 22 (Chen et al. 2019).

The advanced injection policies show substantial influence on the fuel consumption and emissions. Both, these characteristics are extremely relying on the combustion. The study of spray formation is required to optimize the combustion in a diesel engine (Liu et al. 2020).

The diesel engine characteristics can be grouped as follows and also shown in Fig. 3.1.

- Spray parameters
- Injection system
- Combustion phenomenon
- Engine performance

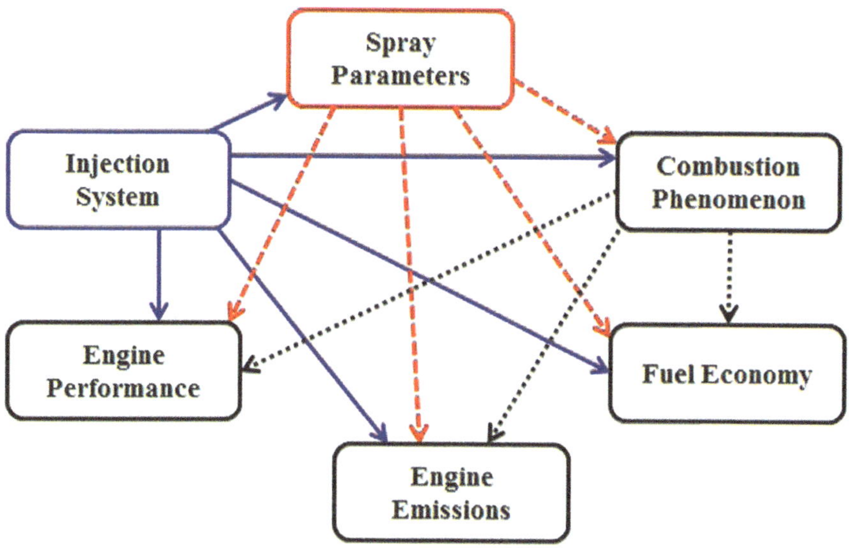

Fig. 3.1 Engine characteristics

- Engine emissions
- Fuel economy

It could be very well interpreted from Fig. 3.1 that all the engine characteristics are depending mainly on fuel injection system. The main outcome of the engine depends on various parameters such as mass of fuel injected, injection timing, fuel rate shaping, type of combustion chamber, number and diameter of nozzle holes, spray pattern, swirl, etc. (Som et al. 2010; Lee et al. 2002). One of the important parameters which influence the performance and emissions of the diesel engine is injection pressure. It is observed that rise in injection pressure tends to longer penetration, better air entrainment, improved mixture formation of fuel and air, and shorter combustion duration (Lee et al. 2002).

The design of fuel injection system and its operating condition are the contributing factors for the spray characteristics. In a diesel engine the combustible mixture form during spray processes such as break-up length, sauter mean diameter (SMD), spray penetration, spray cone angle, air entrainment and vaporization (Lahane and Subramanian 2015). Therefore, the fuel spray characteristics need to be closely paired with the design of combustion chamber and the aerodynamic conditions within the chamber to form a good mixture (Hamidi et al. 1990).

Break-up length and spray cone angle are the important spray characteristics for air entrainment into the spray. Increase of air entrainment takes place with shorter break-up length and higher spray cone angle. Vaporization of the fuel depends on the SMD of the fuel droplet. If SMD is high, it would enhance the vaporization time, and resulting lower vaporization. It is also essential to identify the spray penetration in a direct injection diesel engine since over penetration can lead to fuel impingement while under penetrate tends to less air entrainment (Urbán et al. 2020; Laryea and No 2004).

Significant amount of work has been carried out by various researchers to understand the fuel spray characteristics for diesel fuel. Injection characteristics (injection delay, injection duration, and dynamic injection timing) and spray characteristics such as break-up length, spray cone angle, SMD, spray penetration, air entrainment and vaporization play an important part in combustion phenomena of engine. Therefore, it is required to go through the important work in the area of this field and find out the challenges and more clarity for further development, if possible.

3.1.1 Injection Characteristics

Ever increasing rate of environmental degradation and energy crunch, the optimization of combustion process of a diesel engine appears to be more crucial to the greatest extent so far; consequently, cutting-edge injection schemes are step by step proposed and implemented in recent engines.

Injection system is the foremost importance of any diesel engine for smooth operation. The injection characteristics involve injection delay, static injection timing,

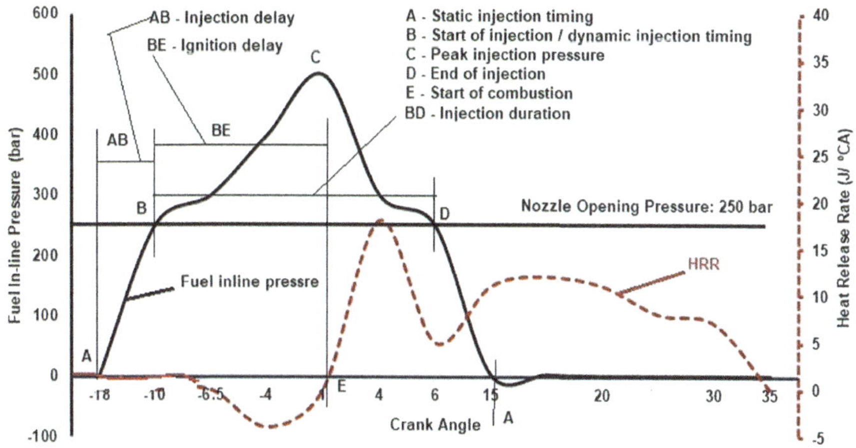

Fig. 3.2 Injection characteristics

dynamic injection timing (DIT) or start of injection (SoI) and injection duration. The overlapping of fuel in-line pressure and heat release rate is shown in Fig. 3.2 to represent the injection characteristics. The duration between start of injection (SoI) and static injection timing is known as injection delay. Dynamic injection timing or start of injection is defined as the real timing where the fuel is beginning to inject into the combustion chamber whereas static injection timing is always constant. Precise control of fuel injection factors such as injection timing, fuel rate shaping, and fuel-line injection pressure is necessary as it influences the atomization process (Lahane and Subramanian 2015).

General requirement for a spray is adequate mixture supply to a compressed air at high pressure and temperature in a cylinder for better combustion. Then, the main purpose of fuel injection is as follows.

- Fuel mass and combustion timing control
- Mixture formation
- Fuel preparation in a combustion chamber.

The essential task of injection system is the transportation of fuel to a desired space at a desired timing (Arai 2012). Wide-ranging computer codes are currently applied in the research of engine instead describing engine processes empirically. CFD modeling is a key facilitating knowledge in the growth of engine. CFD, combined with genetic algorithm optimization techniques, observed smooth operating conditions even with very high injection pressure like in CRDI (Reitz 2013).

3.1.2 Fuel Spray Characteristics

Spray characteristics such as break-up length, spray cone angle, SMD, penetration and air entrainment (shown in Fig. 3.3) are influencing the atomization process and combustion phenomenon. The injected fuel penetrates into the hot air at high pressure and temperature at the end of compression stroke. The fuel is atomized to very fine droplets and vaporized to form an ignitable mixture. The preparation of this mixture depends upon the temperature of the air and pressure of numerous pockets of desired air–fuel ratio. Therefore, spray characteristics show a foremost part in the combustion process for realization of thermal energy (Hamidi et al. 1990).

The description of spray process may be dividing in two categories, macroscopic characteristics such fuel spray penetration and cone angle of spray, and microscopic characteristics such as the spray break-up length and SMD. The unavoidable complex phenomenon in both experimental as well as and theoretical research, the entire phenomenon of some detailed microscopic characteristics like spray break-up length and atomization still remains unknown (Liu et al. 2020). Nevertheless, for the research of macroscopic characteristics, the fuel spray development can be critically streamlined, where the fuel spray can be preserved as mixture, ignoring microscopic characteristics (Liu et al. 2020).

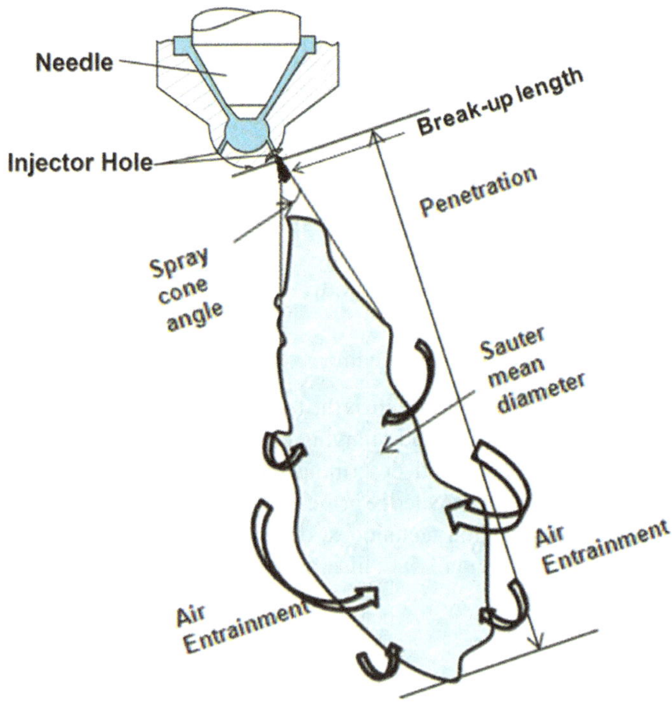

Fig. 3.3 Spray characteristics

The study of fuel spray near the nozzle exit is still difficult due to rich mixture and droplet clouds. The holography, diffraction-based techniques and laser doppler-anemometry (LDA) has been used for drop size and velocity distribution diagnostics (Hamidi et al. 1990). However, these measurement techniques faced the challenges to measure the drop size and velocity distribution near the nozzle exit. The absorption-based measurement techniques 'x-ray radiography' has been developed to overcome the challenges near the nozzle due to dense liquid. But it is very time and resource intensive (Arai 2012). Another experimental measurement technique needs to develop which ensures the less time and resource intensive to accelerate our understanding of spray and benefit as input data for mathematical models of diesel spray (Reitz 2013).

3.2 Models for Spray

The package, parcel and two-phase flow models are used for numerical simulation. Package model is generally used for phenomenological simulation of fuel spray. Package contains of different size of droplets and gaseous mixture. As air entrainment increases, package volume also increases. All the evaporation and combustion phenomena are considered in the package model. However, parcel only contains different size of droplets. Its movement is supposed to be the same as that of a SMD in the parcel. Every parcel means source terms of mass, momentum, and energy in the Navier–Stokes equations. It is also adopted for many CFD simulations of fuel spray. There are still many disagreements between models and actual movement of fuel spray. Two phase flow model is used in a direct simulation of a fuel spray. However, without sufficient theoretical interpretations for fuel droplets, it might be challenging to use for fuel spray simulation (Arai 2012).

3.2.1 Models Based on Spray Equation

O'Rourke (O'Rourke 1984) studied the dynamic and evaporation analysis of spray. The main objective his research was to develop the spray equation and study the effect of vaporizing spray. Spray equation (O'Rourke 1984) can be written as Eq. (3.1)

$$\frac{\partial f}{\partial t} + \nabla_x \cdot (f v_d) + \nabla_{vd} \cdot (f F_d) + \frac{\partial}{\partial r_d}(f R_d) + \frac{\partial}{\partial T_d}(f \dot{T}_d) = \dot{Q} \tag{3.1}$$

where,

v_d, F_d, R_d, and T_d	denote the velocity, force, change rate of the radius, and temperature of the droplets, respectively.
r_d	is the droplet radius.

$f\,dr_d\,dv_d\,dx\,dT_d$	is the possible number of droplets in the radius range dr_d around r_d located in the spatial range dx around x with velocities in the range dv_d around v_d and with temperatures in the range dT_d around T_d.
$\dot{Q}$	represents the rate of increase of f due to collisions, coalescence, and breakup.
$\dot{T}_d$	represents the change in droplet temperature with respect to time.

Williams (1985) generalizes the spray equation and accounts for variations of the droplet temperature. The value of F_d, R_d, and $\dot{T}_d$ and $\dot{Q}$ need to be specified in order to solve Eq. (3.1). These values are frequently taken from experiments for isolated droplets, and they account for the effects of vaporization on F_d, R_d, and dv_d/dt.

The void fraction can be calculated as given in Eq. (3.2).

$$\varphi = 1 - \int \frac{4}{3}\pi (r_d)^3 f\,dr_d\,dv_d\,dx\,dT_d \tag{3.2}$$

where, the integral is to be performed with respect to all the velocities, radii, and temperatures of the droplets.

The effect of the gas phase turbulence on the droplets can be represented by considering that the gas phase velocities are the sum of the time averaged and fluctuating gas phase velocities as given in Eq. (3.3).

$$u = u + u' \tag{3.3}$$

where u' can be obtained from turbulence model and assuming that in a turbulent flow, $f = f\left(t, x, v_d, , r_d, T_{d,}u'\right)$.

If each component of u' is distributed according to a Gaussian distribution with a mean square deviation $\left|u'\right|^2 = 2k/3$ calculated from isotropic turbulence, the distribution function f can be written (Amsden et al. 1985) as given in Eq. (3.4).

$$f\left(t, x, r_d, v_d, T_{d,}u'\right) = \left(\frac{4}{3}\pi k\right)^{-3/2} \exp\left[\frac{-3\left|u'\right|^2}{4k}\right] f^*(t, x, r_d, v_d, T_d) \tag{3.4}$$

The turbulence kinetic energy, 'k' of the gas phase and f* can be obtained from 'f' by integrating over all values of u'.

Under these assumptions, the spray Eq. (3.1) can be written as given in Eq. (3.5).

$$\begin{aligned} &\frac{\partial f}{\partial t} + \nabla_x \cdot (f v_d) + \frac{\partial}{\partial r_d}(f R_d) + \nabla_{vd} \cdot (f F_d) \\ &+ \frac{\partial}{\partial T_d}(f \dot{T}_d) + \nabla_{u'}\left(f\frac{\partial u'}{\partial t}\right) = \dot{Q} \end{aligned} \tag{3.5}$$

A spray has basically two major components such as pocket of droplets has liquid phase and air contains which has gas phase. In the spray equation of Shuai (2009), the velocity axial component of gas phase in the equation of drop is developed by using theory of gas-jet near the exit of nozzle. The momentum of drop can be calculated using Eq. (3.6).

$$\frac{dU}{dt} = \frac{3}{8} C_d \frac{\rho_g}{\rho_l} \frac{1}{D} \left| U - V_{gas} \right| \left(U - V_{gas} \right) \tag{3.6}$$

The axial component of gas phase velocity near the nozzle exit in the x direction is modeled using gas-jet theory (Lippert et al. 2005) and can be calculated by using Eq. (3.7).

$$V_x = V_{gas} = \min \left(U_{inj}, \frac{3U_{inj}^2 d_{eq}^2}{\left[32\vartheta_t x \left(1 + \frac{3U_{inj}^2 d_{eq}^2 r^2}{256\vartheta_t^2 x^2} \right)^2 \right]} \right) \tag{3.7}$$

U_{inj} is the liquid jet velocity also assumed as gas jet velocity.
x is the axial distance between droplet parcel and injector tip.
r is the radial distance between parcel and spray axis.
d_{eq} is the corresponding diameter of gas jet and can be calculated using Eq. (3.8).

$$d_{eq} = D_n \sqrt{\frac{\rho_l}{\rho_g}} \tag{3.8}$$

ϑ_t is turbulent viscosity for jet and can be calculated using Eq. (3.9).

$$\vartheta_t = C_t \pi^{0.5} U_{inj} \frac{d_{eq}}{2} \tag{3.9}$$

C_t is the air entrainment constant and is equal to 0.0161.

Haselman and Westbrook (1978) and Westbrook (1977) solved the spray equation and the gas phase conservation equations of mass, linear momentum, species, and energy; they used a turbulent diffusivity as a function of strain rate of the mean flow field. The solution of the spray equation yields the fuel vaporization rate, but it does not provide resolution of scales of the order of or larger than the droplet spacing. However, the spray equation provides a basis for developing the thin and thick spray models.

3.2.2 Computational/Mathematical Models

The spray process is an essential chunk of the in-cylinder process of advanced engines and influence engine performance. The physical models of spray processes applied in in-cylinder are very complex and critical phenomenon like atomization. The fundamental physical phenomenon in liquid-sprays is demonstrated in Fig. 3.4. The sufficient tenacity of the complex spray within cylinder must account for the spray behavior, primary and secondary atomization, droplet collision and coalescence, drag and vaporization, and surface of the piston bowl (Lippert et al. 2005).

In Lagrangian approach, the study of spray formation is established on the Eulerian–Lagrangian approach wherein the Eulerian approach is used for analyzing the gas phase while the Lagrangian approach is used for analyzing the liquid phase (Sazhin et al. 1096). Another approach recommended by Osiptsov (Sazhin et al. 1096), which is broadly recognized as the Fully Lagrangian (Osiptsov) Approach (FLA). Nowadays, both these approaches are being used in all CFD codes.

The modified WAVE model based on linear stability analysis, Taylor Analogy Break-up (TAB) which gives liquid droplet break-up using spring mass system, gas the Effective Thermal Conductivity (ETC) liquid phase model, and the modified version of the Shell autoignition model have been incorporated in the KIVA 2 CFD code. It was observed that the modified WAVE model shows good agreement between experimental and calculated results compared to other models (Nijdam et al. 2006;

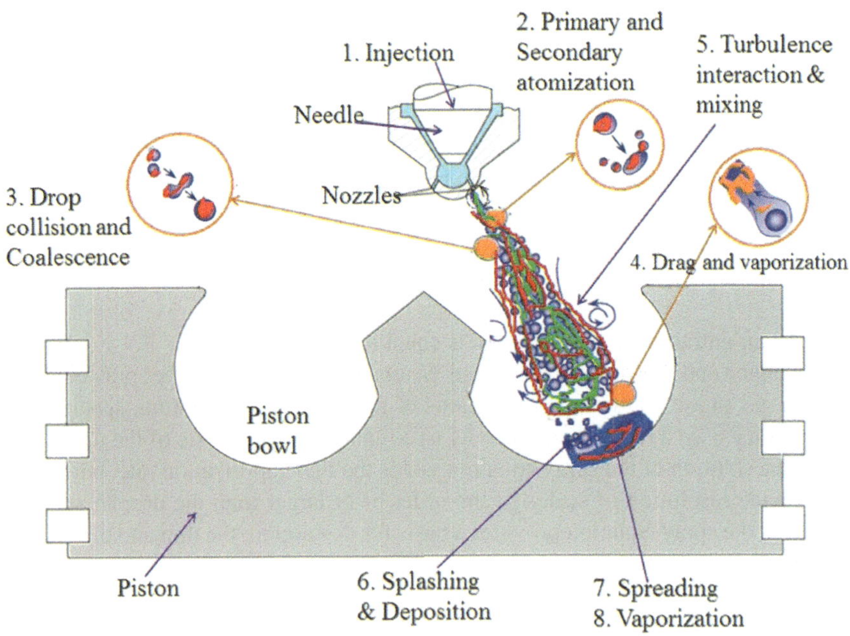

Fig. 3.4 Physical process in liquid sprays

Sazhin et al. 2008). The main drawback of these models is based on unrealistic assumption of single droplet created after break-up and quasi-steady-state flow situation whereas spray is transient process due to which they are not showing good agreement with spray calculations (Nijdam et al. 2006).

Lagrangian-Droplet and Eulerian-Fluid (LDEF) and Eulerian-Liquid-Eulerian-Gas models are generally being used to model the fuel spray in a diesel engine (Abani et al. 2007; Post et al. 2000). These approaches of modeling weakly predict the spray, spray penetration and SMD on low quality meshes of CFD. In both of these models, fine mesh close to nozzle is required to the scale of nozzle hole size for improved spray predictions (Abraham 1997). Therefore, it is necessary to develop the globally applicable grid independent spray model. Time-step independency is another integral modeling issue which is hardly been stated in the earlier study (Abani et al. 2008). Further, this model implements the LDEF method and O'Rourke's collision model (O'Rourke 1984), it is applied in CFD codes, with the KIVA-3V code (Amsden et al. 1985) and a diverse variety of mesh sizes and time-steps is used.

LDEF models has been improved with atomization models and used popularly with assumptions of Kelvin–Helmholtz and Rayleigh–Taylor (KH-RT breakup) instabilities (Beale and Reitz 1999). KH is mainly responsible for primary break-up and RT is takes care of secondary atomization (Nijdam et al. 2006). Another study on sprays near to nozzle was conceded out by Post et al. (2000). They observed that Lagrangian models do not offer the resolution independent results when the spatial dispersal of the drops is extremely deviating. This problem is overcome by improved model of O'Rourke and Bracco in which separate collision mesh has been used which tends to faster collision computations with no-time counter approach (Schmidt and Rutland 2004). An adaptive mesh has been proposed and it solve the problem of close to nozzle region and predict spray structures accurately (Lippert et al. 2005).

Abraham et al. (1997) verified the grid-independency in the calculations of different spray model for steady state injections and suggested that the entrainment of air can be modeled as per gas jet theory (Abraham 1996). It is also proposed that the first order entrainment velocity changes as per the time derivative of the velocity of injection for unsteady conditions (Abani et al. 2008).

The combination of turbulent jet-theory along with superposition integral formulation is used to establish velocity of injection and then penetration. This concept can be used for any random differing injection velocity at the exit of nozzle (Abani and Reitz 2007). The penetration for sprays of evaporating is modeled by scaling parameters. It was observed that the model shows good result for a linearly growing velocity of injection (Abani et al. 2008).

For reducing grid-dependency, the results obtained from jet theory of gas are presented with a Lagrangian model. The enhanced spray model is applied in the engine simulation code KIVA-3V and checked under several conditions and compared with LDEF. It is observed that the improved model shows lesser mesh dependency (Abani et al. 2008).

CFD fuel spray models consider the fuel spray formation, break-up, atomization, evaporation, and turbulence. It facilitates the prediction of both macroscopic as well as microscopic fuel spray characteristics, such as spray cone angle, penetration, atomization, and SMD in time-based and spatial scale. However, CFD fuel spray models are expensive and require more computational time (Shi et al. 2011).

3.2.3 Cavitation in Injector

The physical approaches regarding spray performance such as liquid break-up, penetration, spray volume, velocity scattering, and air entrainment have been analyzed by Arai (2012). Cavitation in an injector, break-up process, development of spray, and atomization phenomenon and then combustion process contribute to the scientific and engineering progresses in hydrodynamics. The physical significance of turbulence initiated because of cavitation bubble disruption. It may be the origin of atomization. However, it is also reported that origin of atomization still unknown physics (Arai 2012). Cavitation used some kinetic energy and might alter some properties of fuel. Cavitation number given in Eq. (3.10) is good index of cavitation but need to find some relation between cavitation and atomization.

The static pressure is defined using Blasius's equation of the wall friction as given in Eqs. (3.10–3.11)

$$P_2 = P_a + \left\{ \frac{1}{2} f \left[\frac{(L - L_c)}{D} \right] \times V_i^2 \right\} \tag{3.10}$$

$$f = \frac{0.316}{Re^{0.25}} \tag{3.11}$$

The velocity V_1 is calculated by using Eq. (3.12)

$$V_1 = \left(\frac{D}{D_c} \right)^2 \times V_i \tag{3.12}$$

And P_1 is calculated using Eq. (3.13)

$$P_1 = P_2 + \left\{ \frac{1}{2} \rho \times V_i^2 \left[1 - \left(\frac{D}{D_c} \right)^4 \right] \right\} \tag{3.13}$$

Then, the cavitation K_c is expressed as given in Eq. (3.14)

$$K_c = \frac{P_1 - P_v}{\frac{1}{2} \rho V_1^2} \tag{3.14}$$

Using this model (Fig. 3.5), cavitation number coupling with nozzle friction f and back pressure Pa can be obtained as given in Eq. (3.15)

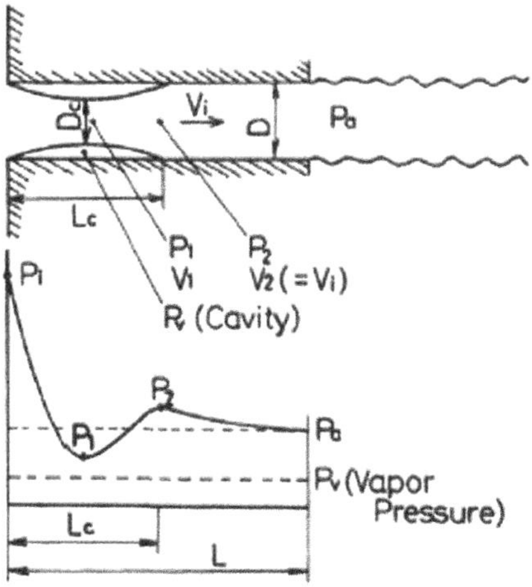

Fig. 3.5 Cavitation model in a nozzle (Arai 2012)

$$K_c = \left(\frac{D_c}{D}\right)^4 \left\{ \frac{P_a - P_v}{\frac{1}{2}\rho V_i^2} + f\frac{(L - L_c)}{D} + 1 \right\} - 1 \tag{3.15}$$

The transparent scale-up nozzle has been used to observe the atomization phenomena related with cavitation (Shimizu et al. 1988). It was observed that smooth jet coupled with no cavitation was hardly atomized. As the velocity increased beyond a certain limit, cavitation stabled at the entrance. Disruption of the cavity produced extreme turbulence that could disturb the jet surface near exit. This condition was generally observed in an atomization process of a high-speed liquid jet (Hiroyasu et al. 1991).

3.3 Thin Spray and Thick Spray

In thin or dilute spray models volumetric displacement effects ($\varphi \neq 1$) can be neglected, as can droplet collisions, coalescence, and breakup; the interactions between individual droplets can be disregarded or highly simplified (Marble 1970). These assumptions correspond to dilute sprays for which the droplet loading, and correlations can be derived to estimate the effects of the droplets on the gas phase equation for the turbulence kinetic energy and its rate of dissipation (Faeth 1987).

The volumetric displacement effects which were considered in thick or dense spray models seem to have originated with Dukowicz's work (Dukowicz 1980), in which a Lagrangian–Eulerian formulation was applied to calculate the behavior of atomized, non-evaporating liquid sprays introduced into a vaporous environment. Since a Lagrangian formulation was employed to calculate the droplet trajectories, numerical diffusion was minimized.

3.4 Empirical Models/Correlations for Injection and Spray Characteristics

The analysis of fuel injection and spray can be done using experimental and models approach. The fuel injection and spray models required less cost and time for the development of new diesel engine. It can also use to simulate the extreme circumstances that could be difficult to achieve using experimentations. Hence, in real-world engineering uses, fuel injection and spray models are commonly favored in the development and further research in engines (Liu et al. 2020).

3.4.1 Injection Delay

The injection delay can be measured from the in-line pressure verses crank angle plot as it is time for pressure wave to reach the nozzle holes for a connecting pipe length of L_p of the injector. The injection delay model has been developed by Rakopoulos et al. (2006, 2007) as shown in Eq. (3.16). The speed of sound (a_s) is vital for calculating the performance of fuel injection for diesel engines (Rakopoulos et al. 2007).

$$\Delta\tau_{injdel} = \left(\frac{L_p}{a_s}\right)6N \tag{3.16}$$

where, speed of sound (a_s) can be calculated by using Eq. (3.17).

$$a_s = \sqrt{\frac{K_{bm}}{\rho_l}} \tag{3.17}$$

3.4.2 Dynamic Injection Timing (DIT) or Start of Injection (SoI)

Dynamic injection timing (Rakopoulos et al. 2006, 2007) can be predicted by adding injection delay to well-known static injection timing as given in Eq. (3.18).

$$DIT = \Delta\tau_{injdel} + \tau_{static} \tag{3.18}$$

The pressure wave magnitude in the nozzle holes of the injector can be predicted as per Eq. (3.19).

$$\Delta p_w = a_s \rho_l c_{pump} \frac{F_{pump}}{F_{nozzle}} \tag{3.19}$$

where, F_{pump}, F_{nozzle} and c_{pump} are pump barrels cross sectional areas, nozzle holes cross sectional areas and variation of plunger speed respectively.

The velocity of jet from each nozzle is given in Eq. (3.20).

$$u_{inj} = c_d \sqrt{\frac{2\Delta p_{inj}}{\rho_l}} \tag{3.20}$$

The wall friction loss (f), and the discharge coefficient (c_d) which closely related to the inlet loss coefficient (K_{inlet}) is calculated by Eqs. (3.21–3.22).

$$c_d = \left(K_{inlet} + f \cdot \left(\frac{L}{D_n}\right) + 1\right)^{-0.5} \tag{3.21}$$

$$f = \max\left(0.316 Re^{-0.25}, \frac{64}{Re}\right) \tag{3.22}$$

The inlet loss coefficient (K_{inlet}) can be tabulated according to the R/D_n ratio. 'L' is length and 'R' is the orifice inlet radius and 'D_n' is the hole diameter.

Then the mean fuel injection flow rate (kg/CA) can be predicted by using Eq. (3.23).

$$m_{finj} = \frac{\pi D_n^2}{4} \frac{\rho_l u_{inj}}{6N} \tag{3.23}$$

The total duration of injection is the time from start of injection to the end of injection. The total duration of injection can be predicted by using Eq. (3.24).

$$\Delta \tau_{inj} = \frac{m_{ftot}}{n} \frac{1}{m_{finj}} \tag{3.24}$$

where, n is the number of holes of nozzle.

3.4.3 *Spray Break-Up Length*

Since the rich fuel is available near the nozzle hole and the injection velocity is also too high so that observation of breakup process is impossible (Hamidi et al. 1990). The detailed mechanism of fuel breakup process is still unclear matter for diesel spray research. However, to satisfy the demand of numerical simulation, some models have been proposed (Arai 2012).

Table 3.1 Available models of spray break-up length

S. No.	Break up length models	References
1	$L_b = 15.8 \times (\rho_l/\rho_a)^{0.5} \times D_n$	Dan et al. (1997)
2	$L_b = 8.14 \times D_n \times (We)^{0.5}$	Siebers (1999)

The difference of pressure across the nozzle dictates the state of the liquid jet emerging from the nozzle. The distance, the jet moves from the nozzle outlet before breaking up in a spray is generally termed as spray break-up length. A lesser spray break-up length specifies better droplet size dispersal, better air entrainment and lower ignition delay. The break-up length is influenced by nozzle diameter, air density and, turbulence and physical characteristics of fuel. Two models are available for the spray break-up length of diesel fuel (Dan et al. 1997; Siebers 1999) as given in Table 3.1.

3.4.4 Spray Cone Angle

Near nozzle due to dense liquid cone angle is very small. The aerodynamic interaction with air and fuel stream, the injected fuel disintegrates results in higher cone angle (Varde and Popa 1983). The spray appears as a cone in shape with vertex of the injector side and the solid angle of the cone is called fuel spray cone angle. Higher cone angle is always advantages for better air utilization and atomization point of view (Araneo et al. 1999). But too large cone angle may have disadvantages like fuel impingement on side wall of the crown. With the increase in the pressure difference across the nozzle, spray break-up length decreases and spray cone angle increases, until the apex of the cone particularly co-insides of the orifice outlet. Available models of spray cone angle are given in Table 3.2.

Table 3.2 Available models of spray cone angle

S. No.	Models for spray cone angle	References
1	$\theta = 0.03*(L_n/D_n)^{-0.3}(\rho a/\rho_l)^{0.1} * Re^{0.7}$	Varde and Popa (1983)
2	$\theta = 83.5 * (L_n/D_n)^{-0.22} (\rho_a/\rho_l)^{0.26} * (D_n/d_n)^{0.15}$	Varde and Popa (1983)
3	$\theta = 0.135 * (L_n/D_n)^{-\beta} * (\rho_l/\rho_a) * Re^{0.46}$ and $\beta = 0.0284 (\rho_l/\rho_a)^{0.39}$	Araneo et al. (1999)
4	$\tan(\theta/2) = (1/A)\, 4\pi(\rho_a/\rho_l)^{0.5} * 3^{0.5}/6$; and $A = 3 + (L_n/D_n)/3.6$	Araneo et al. (1999)
5	$\tan(\theta/2) = A * (\rho_a/\rho_l)^{0.5}$ and $A = 0.49$	Araneo et al. (1999)
6	$\tan(\theta/2) = (1.38 * 10^{-7} * V_i + 3.63 * 10^{-3}) * \rho_a + 2.86 * 10^{-5} * V_i + 1.88 * 10^{-3}$	Ishikawa and Zhang (1999)
7	$2\theta = 0.05 (\rho_a * \Delta P * D_n^2/(\mu_a)^2)^{0.25}$	Arai et al. (1984)

3.4.5 Sauter Mean Diameter (SMD)

The sauter mean diameter is suggestive droplet diameter corresponding to same surface of spray and could show the average evaporation characteristics of fuel spray (Hiroyasu and Kadota 1974). SMD characterize a single droplet with the same ratio of volume to surface area as the respective quantity's ratio integrated over the whole droplet size dispersal as shown mathematically in Eq. (3.25). Typical empirical equations of the SMD of fuel spray are as follows given in Table 3.3.

The injection pressure and injection rate may diverge during the injection duration. Accordingly, drop size distribution at a given situation may also vary with time during the injection duration. The drop size study with continuous spray have been examined in order to find the effects of atmospheric pressure, injection pressure, nozzle diameter and the ratio of the nozzle length to the diameter, and also effects of viscosity and surface tension of the fuel. An optical technique based on Fraunhofer diffraction has been developed for on-line measurement of drop size distribution (Hiroyasu and Arai 1989).

In diesel engine atomization must be completed before the start of combustion. Basically, atomization is the middle stage of injected fuel and its subsequent combustion. The rate of atomization depends upon the relative movement of the droplets to

Table 3.3 Available models of sauter mean diameter (SMD)

S. No.	SMD models	References
1	$X_{32} = A^*(\Delta P)^{-0.135} * (\rho_a)^{0.121} * (Q)^{0.131}$ and A = 25.1 for pintle nozzle, A = 23.9 for hole nozzle, and A = 22.4 for throttling pintle nozzle	Hiroyasu and Arai (1989)
2	$X_{32} = \text{Max}\ (X32LS, X32HS)$ $X_{32}{}^{LS} = 4.12 * D_n * Re^{0.12} * We^{-0.75} * (\mu_l/\mu_a)^{0.54} * (\rho_l/\rho_a)^{0.18}$ $X_{32}{}^{HS} = 0.38 * D_n * Re^{0.25} * We^{-0.32} * (\mu_l/\mu_a)^{0.37} * (\rho_l/\rho_a)^{-0.47}$	Hiroyasu and Kadota (1974)
3	Complete Spray: $X_{32} = 0.14 * D_n * Re^{0.25} * We^{-0.32} * (\mu_l/\mu_a)^{0.37} * (\rho_l/\rho_a)^{0.17}$ $X_{32} = 47 * 10^{-3} * (D_n/Vi) * (\sigma/\rho_a)^{0.25} * g^{0.2} * (1 + (33.1 * 101.5 * \mu_l/(\sigma * \rho_l * D_n)^{0.5}$ Ambient pressure: $X_{32} = 0.38 * D_n * Re^{0.25} * We^{-0.32} * (\mu_l/\mu_a)^{0.37} * (\rho_l/\rho_a)^{-0.47}$	Hiroyasu and Kadota (1974)
4	$X_{32} = ((585 * \sigma^{0.5})/(V_i * \rho_l{}^{0.5})) + 597(\mu_l/(\sigma * \rho_l)^{0.5})^{0.45}$	Lichty and Streeter (1939)
5	$SMR = C * (4 * \Pi * \sigma * 3)/(\rho_a * V_i{}^2 * 2)$, where, C = 0.75	Kuo and Yu (1984)
6	$SMD = 7.3 * \sigma^{0.6} * \nu^{0.2} * m^{0.25} * \Delta p^{-0.4}$	Kamimoto and Matsuoka (1977)
7	$X_{32} = C * 83.2 * (D_n/V_i) * (\sigma * g/\rho_a)^{0.25}\ ((1 + 10.4 * \mu_l)/(\sigma * \rho_l * D_n)^{0.5})$ C = 0.721 for Dn = 0.2 and, C = 0.455 for D_n = 0.3	Kamimoto and Matsuoka (1977)

the surrounding air. It would appear that the fine droplets, larger surface area for the same amount of fuel injected which is subjected to high pressure rise rate during the premixed combustion duration (Hamidi et al. 1990).

$$SMD = \frac{\sum_{i=n}^{N_{drops}} d_i^3}{\sum_{i=n}^{N_{drops}} d_i^2} \quad (3.25)$$

3.4.6 Spray Penetration

The spray penetration signifies the spray depth, and it implies the distance that the spray followed before the vaporization of fuel droplets. The contact area of air and fuel can be used to determine the fuel spray penetration. The fuel spay penetration affects the air entrainment process and the mixture uniformity, and subsequently the combustion process. While analyzing the length of fuel spray penetration, it needs special consideration to avoid the spray wall impingement. Certainly, in some engine's impingement is acceptable due to presence of hot surface and greater swirl effect. Else, the over fuel spray penetration would result in soot deposition, unburned hydrocarbon, and partial combustion. However, lower spray tends to less air entrainment may lead to the undesirable combustion and subsequently, poor engine performance and emission. Therefore, the research on fuel spray penetration is essential phase for the optimization of combustion of a diesel engine (Liu et al. 2020). While designing the combustion chamber, penetration is given higher importance in connect with the size of combustion chamber in diesel engine. Pressure difference across the nozzle, nozzle diameter and air density mainly influence the penetration.

It is also well established that the NOx emission always forms at the fuel spray periphery due to lean air–fuel mixture and soot emission forms at the core of the spray (Hamidi et al. 1990). Therefore, spray penetration may be one of the good tools to optimize engine emissions and achieve the emission norms.

Wakuri et al. (1960) established a model for fuel spray penetration and resolved that the penetration is directly proportional to the square root of time and also depends on the fuel spray cone angle. The empirical correlations for spray break-up and fuel spray penetration developed from the available experimental data (Nijdam et al. 2006). Before and after the break-up has been considered for this model and accordingly two different fuel sprays penetration models are developed. These fuel spray penetration models are extensively used in phenomenological modeling as well as in CFD model approach. Further, Arai (2012) modified these models by incorporating non-dimensional scales of time and fuel spray penetration length. Dent and Mehta (1981) observed the reduction in fuel spray penetration as they have considered gas temperature in their early phenomenological model. Available models are given in Table 3.4.

Table 3.4 Available models of spray penetration

S. No.	Models of spray penetration	References
1	$S = (2 * Ca * \Delta P/\rho_a)^{0.25} * (D_n * t/\tan(\theta/2)^{0.5})$	Wakuri et al. 1960)
2	$S = 0.39 * (2 * \Delta P/\rho_f)^{0.5} * t$ for $0 < t < t_b$ $S = 2.95 * (\Delta P/\rho_a)^{0.25} * (D_n * t)^{0.5}$ for $t > t_b$ And $t_b = 28.65\{(\rho_f * D_n)/(\Delta P * \rho_a)^{0.5}$	Varde and Popa (1983)
3	$S^* = t^*$ for $0 < t^* \leq 1$ $S^* = (t^*)^{0.5}$ for $1 < t^*$ Where $t^* = (t/t_b)$ and $S^* = (S/L_b)$ and $L_b = 15.8\, D_n\, (\rho_f/\rho_a)^{0.5}$	Arai (2012)
4	$S = 3.07 * (\Delta P/\rho_a)^{0.25} * (D_n * t)^{0.5} * (294/T_g)$	Dent and Mehta (1981)

3.4.7 Air Entrainment

The phenomenon of air entrainment involves drawing of air into the spray of fuel, which helps to maintain the homogeneity of the mixture into the chamber (Lippert et al. 2005). In the mixing process of air fuel, air entrainment has a crucial role, and it primarily function of cone angle of spray, and the distance traveled by fuel spray (Ishikawa and Zhang 1999). The injection system affects the mixture formation and air entrainment process. The mixture formation process could significantly improve by fine-tuning of injection system, and optimization of spray characteristics, such as break-up length, spray penetration, and spray cone angle (Chikahisa and Murayama 1995). The soot emission is controlled by air entrainment process because rich core of fuel is mainly responsible for generation of soot in diesel engine (Lahane and Subramanian 2015). The available models for air entrainment are given in Table 3.5.

As the fuel spray development with constant fuel spray cone angle and conservation of fuel spray momentum are maintained, deceleration of fuel spray tip velocity (= dS/dt) was produced by mass of injected fuel spray by air entrainment. Then, following equations (Eqs. 3.26–3.27) can be derived to predict the air entrainment (Hiroyasu et al. 1983).

$$\int_t \frac{dM_f}{dt} V_{inj} dt = \int_t \left(\frac{dM_f}{dt} + \frac{dM_a}{dt} \right) \frac{dS}{dt} dt \tag{3.26}$$

Table 3.5 Available models of air entrainment

S. No.	Models of air entrainment	References
1	$ma = 0.26 * (\rho_a * \rho_l)^{0.5} * D_n * S * V_i * \tan(\theta/2)$	Siebers (1999)
2	$ma = 40 * V^{0.5} * D_n^{-0.5} * t^{1.5} * mf$	Shigeru et al. (1999)
3	$ma = -(mf * V_i/(dS/dt)^2) * d^2S/dt^2$	Chikahisa and Murayama (1995)
4	$ma = (\pi/3) * (\tan(\theta/2))^2 * S^3 * \rho_a$	Rakopoulos et al. (2004)

$$\int_{M_f} \left(\frac{V_{inj}}{dS/dt} - 1 \right) dM_f = M_a \tag{3.27}$$

The Eq. (3.27) is an estimation base of air entrainment and used as a principal concept of phenomenological diesel combustion modeling (Hiroyasu et al. 1983).

3.4.8 Vaporization

The injected fuels particles vaporize due to existence of hot air ensure the formation of potential flammable mixture (Siebers 1999). The vaporization process mainly depends on droplet size, and precisely the vaporization percentage is directly proportional to square of diameter of fuel droplet (SMD) (Abraham 1996). At higher values of SMD, the time for vaporization process would increase, causing low percentage of vaporization (Lahane and Subramanian 2015). Therefore, SMD is a major factor for process of vaporization. Vaporization time is the main influencing factor for physical ignition delay period. Physical characteristics of the fuel and temperature and state of air greatly influence the vaporization process (Urbán et al. 2020; Laryea and No 2004). The vaporization rate can be obtained by using Eqs. (3.28–3.30).

$$\frac{dD}{dt} = -\frac{K}{2} D \tag{3.28}$$

$$K = 8 \times \lambda_g \times \frac{\ln(BT + 1)}{\rho_l \times C_{pg}} \tag{3.29}$$

$$BT = C_{pg} \times \frac{(T - T_s)}{L} \tag{3.30}$$

3.4.9 Spray Models with Swirl

Induction swirl can be produced by introducing a tangential directional movement to the air into the cylinder based on either by blend of the tangential flows with a pre-swirl flow or inducing a pre-swirl in the induction port (Laryea and No 2004; Shigeru et al. 1999). The angular rotational speed of air about the cylinder axis is called air swirl (Faeth 1987). The crankshaft angular rotational speed can be conveniently related to this air rotational speed as swirl ratio (Eq. 3.31).

$$Swirl\,ratio = \frac{Air\,rotational\,speed}{Crankshaft\,rotational\,speed} \tag{3.31}$$

Table 3.6 Spray models with swirl

Characteristics	Models	References
Air-entrainment	$m_S = (\pi/3) * (\tan(\theta_S/2)) * \rho_a * X_S{}^3$	Faeth (1987)
Penetration distance	$X_S = X * (1 + \pi * R_S * N * X/30 * U_{inj})^{-1}$	Faeth (1987)
Cone angle of spray	$\theta_S = \theta\ (X/X_S)$	Rakopoulos et al. (2004)

Faeth (1987) has developed models for cone angle of spray, air entrainment with swirl, and penetration distance presented in Table 3.6.

Phenomenological spray models streamline the physical process of the fuel spray formation and critical relationship for prediction of macroscopic spray characteristics. It performs instantaneous prediction, and further recognizes the impact of key parameters on the fuel spray formation. However, it may not offer the facts of the microscopic fuel spray characteristics such as spray break-up length, SMD and atomization process.

3.5 Selection of Suitable Empirical Spray Models

It has clearly been seen from Sect. 3.4 that numerous models are presented for the estimation of diesel fuel spray characteristics. It is important to note that no model is available to estimate the spray characteristics of any other alternative fuel such as biodiesel or its blends. Hence, the biodiesel blend spray characteristics are essential and it is anticipated using the reported correlations/models related to base diesel fuel. Lahane and Subramanian (2013a) validated these available models with experimental data available in literature at same condition and assumptions. Those model results are matching with experimental data, that models are selected for other alternative fuels (Som et al. 2010; Lahane and Subramanian 2015; Rakopoulos et al. 2006, 2007; Subramanian and Lahane 2013a). The selected models are shown in Table 3.7.

It is stated that the models mentioned in Table 3.7 can be used for any other alternative fuels for predicting the spray characteristics (Rakopoulos et al. 2006, 2007; Subramanian and Lahane 2013b). The measured experimental data such as fuel density, air density, fuel viscosity and surface tension, air viscosity, measured fuel pressure and combustion pressure have been incorporated in the selected spray models mentioned in Table 3.7 for predicting the spray for biodiesel and pure diesel (Lahane and Subramanian 2015; Hamidi et al. 1990). The measured injection pressure with respect to crank angle and in-cylinder pressure with respect to crank angle along with other required engine parameters and measured fuel properties have been used to predict the injection and spray characteristics with respect to crank angle using selected models shown in Table 3.7 of a diesel engine (Lahane and Subramanian 2015; Subramanian and Lahane 2013b).

Table 3.7 Selected spray models based on experimental data

Characteristics	Models
Break-up length	$L_b = 15.8 \times (\rho_l/\rho_a)^{0.5} \times D_n$
Spray cone angle	$\tan(\theta/2) = (1/A) \times 4\pi \times (\rho_a/\rho_l)^{0.5} \times (3^{0.5}/6)$ where, $A = 3 + ((L_n/D_n)/3.6$
Sauter mean diameter	$X_{32} = \text{Max}(X_{32}^{LS}, X_{32}^{HS})$ $X_{32}^{LS} = 4.12 \times D_n \times Re^{0.12} \times We^{-0.75} \times (\mu_l/\mu_a)^{0.54} \times (\rho_l/\rho_a)^{0.18}$ $X_{32}^{HS} = 0.38 \times D_n \times Re^{0.25} \times We^{-0.32} \times (\mu_l/\mu_a)^{0.37} \times (\rho_l/\rho_a)^{-0.47}$
Spray penetration	$S = 2.95 \times (\Delta P/\rho_a)^{0.25} \times (D_n \times t)^{0.5}$ for $t > t_b$ and $S = 0.39 \times (2 \times \Delta P/\rho_a)^{0.5} \times t$ for $t < t_b$
Air entrainment	$ma = (\pi/3) \times (\text{Tan}(\theta/2))^2 \times S^3 \times \rho_a$
Vaporization	$dD/dt = -K/2 \times D$ where, $K = 8 \times \lambda_g \times \ln(BT + 1)/(\rho_l \times C_{pg})$, $BT = C_{pg} \times (T - T_s)/L$

3.6 Effect of Injector Parameters/Fuel Properties on Injection and Spray Characteristics

3.6.1 Effect of Injector Parameters

It is well established that to reduce the emissions, injection system plays vital role. There is a rising inclination towards the electronic injection, higher injection pressures like in CRDI and smaller nozzle sizes (Gupta et al. 2000). It was observed that increase in plunger diameter would increase injection rate, injection pressure and advance the dynamic injection timing (Reddy and Ramesh 2006). It can be clearly seen from the available spray models that higher injection pressure would give better fuel injection and spray characteristics (Hiroyasu and Kadota 1974; Hiroyasu and Arai 1989; Lichty and Streeter 1939; Kuo and Yu 1984; Kamimoto and Matsuoka 1977). It was observed that increase in injection pressure results in decrease in droplet diameter (Hiroyasu and Kadota 1974; Hiroyasu and Arai 1989). On the other hand, higher injection pressure some time may lead to fuel wall impingement results in CO, HC, and smoke emission formation (Urbán et al. 2020; Laryea and No 2004). But this problem of wall impingement can be weakened by increasing the nozzle holes number, decreasing the diameter of holes and by implementing swirl in case of higher injection pressure such as CRDI system (Reddy and Ramesh 2006; Subramanian and Lahane 2013b; Lahane and Subramanian 2012). The increase in nozzle holes number would result in reduction in fuel quantity per hole. The reduction in fuel quantity results in less fuel momentum which further result in lower spray penetration and hence no wall impingement problem (Lahane and Subramanian 2012).

Increase in nozzle hole number increases the number of sprays in the combustion chamber which may ensures the more homogeneity in the combustion chamber results in better combustion and higher performance of the engine. However, larger droplet diameter observed with groped-hole nozzle due to collision and coalesces of small droplets which further may reduce the mixing process (Suh et al. 2009).

Decrease in nozzle hole diameter would give lower sauter mean diameter which ensures the lower vaporization time and early start of combustion (Bergstrand and Denbratt 2001).

3.6.2 *Effect of Fuel Properties*

It is well known that the fuel properties like viscosity, surface tension, density, bulk modulus, latent heat of vaporization, thermal conductivity, specific heat capacity, boiling point and heat of combustion influence the atomization process in a diesel engine (Deeparaj et al. 2016). Fuel mass flow rate and injection velocity at the nozzle is intensely influenced by the change in fuel properties. Fuel properties are also intensely influence the cavitation region in nozzle than injection parameters (Deeparaj et al. 2016). The higher fuel density, fuel viscosity, bulk modulus and surface tension would increase the break-up length, sauter mean diameter, spray penetration, vaporization time and decrease the spray cone angle results in lower air entrainment (Rakopoulos et al. 2006, 2007; Lahane et al. 2010). It was observed that break-up length, SMD and penetration is proportional to the fuel viscosity and density. Therefore increase in fuel viscosity and density tends to increase in break-up length, SMD and penetration (Callahan et al. 1987). It is stated that 100% biodiesel has poor fuel injection and spray characteristics because of its higher density (about 900 kg/m^3), viscosity (about 5.8 cSt) and surface tension (about 0.035 N/m) as compared to diesel fuel (820 kg/m^3, 2.64 cSt, 0.02 N/m respectively) (Rakopoulos et al. 2006, 2007). The biodiesel can be used in a better way injection and spray characteristics point of view if it used with diesel in lower blend ratio (up to B20) (Subramanian and Lahane 2013a; Reddy and Ramesh 2006; Lahane et al. 2010). The higher biodiesel blend can also be used in a diesel engine with little modification in injection system (injection pressure and plunger size) (Gupta et al. 2000) and injector parameters such nozzle hole size, number of nozzle holes and L/D ratio (Subramanian and Lahane 2013b). The problem poor atomization characteristics with higher density, viscosity, surface tension and bulk modulus fuel can be overcome by adding alcohols, ethers, esters, and dimethyl ether (Zhang et al. 2020).

3.7 Conclusions

The advanced injection system such as high injection pressure, injection timing, injection rate shaping, and split injections results in shorten the injection duration may lead to end of injection before ignition. The split or multiple injections may lead to interaction between adjacent sprays changes the spray formation. Therefore, it is required to simulate and study the injection system thoroughly before implementation in actual system.

It is very difficult task to notice the actual fuel spray in a diesel as it is high-speed transient occurrences in high temperature and pressure. However, fuel spray formation partially clarified with its fundamental studies. The good combustion chamber is always designed by keeping in mind the fuel spray characteristics. The practical approach for break-up and atomization of liquid fuel injected are still unclear phenomena.

Liquid surface immovability problem in breakup process has leading many liquid disintegrating concepts and their numerical simulation methods. Measurement of a fuel spray encouraged new scientific accomplishment of laser diagnostics. These accomplishments usually combine with physical concerns of fuel spray. Then finding of unknown physics behind fuel spray is the necessary of next approach of spray.

The study of liquid beam of fuel near the nozzle exit is still difficult due to rich mixture and droplet clouds. It is required to use the CFD models with high end technology to evolve more about microscopic and macroscopic spray characteristics. Based on the literature review on injection and spray characteristics, following conclusions are drawn.

- It is observed that many models of injection and spray characteristics are available for diesel fuel only.
- Most of the spray characteristics models are based on injector parameters and fuel properties.
- Higher density (about 900 kg/m^3), viscosity (about 5.8 cSt) and surface tension (about 0.035 N/m) would lead to poor injection and spray characteristics.
- Higher break-up length (about 14 mm near TDC) and lower spray cone angle (about 12° near TDC) results in poor air entrainment which leads to poor combustion of fuel.
- Over penetration may lead to wall impingement problem and lower penetration may lead to poor air utilization. Therefore, it is necessary to optimize the spray penetration for better utilization of air and hence combustion.
- There is no single model available of spray, and injection characteristics for biodiesel or any other alternative fuels.

On the whole, it may be concluded that the generalized version of the FLA need to be developed which may be applicable to a wide range of turbulent phase. Traditionally the development of random droplet shapes has been modeled using Volume of Fluid (VOF) methods, but the application of these methods in CFD codes would be impractical. Even in case of slightly deformed spheroidal droplets the model available is not ideal for implementation in CFD codes. The available models are suitable only for heating and evaporation of droplet in sub-critical condition. However, the interface between liquid and gas in supercritical condition is still challenge.

References

Abani N, Reitz RD (2007) Unsteady turbulent round jets and vortex motion. Phys Fluids 19(12):125102

Abani N, Munnannur A, Reitz RD (2007) Reduction in numerical parameter dependencies in diesel spray models. ICEF2007-1667. In: Proceedings of ASME internal combustion engine division, Charleston, South Carolina, USA

Abani N, Kokjohn S, Park SW, Bergin M, Munnannur A, Ning W, Sun Y, Reitz RD (2008) An improved spray model for reducing numerical parameter dependencies in diesel engine CFD simulations. In: 2008 world congress Detroit, Michigan, SAE 2008-01-0970

Abraham J (1996) Entrainment characteristics of transient jets. Numer Heat Transf Part A 30:347–364

Abraham J (1997) What is adequate resolution in the numerical computation of transient jets. SAE paper 970051

Amsden AA, Ramshaw JD, O'Rourke PJ, Dukowicz JK (1985) KIVA: a computer program of two- and three-dimensional fluid flows with chemical reactions and fuel sprays. Report No. LA-10245-MS, Los Alamos National Laboratory, Los Alamos

Arai M (2012) Physics behind diesel sprays. ICLASS 2012. In: 12th triennial international conference on liquid atomization and spray systems, Heidelberg, Germany, 2–6 Sept 2012

Arai M, Tabata M, Hiroyasu H (1984) Disintegrating process and spray characterization of fuel jet injected by a diesel nozzle. SAE J Engines 2:358–371

Araneo L, Coghe A, Brunello G, Cossali GE (1999) Experimental investigation of gas density effect on diesel spray penetration and entrainment. In: International congress and exposition, Detroit, Michigan, SAE No. 1999-01-0525, pp 679–693

Beale JC, Reitz RD (1999) Modeling spray atomization with the Kelvin-Helmholtz/Rayleigh-Taylor hybrid model. Atom Sprays 9:623–650

Bergstrand P, Denbratt I (2001) Diesel combustion with reduced nozzle orifice diameter. SAE technical paper 2001-01-2010

Callahan T, Ryan T, Dodge L, Schwalb J (1987) Effects of fuel properties on diesel spray characteristics. SAE Technical Paper 870533

Chen H, Su X, He J, Xie B (2019) Investigation on combustion and emission characteristics of a common rail diesel engine fueled with diesel/n-pentanol/methanol blends. Energy 167:297–311

Chikahisa T, Murayama T (1995) Theory and experiments on air entrainment in fuel sprays and their application to interpret diesel combustion processes. In: International congress and exposition, Detroit, Michigan, SAE No. 950447, pp 787–797

Dan T, Yamamota T, Senda J, Fujimoto H (1997) Effect of nozzle configurations for characteristics of non-reacting diesel fuel spray. In: International congress and exposition, Detroit, Michigan, SAE No. 970355, pp 581–596

Deeparaj E, Vivek B, Gunasekaran D, Satheeshkumar N, Magudeswaran M, Srinivasan SA (2016) Spray characteristic analysis of diesel injector with biodiesel. South Asian J Eng Technol 2(23):195–205

Dent JC, Mehta PS (1981) Phenomenological combustion model for a quiescent chamber diesel engine, SAE No. 811235, pp 3884–3902

Dukowicz JK (1980) A particle fluid numerical model for liquid sprays. J Comput Phys 35:229–253

Faeth GM (1987) Mixing, transport, and combustion in sprays. Prog Energy Combust Sci 13:293–345

Gupta SE, Sekar R, Poola R (2000) Effect of injection parameter on diesel spray characteristics. SAE technical paper 2000-01-1600

Hamidi AA, Swithenbank J (1990) Diesel engine fuel injection processes and sprays diagnostic methods. In: Weaving JH (ed) Internal combustion engineering: science and technology. Springer, Dordrecht, pp 213–242

Haselman LC, Westbrook CK (1978) A theoretical model for two phase fuel injection in stratified charge engine, SAE 780318

Hiroyasu H, Kadota T, Arai M (1983) Development and use of a spray combustion modelling to predict diesel engine efficiency and pollutant emissions, part 1 combustion modelling. Bullet JSME 26(214):569–575
Hiroyasu H, Arai M (1989) Empirical equations for the Sauter mean diameter of a diesel spray. SAE J Eng 3:868–877
Hiroyasu H, Kadota T (1974) Fuel droplet size distribution in diesel combustion chamber. SAE J Eng 3:2615–2624
Hiroyasu H, Arai M, Shimizu M (1991) Break-up length of a liquid jet and internal flow in a nozzle. In: 5th international conference on liquid atomization and spray systems (ICLASS-91, Gaithersburg, MD, USA), pp 275–282
Ishikawa N, Zhang L (1999) Characteristics of air entrainment in a diesel spray. SAE J Eng 3:644–651
Kamimoto T, Matsuoka S (1977) Prediction of spray evaporation in reciprocating engines, SAE No. 770413, pp 1792–1802
Kozina A, Radica G, Nižetić S (2020) Analysis of methods towards reduction of harmful pollutants from diesel engines. J Cleaner Prod 262:121105
Kuo W, Yu RC (1984) Modeling of transient evaporating spray mixing processes effect of injection characteristics, SAE No. 840226, pp 16–29
Lahane S, Subramanian KA (2012) Modelling and CFD simulation of effects of spray penetration on piston bowl impingement in a DI diesel engine for bio-diesel-diesel blend (B20). ICES2012e81171. In: Proceeding of ASME 2012 ICED spring technical conference, ICES 2012, pp 163–170, Torino, Piemonte, Italy
Lahane S, Subramanian KA, Babu MKG (2010) Comparative evaluations of injection characteristics of a diesel engine for use of biodiesel-diesel blends. ISFL 2010, SAE Number 2010-28-0008, New Delhi, India, pp 35–40
Lahane S, Subramanian KA (2015) Effect of different percentages of biodiesel–diesel blends on injection, spray, combustion, performance, and emission characteristics of a diesel engine. Fuel 139:537–545
Laryea GM, No SY (2004) Spray angle and breakup length of charge-injected electrostatic pressure-swirl nozzle. J Electrostat 60(1):37–47
Lee S, Tanaka D, Kusaka J, Daisho Y (2002) Effects of diesel fuel characteristics on spray and combustion in a diesel engine. JSAE Rev 23:407–414
Lichty LC, Streeter RL (1939) Internal combustion engine, 5th edn. McGraw-Hill Book Company, Inc. Textbook
Lippert AM, Chang S, Are S, Schmidt DP (2005) Mesh independence and adaptive mech refinement for advanced engine spray simulations. SAE technical paper 2005-01-0207
Liu L, Peng Y, Liu D, Han C, Zhao N, Ma X (2020) A review of phenomenological spray penetration modeling for diesel engines with advanced injection strategy. Int J Spray Combustion Dyn 12(1):1–22
Marble FE (1970) Dynamics of dusty gases. Annu Rev Fluid Mech 2:397–446
Nijdam JJ, Guo B, Fletcher DF, Langrish TAG (2006) Lagrangian and Eulerian models for simulating turbulent dispersion and coalescence of droplets within a spray. Appl Math Model 30(11):1196–1211
O'Rourke PJ (1984) Collective drop effects on vaporizing liquid sprays. PhD Thesis, Princeton University, Princeton, NJ
Post S, Iyer V, Abraham J (2000) A study of the near-field entrainment in gas jets and sprays under diesel conditions. ASME J Fluids Eng 122:385–395
Rakopoulos CD, Rakopoulos DC, Giakoumis EG, Kyritsis DC (2004) Validation and sensitivity analysis of a two zone Diesel engine model for combustion and emissions prediction. Energy Convers Manage 45:1471–1495
Rakopoulos CD, Antonopoulos KA, Rakopoulos DC, Hountalas DT (2006) Multi-zone modeling of diesel engine fuel spray development with vegetable oil, biodiesel or diesel fuels. Energy Convers Manage 47:1550–1573

Rakopoulos CD, Antonopoulos KA, Rakopoulos DC (2007) Development and application of multi-zone model for combustion and pollutants formation in direct injection diesel engine running with vegetable oil or its biodiesel. Energy Convers Manage 48:1881–1901
Reddy JN, Ramesh A (2006) Parametric studies for improving the performance of a Jatropha oil-fuelled compression ignition engine. Renew Energy 31:1994–2016
Reitz R (2013) Directions in internal combustion engine research. Combust Flame 160(1):1–8
Sazhin S, Shchepakina E, Sobolev V (2018) Modelling of sprays: recent results and future challenges. IOP Conf Ser J Phys Conf Ser 1096
Sazhin SS, Martynov SB, Kristyadi T, Crua C, Heikal MR (2008) Diesel fuel spray penetration, heating, evaporation and ignition: modelling versus experimentation. Int J Eng Syst Model Simul 1(1):1–19
Schmidt DP, Rutland CJ (2004) Reducing grid dependency in droplet collision modeling. J Eng Gas Turbines Power 126:227–233
Shi Y, Ge H, Reitz R (2011) Computational optimization of internal combustion engines. Springer
Shigeru S, Kakegawa T, Tsujimura K, Kobayashi S (1999) The effect of injection parameters and swirl on diesel combustion with high pressure fuel injection. SAE No. 910489, pp 793–805
Shimizu M, Arai M, Hiroyasu H (1988) Disintegrating process of a high-speed liquid jet. J Japan Soc Mech Eng (JSME):2236–2244
Shuai S, Abani N, Yoshikawa T, Reitz RD, Park SW (2009) Simulating low temperature diesel combustion with improved spray models. Int J Therm Sci 48:1786–1799
Siebers DL (1999) Scaling liquid phase fuel penetration in diesel sprays based on mixing limited vaporization. In: international congress and exposition, Detroit, Michigan, SAE No. 1999-01-0528, pp 703–728
Som S, Longman DE, Ramírez AI, Aggarwal SK (2010) A comparison of injector flow and spray characteristics of biodiesel with petrodiesel. Fuel 89:4014–4024
Subramanian KA, Lahane S (2013a) Comparative assessment of injection, combustion, performance and emission characteristics of a diesel engine for biodiesel–diesel blends. Int J Renew Energy Technol 3(4):410–429
Subramanian KA, Lahane S (2013b) Comparative evaluations of injection and spray characteristics of a diesel engine using Karanja biodiesel–diesel blends. Int J Energy Res 37:582–597
Suh HK, Park SW, Lee CS (2009) Effect of grouped-hole nozzle geometry on the improvement of biodiesel fuel atomization characteristics in a compression ignition engine. Proc IMechE Part D J Automob Eng 223(12)
Urbán A, Katona B, Malý M, Jedelský J, Józsa V (2020) Empirical correlation for spray half cone angle in plain jet airblast atomizers. Fuel 277:118197
Varde KS, Popa DM (1983) Diesel fuel spray penetration at high injection pressures. SAE J Eng 2:265–278
Wakuri Y, Fujii M, Amitani T (1960) Studies on the penetration of fuel spray in a diesel engine. Bull JSME 3:123–130
Westbrook CK (1977) Three dimensional numerical modeling of liquid sprays. In: 16th symposium on combustion. The combustion Institute, pp 1517–1526
Williams FA (1985) Combustion theory, 2nd edn. Benjamin-Cummings, Menlo Park, Calif, pp 446–484
Zhang P, Su X, Yi C, Chen H, Xu H, Geng L (2020) Spray, atomization and combustion characteristics of oxygenated fuels in a constant volume bomb: a review. J Traff Transp Eng (english EdN) 7(3):282–297

Chapter 4
Spray Breakup Modelling for Internal Combustion Engines

Utkarsha Sonawane and Avinash Kumar Agarwal

Abstract Rising concerns about emissions have led to a significant tightening of pollution norms for internal combustion (IC) engines. High-pressure direct injection (HPDI) technologies have been adopted for most on-road and off-road engines to meet the global demand for clean and efficient powertrains. Higher fuel efficiency, superior combustion, and lower pollutant formation are the characteristic features of the HPDI. The introduction of alternative fuels, modified combustion geometry, and novel combustion concepts demand continuous improvement in fuel injection equipment (FIE). The complicated physics of HPDI and its modelling is an active area of research among researchers and engine developers. Fuel-injected in the combustion chamber breaks up into a spray of fine droplets, evaporating, mixing with ambient air, and forming a fuel–air mixture, greatly affecting the engine combustion and emission characteristics. Therefore, it is necessary to study the fuel breakup phenomenon under different engine conditions comprehensively. Detailed understanding of the spray breakup phenomenon is unavailable due to difficulties in optical access, highly dense sprays, complex processes, etc. However, recent advances in measurement technologies and computational tools have made it feasible for researchers. This chapter attempts to capture widely used spray breakup models and research studies involving IC engines. Fundamental spray breakup and atomization have been discussed at the beginning of the chapter. Subsequently, the basis and fundamentals of popular spray models have been discussed. Finally, the authors have comprehensively discussed the key contributions in sprays to provide an overall idea about the spray models and their application for IC engine studies. Various spray breakup models such as Blob Model, Linear Instability Sheet Atomization (LISA) Model, Kelvin–Helmholtz (KH) Model, Kelvin–Helmholtz-Aerodynamics Cavitation Turbulence (KH-ACT) Model, RT (Rayleigh–Taylor) Model, Hybrid/Modified Kelvin–Helmholtz Rayleigh–Taylor (KH-RT) Model, Taylor Analogy Breakup (TAB) Model, Enhanced TAB breakup model (ETAB) are discussed briefly in this chapter. Towards the end, a summary of the contents of the chapter is provided, which covers highlights and significant observations.

U. Sonawane · A. K. Agarwal (✉)
Design Programme, Indian Institute of Technology Kanpur, Kanpur 208016, India
e-mail: akag@iitk.ac.in

A. K. Agarwal et al. (eds.), *Engine Modeling and Simulation*, Energy, Environment, and Sustainability, https://doi.org/10.1007/978-981-16-8618-4_4

Keywords Spray Breakup model · Eulerian–Lagrangian Method · Atomization · Engine

4.1 Introduction

Increased goods transport to cater to ever-increasing population leads to harmful emissions from the transport sector. A major portion of transport is based on IC engines. IC engines emit harmful pollutants such as NOx, CO, CO_2, HC, and PM. These pollutants degrade the ambient air quality, leading to serious health issues among the people exposed to these pollutants. Therefore, stringent emission regulations have been implemented on the IC engines globally. The emission norms are becoming increasingly stricter with time. Complying with these emission norms is the driving force for the automotive industry and researchers to develop cleaner and more efficient combustion systems. The fuel injection equipment (FIE) is an integral part of engine combustion architecture. Therefore, fundamental understating and knowledge of the fuel injection process would help to achieve cleaner combustion. Spray formation and atomization are crucial processes controlling the IC engine emissions. The fuel-injected in the combustion chamber forms sheet/jet/ligaments (Smirnov et al. 2012; Magnotti and Genzale 2017). Interactions between the surrounding ambient air and fuel jet spray lead to interfacial instabilities and momentum diffusion. The mechanisms involved in these phenomena are complex and influenced by ambient conditions in which fuel is injected. The fuel spray is dense near the nozzle, which dilutes out in the far-field, resulting in multiphase, multiscale, turbulent physical processes in the engine cylinder. Aerodynamic interactions affect the liquid core column. Liquid ligaments form when the liquid surface becomes unstable, creating parent droplets followed by children droplets. Therefore, modelling of spray and jet breakup is one of the most critical steps in the simulation of IC engines (Arcoumanis et al. 1997). The study of droplet deformation is vital as it significantly affects droplet drag and hence trajectory. The instantaneous drag force, which leads to droplet deformation, is given as:

$$F_D = \frac{1}{2}\rho_a U_0^2 C_D \frac{\pi d_{cro}^2}{4}$$

Here, C_D. is the instantaneous drag coefficient, and d_{cro} is deformed droplet diameter. The equation requires information of C_D as a function of deformation and d_{cro} as a function of time.

Spray modelling includes macroscopic and microscopic events, including the spatial distribution of fuel vapour and droplets. The total surface area of fuel droplets contributes to the fuel evaporation rate, which further influences the combustion reactions. Misinterpretation of droplet size distribution may lead to inaccurate anticipation of fuel evaporation rate, fuel vapour distribution, and mixture formation, resulting in an increct prediction of combustion reactions. The complexity in understanding spray breakup is due to the interdependency of multiple processes and mechanisms (Stiesch 2003). These processes include spray jet breakup, heat and mass transfer, evaporation, droplet to droplet interactions, ambient gas forces, fluid properties, droplet deformation, collision, coalescence, etc. Failure to predict any of the above processes may lead to misinterpretation of successive processes. CFD simulation is a powerful tool for investigating such complex processes and mechanisms. The simulation assumes boundary conditions based on the experiments and simplifies the actual physics behind these phenomena. CFD evaluates complex equations and solves them numerically to capture complex flow phenomena. Also, improvement in computation power and capability opens up doors for new research and technology development (Helldorff and Micklow 2019). CFD analysis of spray is an integral part of combustion research and improves the interpretation of complicated in-cylinder processes. Practical and effective application of modelling is related to the present understanding of the physical process involved.

Numerical modelling and simulations can be used to explore the grey and less understood areas of engine research. This chapter aims to provide an overview of CFD spray breakup models used in IC engines. An attempt has been made to present the fundamental concepts and assumptions in various spray breakup models in a simplified manner. This chapter is mainly divided into three parts. First, detailed information of primary and secondary breakup phenomena has been presented. Then the authors briefly discussed primary and secondary breakup models in the second part. The third part is dedicated to spray breakup modelling literature, covering the application and accuracy of the spray models for various applications. Finally, a summary is presented towards the end of this chapter.

4.2 Fundamentals of Liquid Spray Atomization

Spray atomization is categorized into two: primary and secondary spray breakup. The combined effect of both breakups leads to the formation of fine droplets from the liquid jets emanating from the injector nozzle. Both types of breakups happen in different ways, depending on the ambient conditions and droplet properties. These properties, along with geometrical parameters, are grouped in the form of non-dimensional numbers such as Reynold Number (*Re*), Weber number (*We*), and Ohnesorge number (*Oh*). The critical points in these breakup mechanisms are governed by the specific value of these non-dimensional numbers. Figure 4.1 shows a classification of primary and secondary breakup models.

$$Re_l = \frac{Inertia\ Force}{Viscous\ Force} = \frac{\rho_l u D}{\mu_l}$$

$$We_l = \frac{Inertia\ Force}{Surface\ tension\ Force} = \frac{\rho_l u^2 D}{\sigma}$$

$$Oh = \frac{Viscous\ Force}{Surface\ Tension\ Force} = \frac{\mu_l}{\sqrt{\sigma \rho_l D}}$$

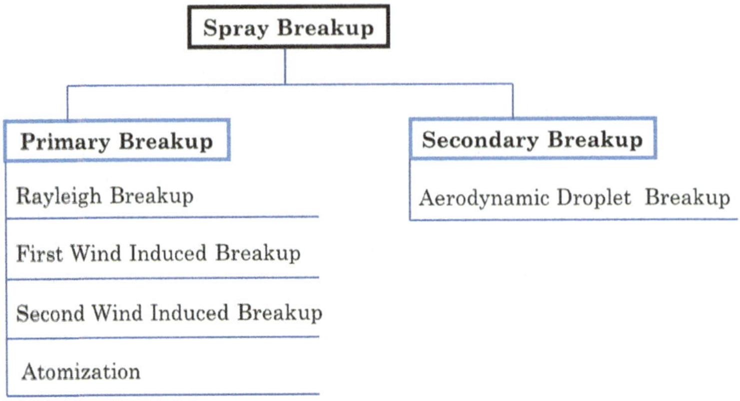

Fig. 4.1 Spray breakup type [Adapted from Web source (2021)]

Here,

ρ_l liquid density,
σ surface tension,
μ_l dynamic viscosity,
D nozzle diameter,
u jet velocity,
l represents the liquid phase.

A higher weber number generates a higher number of fine droplets through the primary and secondary breakups. Therefore, spray with a higher weber number is of specific interest for IC engines since they have fast and superior atomization and faster fuel evaporation.

4.2.1 Primary Spray Breakup

The primary breakup of sprays is related to the breakup of the intact liquid core. Therefore, it is a crucial breakup process since it determines the sizes of droplets formed from the liquid core.

The cylindrical liquid column emerging from the injector nozzle breaks into discrete ligaments/blobs/droplets in the primary spray breakup, primarily because of surface instabilities. The surface instabilities depend upon the interactions of competing surface forces (surface tension, aerodynamic drag) and initial perturbations from the internal nozzle surfaces. Surface instabilities form on the liquid sheet surface and interact with the ambient high-speed gas. These instabilities grow in magnitude downstream, leading to a breakup. When a liquid sheet is injected from the nozzle, it may encounter three instabilities: (a) Kelvin–Helmholtz (K–H) instability, (b) Rayleigh–Taylor (R–T) instability, and (c) capillary (Rayleigh) instability.

During atomization of the liquid sheet, shear-induced capillary waves grow at the vapour-liquid interface, known as KH instabilities, resulting in ligament growth and breakup. RT instabilities are due to the density difference between the vapour and liquid interface, which improves the disintegration of droplets (Rao et al. 2017).

Figure 4.2 shows the actual and simulated spray liquid core differences close to the nozzle (Helldorff and Micklow 2019). Near nozzle visualization of spray remains an active area of research. The complicated physics near the nozzle is not well understood and is often modelled using simplified assumptions, matching subsequent processes (downstream droplet dispersion) with reasonable accuracy.

Figure 4.3a shows different primary breakup regimes as a function of *Re* and *Oh* numbers. Rayleigh breakup happens at low spray jet velocity because of axisymmetric oscillations due to inertial and surface tension forces. The breakup length is higher in this case, which decreases with increasing spray jet velocity. Here, droplets of size equal to jet diameter are formed. Aerodynamic forces increase the inertial and surface tension forces in the first wind-induced breakup (Fig. 4.3).

In contrast, short unstable surface waves initiate in the second wind-induced breakup, further amplified by the aerodynamic forces. Atomization takes place at the nozzle; therefore, intact liquid core length is zero. Here, droplets of much smaller

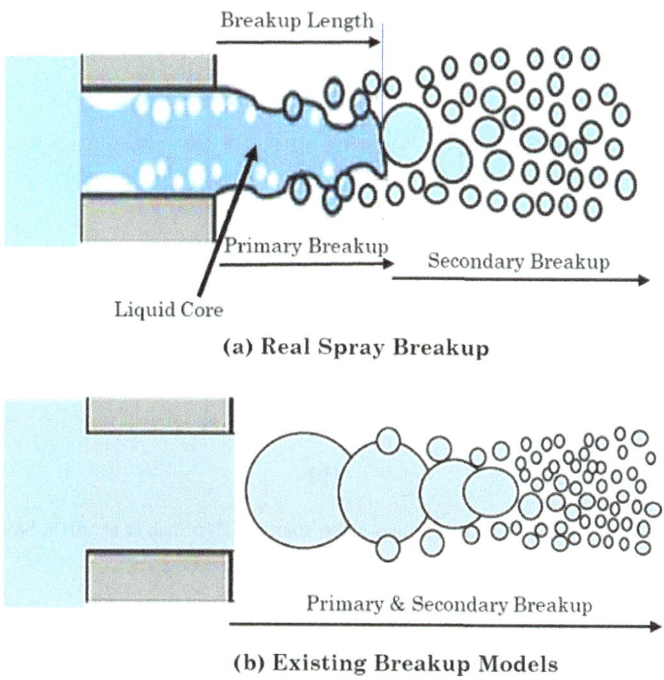

Fig. 4.2 Differences in the actual and simulated breakup of the sprays [Adapted from Helldorff and Micklow (2019)]

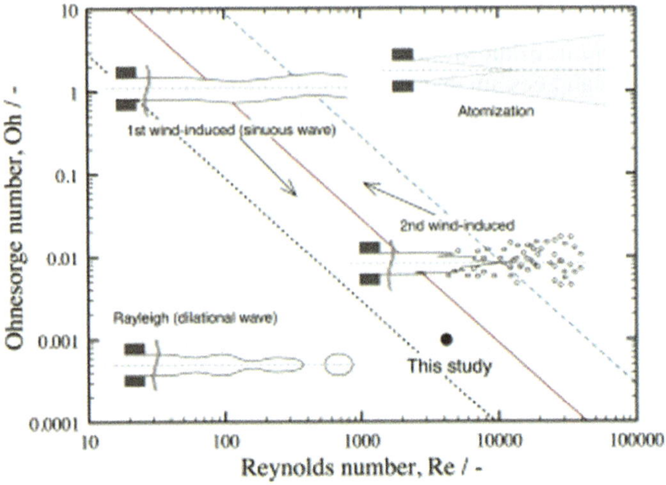

(a) [8] [9]

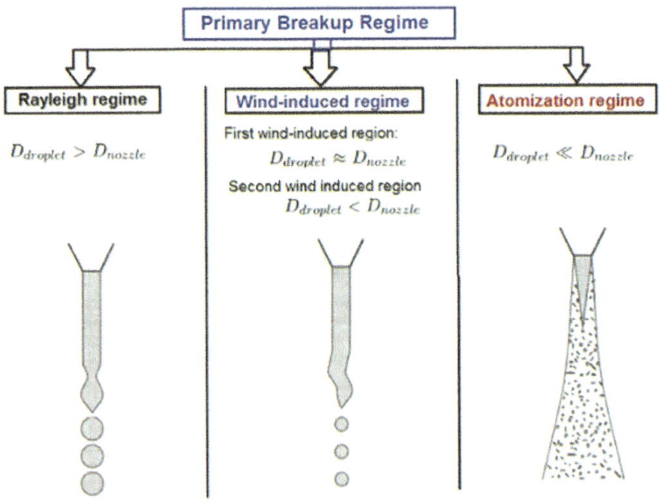

(b)

Fig. 4.3 Different primary breakup regimes (Web source 2021; Shao et al. 2018; Shimasaki and Taniguchi 2011)

size compared to nozzle jet diameter are formed (Shao et al. 2018b). This is a very relevant zone for direct fuel injection diesel engines.

4.2.2 Secondary Spray Breakup

The secondary breakup process disintegrates the primary spray droplets to form smaller droplets due to aerodynamic forces. The aerodynamic forces are proportional to the relative velocity of droplets w.r.t. the ambient air. Therefore, the size of droplets is a characteristic parameter for the secondary spray breakup. Higher We facilitate smaller droplets because of increased aerodynamic forces and faster breakup with increasing *We* (Pilch and Erdman 1987). As *We* increase, various kinds of breakup morphologies begin to happen, such as vibrational, bag, bag-and-stamen, multimode, sheet-thinning, and catastrophic breakup modes, as shown in Figs. 4.4 and 4.5 (Yang et al. 2016).

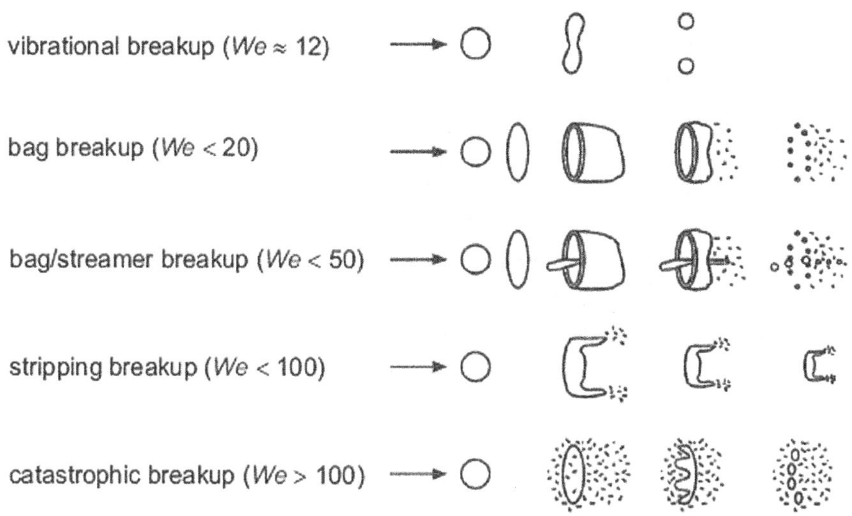

Fig. 4.4 Secondary spray breakup morphology (Web source 2021)

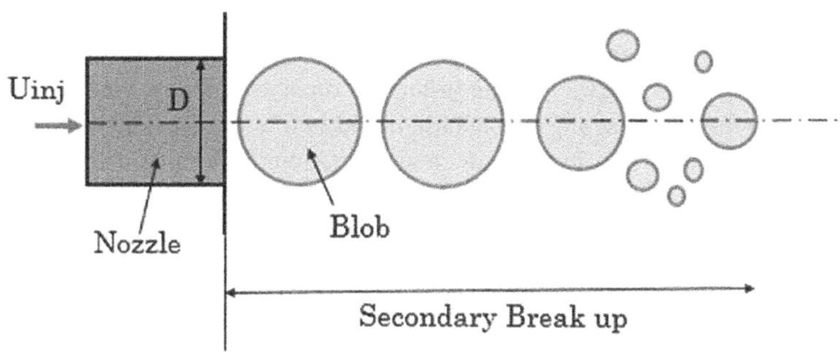

Fig. 4.5 Blob model

Vibrational breakup breaks down the liquid column into coarse droplets. Bag breakup forms a single bag-like shape, whereas multimode breakup leads to the formation of more than one bag-like shape. The striping and catastrophic breakup occur when the outer edges of the liquid column accelerate and break up before the center. Each morphological breakup forms secondary droplets of different sizes. All droplet breakups happen at the same time for the high-velocity sprays in the IC engines. The breakup is most intense near the nozzle due to higher *We*, which decays with *We* downstream of the spray. The lower We downstream of the nozzle is due to the smaller droplets and lower relative velocities (Baumgarten 2006). Depending on the ambient and initial conditions and injected fluid properties, the droplets reach a minimum diameter, which can not be disintegrated further. This diameter is called the 'critical diameter' of the droplets. Conditions favourable for lower critical diameters are desirable for rapid evaporation of sprays in IC engine applications. However, a rigorous breakup may lead to slower and lower spray penetration, leading to inferior air utilization.

4.3 Spray Breakup Models

Unlike real spray ligaments, the spray is treated as a collection of discrete droplet parcels and tracked using the Lagrangian approach in simulations. Generally, spray droplets are modelled using primary breakup models until droplets achieve model switching criteria. Once the switching criteria are met, the secondary breakup modelling is initiated for the particular droplet/droplet group. Droplet groups are formed by taking droplets of a certain range to save computation time. The primary breakup phenomenon determines subsequent air–fuel mixture formation in the combustion chamber. Therefore, it is essential to perform detailed modelling of dense sprays (Bekdemir et al. 2008). Primary spray breakup modelling is meant for predicting the starting conditions of spray droplets, like initial droplet radius and velocity parameters (spray angle), which depend on the flow characteristics in the nozzle and near nozzle instabilities. The relative velocity between the liquid film and ambient air generates instability, which causes the break up of liquid film to ligaments. These ligaments create droplets that further disintegrate into secondary droplets (Duronio et al. 2020). A linear instability theory is implemented to determine the most unstable waves acting on the bulk liquid before the breakup. Any variation in instabilities near the nozzle exit will create variations in dominant unstable wavelength, hence the resulting droplet diameter. The flow conditions near nozzle exit fluctuate under practical conditions, influencing the droplet sizes. Therefore, discrete probability function (DPF) and maximum entropy principle (MEP) were applied by Sovani et al. (Davanlou et al. 2015) and Jaynes (Sovani et al. 1999) to model droplet size in the spray.

4.3.1 Blob Model

This model introduces a set of computational parcels into the multidimensional domain, presented as 'blobs' or portions of the injected liquid column. The blob is defined by a sphere and characterized by an initial diameter same as the effective nozzle diameter. The blob model is the simplest and most widely accepted approach for defining initial conditions for the first spray droplets at the nozzle exit. However, this approach assumes that spray atomization and droplet breakup in the dense spray near the nozzle are not separate phenomena. Therefore, an exhaustive simulation can be replaced by a simplified approach injecting coarse droplets with the same diameter as the nozzle. These droplets are further exposed to secondary aerodynamic induced breakup (Fig. 4.5) (Jaynes 1957).

Spray droplet diameter is equal to the nozzle diameter (D), and droplet generation rate is calculated from the mass flow rate of the fuel. The blob breaks up after interacting with the surrounding gas, but a domain with large discrete particles near the nozzle represents the dense spray core. It is assumed that flow inside a nozzle is slug and injection velocity U_{inj} of the blobs is given by using conservation of mass as given below:

$$U_{inj}(t) = \frac{m_{inj}(t)}{A_{nozzle} \times \rho_l}$$

Here,

$\mathbf{A_{nozzle}}$ is the nozzle cross-section area,
$\boldsymbol{\rho_l}$ is the liquid density, and
$\mathbf{m_{inj}}$ **(t)** is the mass flow rate.

The flow is generally turbulent, and the blob with nozzle diameter (D) size is injected with injection velocity (U_{inj}) at the start of injection. The cavitation is observed in the flow during the main injection of fluid, as shown in Fig. 4.6. Due to

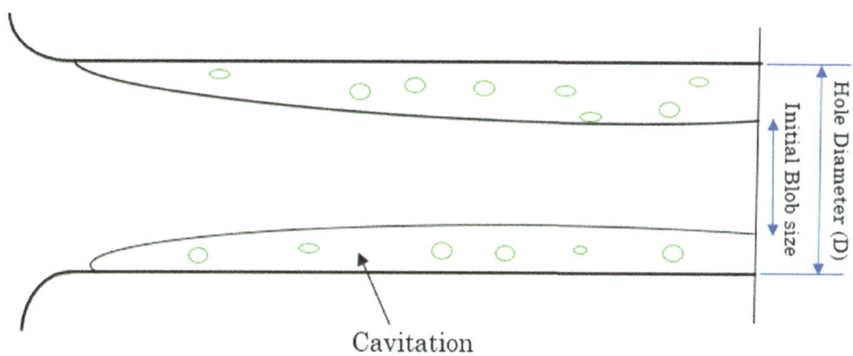

Fig. 4.6 Cavitation inside the nozzle hole

cavitation, an effective cross-sectional area (A_{eff}) of the nozzle decreases, reducing the blob diameter. Therefore, the highest injection velocity and the lowest blob diameter are anticipated during the main injection event.

$$D_{eff} = \sqrt{\frac{4A_{eff}}{\pi}}$$

A more practical value of initial injection velocity and blob diameter can be obtained from the dynamic calculation of both during the whole injection event, including the influence of cavitation. However, only a reduction in cross-sectional area due to cavitation is taken into consideration. Rise in turbulence and break up energy because of the bursting of cavitation bubble is not involved. Blob is a fundamental and widely used approach for investigating the primary breakup in simulation. It is the most suitable method for defining initial conditions of liquid injected in the combustion chamber if there is a lack of knowledge about the composition of primary spray; however, the value of spray angle is obtainable (Baumgarten 2006).

This model does not characterize comprehensive physical and modelling of the related phenomenon during the primary break up. Also, mapping spatial spray angle and droplet size distribution from the three-dimensional nozzle flow cannot be done. Another disadvantage of this approach is excluding the evolution of primary breakup due to turbulence and bursting of bubbles downstream of the nozzle.

4.3.2 Linear Instability Sheet Atomization (LISA) Model

LISA is used to model the liquid sheet breakup. This model requires the spray cone and dispersion angles as input. The spray cone angle represents plume development, and the dispersion angle represents liquid sheet fluctuations from the mean cone angle. Dispersion angle is a critical parameter since it leads to the radial distribution of spray droplets from the mean cone angle (Shrestha et al. 2020). This model has a general liquid sheet breakup and a liquid injection methodology, especially pressure swirl atomizers. Hollow cone sprays produce fine droplets, leading to superior fuel–air mixture formation, lower penetration, and improved spray atomization. The conversion from injector flow to fully evolved spray is modelled in three stages: film formation, sheet breakup, and droplet disintegration (Fig. 4.7).

Figure 4.7 shows that initial perturbations result in the onset of optimal wave mode (λ_{optimal}). These are unstable waves governed by linear stability theory. The unbroken liquid sheet is assumed to fragment into segments ($\lambda_{\text{optimal}}/2$). These broken segments further roll up to form cylindrical ligaments due to surface tension, followed by the breakup of these ligaments into droplets because of Plateau–Rayleigh instability (Deshpande et al. 2015). Uniform velocity at the injector exit is assumed. Total sheet velocity (U) is given by:

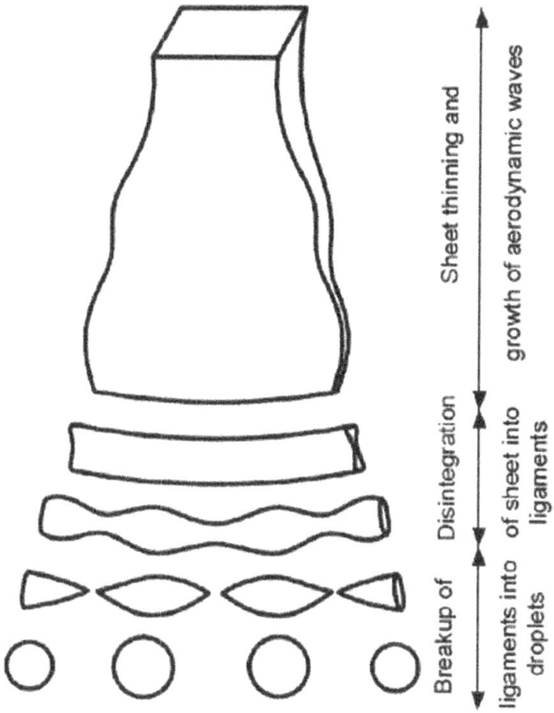

Fig. 4.7 LISA breakup model (Dcshpande et al. 2015; Asgarian et al. 2020)

$$U = k_v \sqrt{\frac{2\Delta p}{\rho_l}}$$

Here, $\boldsymbol{\Delta p}$ is the injection velocity, $\boldsymbol{k_v}$ is the coefficient of velocity and is given by:

$$k_v = max\left[0.7, \frac{4\dot{m}}{\pi d_0^2 \rho_l cos\theta}\sqrt{\frac{\rho_l}{2\Delta p}}\right]$$

Here, $\dot{\boldsymbol{m}}$ is the mass flow rate, $\boldsymbol{d_o}$ is the orifice diameter, $\boldsymbol{\rho_l}$ is the liquid density. The axial (u) and angular (v) components of the injector exit velocity can be calculated from.

$$\mathrm{u} = \mathrm{U}\ \cos\theta,\ \mathrm{v} = \mathrm{U}\ \sin\theta$$

The mass flow rate $\dot{\boldsymbol{m}}$ is given by

$$\dot{m}\rho_l u h_o (d_o - h_o)$$

Here,

u is injector exit axial velocity component,
h_o is initial liquid sheet thickness.

It is assumed that there is no drag on the liquid sheet and no air displacement by the sheet volume due to no direct relationship between the sheet and the gas phase. Hence, there will be no droplet breakup, collision, and drag until the droplet reaches breakup length L_b (Schmidt et al. 1999) given by:

$$L_b = \frac{U}{\Omega} ln\left[\frac{\eta_b}{\eta_0}\right]$$

Here, $ln\left[\frac{\eta_b}{\eta_0}\right]$ is proposed by Dombrowski et al. (1963). Ω is the maximum growth rate. η_b and η_o are critical and original amplitudes. The new diameter after the breakup at the breakup length L_b is calculated by:

$$d_D^3 = \frac{3\pi d_L^2}{K_L}$$

Here,

d_L is the ligament diameter and
K_L is the most unstable wave number.

4.3.3 Kelvin–Helmholtz (KH) Model

Kelvin–Helmholtz (KH) and Rayleigh–Taylor (RT) breakup models are extensively used. It includes two kinds of instabilities, which arise in multiphase flows and can lead to perturbation growth resulting in disruption of spray droplets (Fig. 4.8).

The growth rate of perturbation depends on the conditions of ambient gas and droplet itself. It is assumed that the growth of KH instabilities on the surface of a cylindrical liquid jet causes its disintegration into droplets. These instabilities are due to high shear at the interface of two fluids (Som and Aggarwal 2010). This cylindrical jet penetrates the gaseous environment with a relative velocity of $u_{rel.}$ The liquid and gas are assumed as incompressible. It is assumed that gas has negligible viscosity. The diameter of newly formed droplets is directly proportional to the wavelength of unstable KH wave, which has a fast growth rate on the parent droplets (Nsikane et al. 2017).

$$d_{new} = 2B_0\Lambda$$

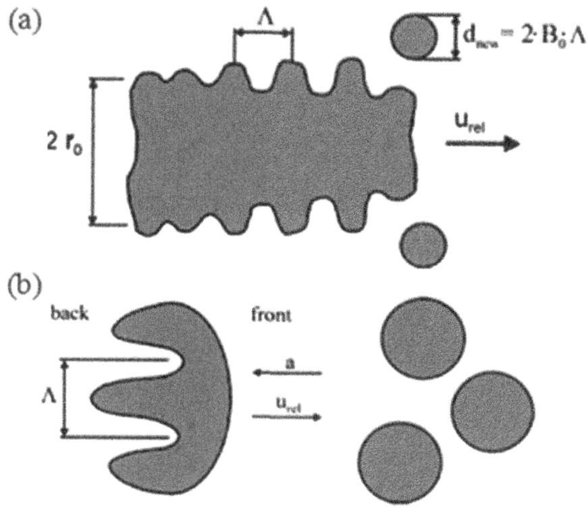

Fig. 4.8 Secondary spray breakup: **a** KH, **b** RT (Nowruzi 2019)

Here,

$\boldsymbol{B}_0$ is constant

Λ is wavelength of the fastest-growing unstable wave having growth rate (Ω_{KH}).

$$\Omega_{KH} = \frac{0.34 + 0.38 We_g^{1.5}}{(1 + Oh)(1 + 1.4T^{0.6})}\sqrt{\frac{\sigma}{\rho_l r^3}}$$

$$\Lambda_{KH} = \frac{9.02r\left(1 + 0.45\sqrt{Oh}\right)(1 + 0.4T^{0.7})}{\left(1 + 0.865 We_g^{1.67}\right)^{0.6}}$$

Here, $\boldsymbol{\sigma}$ = surface tension, **T** is the Taylor number $= Oh\sqrt{We_g}$, **r** is the parent droplet radius, which reduces continuously during the breakup. The rate at which the parent droplet radius reduces until it reaches a stable radius ($r_{KH} = \frac{d_{new}}{2}$) is given by:

$$\frac{dr}{dt} = \frac{r - r_{KH}}{\tau_{KH}}, r_{KH} \le r$$

$$\tau_{KH} = \frac{3.276 B_1 r}{\Omega_{KH}\Lambda_{KH}}$$

Here, $\boldsymbol{\tau}_{KH}$= breakup time, $\boldsymbol{B}_1$ = KH constant. $\boldsymbol{B}_0$ and $\boldsymbol{B}_1$ are the tuning constants that affect the rate at which parent droplet contracts define the child droplet size (Beale and Reitz 1999). Mass is collected from the parent droplet till the shed mass is 5% of the initial parcel mass. Here, a new parcel is formed having a diameter given by $d_{new} = 2B_0\Lambda$. The new parcel provided similar properties as the parent parcel, excluding radius and velocity. The secondary breakup of child droplets happens because of the competing effects of KH and RT models (Kawaharada et al. 2020). Cavitation and turbulence inside the nozzle greatly influence primary breakup; hence primary breakup should incorporate these effects.

4.3.4 Kelvin–Helmholtz-Aerodynamics Cavitation Turbulence (KH-ACT) Model

This model is a remodelled conventional KH model with aerodynamics, cavitation, and turbulence effects in the primary breakup. Turbulence inside the nozzle strongly influences the spray breakup, liquid penetration length, and the diameter of child droplets (Hwang et al. 1996). KH-ACT is quasi dynamically linked to the internal nozzle flow simulations and produces cavitation and turbulence transients at the nozzle exit, acting as starting and boundary conditions (Nsikane et al. 2017).

Turbulence Induced Breakup Model: It calculates turbulent time and length scales by assuming isotropic internal flows and using decay laws:

$$L_t = C_\mu\left(\frac{K(t)^{1.5}}{\epsilon(t)}\right);\ \tau_t = C_\epsilon\left(\frac{K(t)}{\epsilon(t)}\right)$$

Here, $\boldsymbol{K(t)}$ and $\boldsymbol{\epsilon(t)}$ are instantaneous kinetic energy and turbulent dissipation rate, $\boldsymbol{C_\mu}$ and $\boldsymbol{C_\epsilon}$ are model constants. Assuming isotropic turbulence and no diffusion, convection and production terms in K-ϵ equation, $K(t)$ and $\epsilon(t)$ can be estimated as below:

$$K(t) = \left[\frac{(K_0)^{C_\epsilon}}{K_0\left(1 + C_\mu - C_\mu C_\epsilon\right) + \epsilon t(C_\epsilon - 1)}\right]^{1/(1-C_\epsilon)}$$

$$\epsilon(t) = \epsilon_0\left[\frac{K(t)}{K_0}\right]^{C_\epsilon}$$

Here, $\boldsymbol{K_0}$ and $\boldsymbol{\epsilon_0}$ are the starting value of kinetic energy and turbulent dissipation rate at nozzle exit (Nsikane et al. 2017).

Cavitation-Induced Breakup Model: This model considers the cavitation formed inside the nozzle, which may move downstream and reach the nozzle exit. The implosion of bubble improves spray atomization and reduces characteristics breakup timescale. The timescale is decided by the fastest of the two physical phenomena, i.e., the burst of cavitation bubble at the spray periphery or collapse before arriving at:

$$\tau_{CAV} = \text{minimum}(\tau_{collapse}, \tau_{burst})$$

The bubble collapse time is estimated as the time taken to reduce the bubble's radius from 'r' to 'zero.'

$$\tau_{collapse} = 0.9145 R_{CAV} \sqrt{\frac{\rho_l}{p_v}}$$

Here,

ρ_l ***and*** p_v	are the density of fuel and fuel vapour pressure, respectively,
R_{CAV}	is the effective radius of an equivalent bubble from the nozzle and is given by, $R_{CAV} = r_{hole}\sqrt{1 - C_a}$;
r_{hole}	is the exit radius and
C_a	is the coefficient of area reduction.

The time required for the bubble to reach the jet periphery is given by:

$$\tau_{burst} = \frac{r_{hole} - R_{CAV}}{u'_{turb}}$$

Here,

u'_{turb} is the turbulent velocity and is given by $u'_{turb} = \sqrt{\frac{2K(t)}{3}}$

The length scale for this breakup model is given as $L_{CAV} = R_{CAV}$

Aerodynamically Induced Breakup Model: This model calculates the instantaneous length and timescales for each parcel:

$$L_{KH} = r - r_{KH}$$

$$\tau_{KH} = \frac{3.276 B_1 r}{\Omega_{KH} \Lambda_{KH}}$$

The dominant breakup process is decided by the largest value of the ratio of length and the timescale for each process and is presented mathematically as below:

$$\frac{L_A}{\tau_A} = max\left[\frac{L_{KH}(t)}{\tau_{KH}(t)}; \frac{L_{CAV}(t)}{\tau_{CAV}(t)}; \frac{L_t(t)}{\tau_t(t)}\right]$$

If aerodynamically-induced breakup dominates, then the KH model from the previous section is applied for the primary atomization. However, for cavitation or turbulence-induced breakup, below mentioned breakup law is used:

$$\frac{dr}{dt} = -C_{T,CAV} \frac{L_A}{\tau_A}$$

Here,

$C_{T,CAV}$ is the model constant.

4.3.5 RT (Rayleigh–Taylor) Model

KH instabilities occur when velocity gradients are normal to the interface. However, RT instabilities are initiated due to the density difference of the two fluids and can arise at the circumference. These instability waves originate due to the rapid deceleration of droplets because of the drag force (Fig. 4.8). It is based on investigating the stability of the liquid-ambient air interface when liquid accelerates in a direction normal to the plane (Hossainpour and Binesh 2009). Generally, RT instabilities form when acceleration is in the opposite direction to the density gradient. When the droplet decelerates due to ambient gas drag, it may build up instability at the trailing edge of the droplet. The mathematical equation for the acceleration of droplet governed by the gas drag is given as:

$$\left|\vec{F}\right| = \frac{3}{8} C_D \frac{\rho_g v_{rel}^2}{\rho_l r}$$

Here,

v_{rel} is the relative velocity of the droplet w.r.t. the ambient gas and
r is the droplet radius.

The frequency and wavelength of the fastest growing wave are given below by assuming a linear disturbance growth rate and negligible viscosity.

$$\Omega_{RT} = \sqrt{\frac{2\left|\vec{F}\right|}{3} \left[\frac{\left|\vec{F}\right| (\rho_l - \rho_g)}{3\sigma} \right]^{1/4}}$$

$$\Lambda_{RT} = 2\pi \sqrt{\frac{3\sigma}{\left|\vec{F}\right| (\rho_l - \rho_g)}}$$

It was found that acceleration creates the fastest growth of RT instabilities; however, the surface tension of liquid opposes the breakup. Standard RT model defines the growth rate as below:

$$\omega_{RT} = \sqrt{k_{RT} \frac{(\rho_l - \rho_g)}{(\rho_l + \rho_g)} a - \frac{{k_{RT}}^3 \sigma}{\rho_l + \rho_g}}$$

And, the breakup time (t_{bu}) is given by:

$$t_{bu} = \frac{1}{\Omega_{RT}}$$

The diameter of child droplet is evaluated from the RT wavelength (Λ), and breakup occurs when $\Lambda < d_{parent}$. Generally, the RT model is used with the KH model and is known as the hybrid KH-RT model.

4.3.6 Hybrid/Modified Kelvin–Helmholtz Rayleigh–Taylor (KH-RT) Model

The single breakup model cannot completely capture the spray breakup phenomenon as various types of instabilities act on liquid jets simultaneously. Therefore, a hybrid KH-RT model is used, in which KH and RT models are used to capture breakup near the nozzle and at a particular distance downstream of the nozzle, respectively (Beale 1999). KH instabilities occur when velocity gradients are normal to the interface. However, RT instabilities are initiated due to the density difference of the two fluids and can arise at the circumference. Taking account of both instabilities provides a breakup phenomenon approaching to practical. In this model, KH-RT instability theory is applied on the droplet surface to resemble a breakup. Here, spray droplets are injected using the 'blob' model, where the liquid core is represented by defining droplet size equal to the nozzle hole diameter (Fig. 4.9).

In this model, RT and KH breakup models are used in competing modes. The breakup time for both models is calculated for the given set of conditions. The shorter breakup time models are used for further simulation of the breakup (Senecal et al. 2007). RT model is used at the nozzle region as droplet velocities are higher; whereas, the KH model is applied further downstream. These droplets are traced

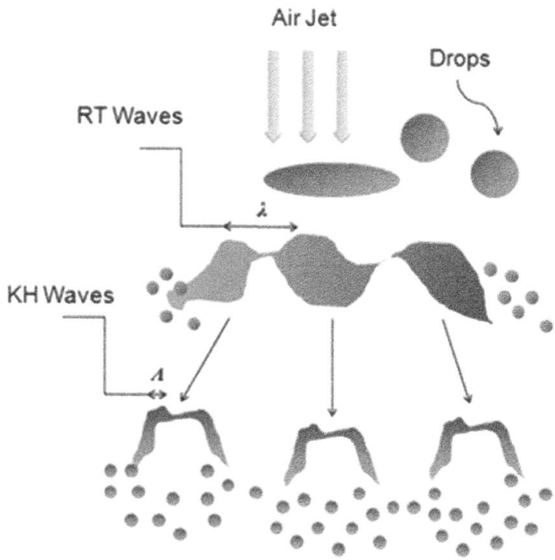

Fig. 4.9 KH-RT model (Som and Aggarwal 2010)

and applied with the KH instability. Calculations of the diameter of parent droplets at each time interval and diameter reduction rate are done in this model. Primary breakup is modelled using KH instabilities, while secondary breakup is modelled using KH and RT models (Sula et al. 2020).

4.3.7 Taylor Analogy Breakup (TAB) Model

TAB model represents an analogy between a spring-mass system with forced oscillations and an oscillating droplet passing through a gaseous atmosphere with a relative velocity of u_{rel} (Fig. 4.10). As per the analogy, the force (F) producing oscillations of mass M, which are like aerodynamic forces, deform the droplet. On the other hand, the restoring spring force ($F_{spring} = - kx$) is similar to a surface tension force, which resists the droplet deformation and maintains its shape (Jaynes 1957). Thus, it is a model to measure the drop distortion and droplet breakup (O'Rourke and Amsden 1987).

Damping force ($F_{damper} = c\,\ddot{x}$) represents friction force due to liquid dynamic viscosity. Therefore, the 2nd order differential equation of motion for the spring-mass system is given as (Table 4.1):

$$\ddot{x} = \frac{F}{M} - \frac{k}{M}x - \frac{c}{M}\dot{x}$$

where

x is displacement of droplet equator from equilibrium,
k is the spring constant,
c is the damping coefficient.

TAB model can track one oscillation mode, whereas there are several such modes. Several investigators have shown that the TAB model is most appropriate for spray

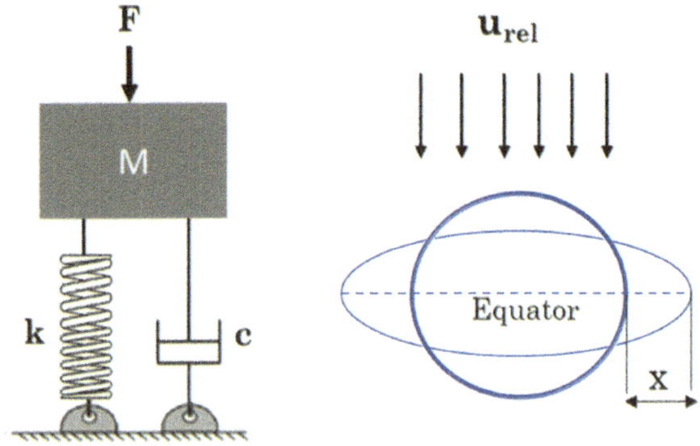

Fig. 4.10 TAB model [Adapted from O'Rourke and Amsden (1987)]

Table 4.1 TAB breakup model analogy

TAB breakup model analogy	
Spring mass system	Oscillating droplet
External force = $\frac{F}{M}$	Droplet drag = $\frac{\rho_a u^2}{\rho_l d}$
Restoring force = $\frac{k}{M}$	Surface tension = $\frac{\sigma}{\rho_l d^3}$
Damping force = $\frac{c}{M}$	Viscous force = $\frac{u_l}{\rho_l d^2}$

under high fuel injection pressures, where aerodynamic effects are predominant, like CI engines.

4.3.8 Enhanced TAB Breakup Model (ETAB)

This model includes the droplet deformation dynamics in the TAB model with a new approach for defining the droplet breakup process. In the TAB model, the radius of product droplets is calculated by the energy balance equation, which generally results in an underprediction of droplet diameter under CI engine conditions. However, the ETAB model utilizes the breakup conditions of the standard TAB model but manages droplet breakup differently. In ETAB, the product droplet formation rate is assumed to be proportional to the number of product droplets, with a proportionality constant depend on the breakup regime. This combines the mass conservation principle and exponential decay law, which relates the mass of product droplets and the breakup time of the parent droplet.

The rate of change of droplet size is calculated from

$$\frac{r}{a} = e^{K_{br}t}$$

Here, a and r are the radius of parents and product droplets. K_{br} is the breakup constant, which depends on the breakup regime as per We as follows,

$$k_{br} = \begin{cases} k_1\omega, if We \leq We_t \\ k_2\omega\sqrt{\mathrm{We}}, if We > We_t \end{cases}$$

Here, k_1 and k_2 are constant, ω is droplet oscillation angular velocity, $We_t = 80$, known as transitional Weber number.

The deformation of the droplet surface is calculated as follows:

$$y(t) = \frac{We}{12} + e^{\frac{-t}{t_d}}\left\{\left[y(0) - \frac{We}{12}\right]cos\omega t + \left[\frac{\dot{y}(0)}{\omega} + \frac{y(0) - \frac{We}{12}}{\omega t_d}\right]\sin\omega t\right\}$$

The breakup happens when the normalized drop distortion, y(t), exceeds the critical value 1. Details of each parameter are given in reference (O'Rourke and Amsden 1987). The product droplets are initially supplied with velocity components perpendicular to the path of the parent droplet with a value $v_{\perp} = A\dot{x}$, Here, $\dot{x}$ is the radial velocity of droplet surface, and A is a constant, which can be calculated from the energy conservation criterion and given as follows:

$$A^2 = 5C_D/4 + 18\left[1 - \frac{a}{r_{32}}\right]/We$$

Here, C_D = aerodynamic drag coefficient, r_{32} = Sauter mean radius. In the standard TAB model, A = 1; whereas, for high fuel injection conditions, A = 0.7. To take into account droplet surface stripping at the nozzle exit initially injected parcels have been equipped with a power-law size distribution as per the following equation:

$$g(r) = \frac{n+4}{r_0}\left(\frac{r}{r_0}\right)^{n+3}$$

Here, r_0 = radius of nozzle, n = 0.5 (Tanner and Weisser 1998).

4.4 Overview of Research Studies

Spray atomization depends on many interdependent parameters such as fuel injection pressure, spray jet velocity, ambient pressure, fuel temperature, ambient temperature, nozzle geometry, and liquid properties. Many researchers investigated the spray phenomena by simulating engine-like conditions in a constant volume spray combustion chamber (Sonawane et al. 2020; Bardi et al. 2012; Pickett 2008). The engine combustion network (ECN) provides high fidelity experimental results of engine-like conditions for diesel and gasoline sprays. For this purpose, ECN has collaborated and worked with many leading universities globally. The database has been generated from the experiments conducted under precisely controlled experimental conditions as per ECN guidelines. This experimental database can validate spray models, the most crucial and mandatory step in the simulations. The accuracy of different spray models matching these experimental results indicates the suitability of models for a particular application. The controlled boundary conditions are extremely helpful for model initialization and setup. The uncertainty in the simulations reduces with the increasing quality of validated results. Some widely accepted CFD tools are CONVERGE, KIVA, OpenFoam, AVL FIRE, and AVBP. In addition, numerous researchers have conducted research studies on spray breakup, which are available in the open literature.

The droplet formation from a primary breakup happens due to shear instabilities like KH instabilities, whereas secondary breakup happens due to RT instabilities. The hybrid KH-RT has been a preferred model used for fuel injection simulations

(Duret et al. 2013). Som et al. (Nsikane et al. 2017) examined the role of primary breakup modelling on the characterization of compression ignition (CI) engines using the Eulerian–Lagrangian framework. KH and KH-ACT models were used for three-dimension simulations of CI engine-like conditions. Simulated results were compared with X-ray radiography results. They reported finer droplets, reduced liquid penetration length, and more radially dispersed spray with the KH-ACT model due to improved primary breakup resulting from nozzle cavitation and turbulence effects. Simulation results showed good agreement with the experimental results of evaporating and non-evaporating sprays. KH-ACT model has predicted experimental results with minimum deviation for liquid and vapour penetration lengths, spray cone angle, axial velocity, droplet dispersion, and flame lift-off location. Improved primary breakup modelling helped superior prediction in the case of the KH-ACT model. However, most diesel engine simulations employ KH-RT atomization models. Many simulation studies employed KH and KH-RT models for primary and secondary breakup (Kawaharada et al. 2020; Reitz and Bracco 1982). This modelling methodology is computationally cost-effective and replicates global spray characterization closely. However, studies have shown that cavitation and turbulence inside the nozzle greatly influence near nozzle spray breakup (Arcoumanis et al. 1997).

Figure 4.11 exhibited liquid penetration at a fixed condition. It can be observed that internal nozzle turbulence has a major influence on spray atomization. However, Sauter mean diameter (SMD) was less sensitive (Bravo and Kweon 2014). A similar conclusion can be drawn from Fig. 4.12. This experimental study was conducted in a constant volume combustion chamber, where operating conditions were maintained constant: fuel injection pressure (FIP) and fuel temperatures were 1420 bar and 438 K, respectively. The KH and KH-ACT models predicted the overall nature of spray; however, the KH-ACT model predicted higher degree atomization with reduced liquid penetration and SMD (Bravo and Kweon 2014). KH model showed higher

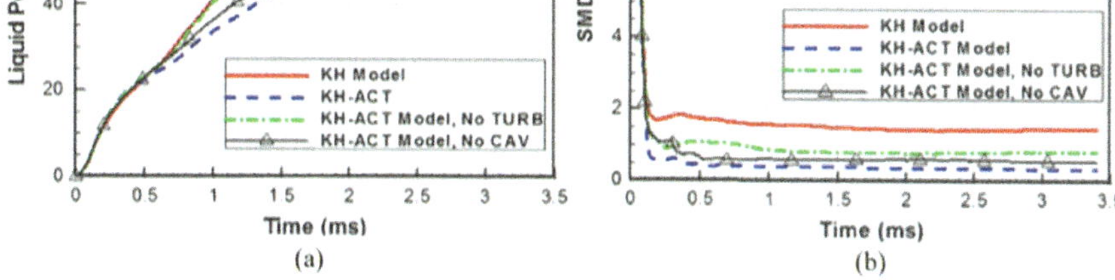

Fig. 4.11 KH and KH-ACT model effect on **a** liquid penetration and **b** SMD (Bravo and Kweon 2014)

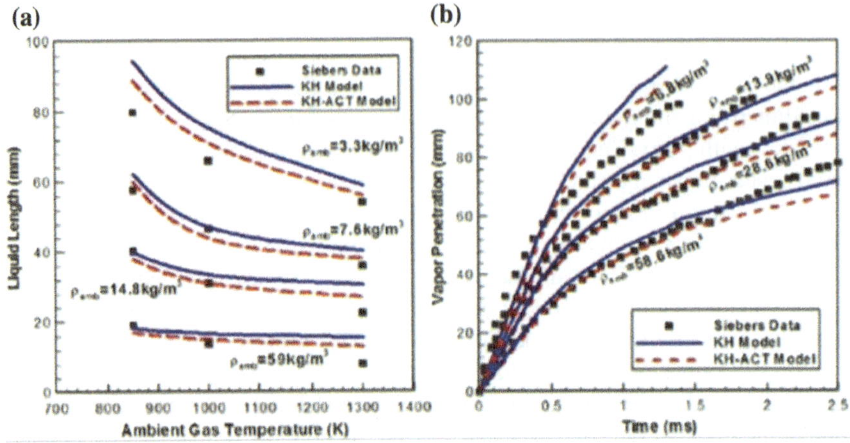

Fig. 4.12 Comparison of hybrid KH model and KH-ACT model for **a** liquid penetration, **b** vapour penetration length (Bravo and Kweon 2014)

penetration as it has not included aerodynamic, cavitation, and turbulence effect on the breakup phenomenon, which is included in KH-ACT.

Rostami and Moghaddam (2020) predicted primary spray breakup length, droplet diameter, and ligaments for diesel, biodiesel, and water using a linear instability model. They reported faster atomization due to increased disturbance by lowering the fuel viscosity and increasing the axial and swirl velocity. Such numerical studies provide a better understanding of spray atomization and how it behaves for various fluids. Generally, the secondary spray breakup modelling assumes that the droplet is moving with the laminar flow. However, the presence of turbulence affects the critical We. Therefore, it results in an inaccurate prediction of the breakup phenomenon. Khaleghi et al. (2021) investigated the influence of turbulence on the secondary breakup using a modified model and found good agreement between simulation results with the experimental results. They concluded that turbulence resulted in an earlier breakup, higher evaporation rate, and lower penetration. However, the difference between the models with and without the turbulence was negligible as the ambient gas pressure increased. This could be a probable reason for the acceptable performance of the KH-RT model in diesel engine simulations where the fuel is injected under high-density ambient.

However, simplified models (turbulence neglected) may lead to inaccuracies in low-density ambient, as commonly seen in SI engines. Chen et al. (2020) numerically investigated the effect of injection angle on the spray structure and droplet gas mixing using the Eularian-Lagrangian framework. The KH-RT model is used for the secondary breakup of droplets and helps understand the mixing processes. Niu et al. (2020) simulated droplet atomization using the KH-RT breakup model. Simulation results showed good agreement with the experimental results. Zhao et al. (2021) performed experiments and simulations to investigate the spray, flow structure and droplet size distribution. Eulerian–Lagrangian approach with KH-RT breakup model was used for the simulations. The simulated results of spray and flow showed good agreement with the experiments. Badra et al. (2016) developed a predictive ability for gasoline compression engines with accurate spray characterization using the Eulerian–Lagrangian framework. For this, they used LISA and KH-RT breakup models, which were found to be most appropriate. They studied the effect of different spray parameters by comparing various available spray models. Results showed a comparison of modified KH-RT and LISA models for various chamber pressure (5, 10 bar) and temperature (300, 600 K) conditions. TAB was used as a secondary breakup model with LISA. The liquid penetration length predicted by both models matched very well with the experimental values. However, the modified KH-RT model showed a longer liquid penetration length than the LISA model. However, trends matched qualitatively well with the experimental results. For better understating of spray modelling, further effects of secondary breakup models of KH-RT and TAB have to be examined.

Liu et al. (2018) studied diesel spray impingement characteristics in small-bore IC engines. LISA was used as a spray breakup model, which resulted in an error of 3% between the experimental and simulation results. Park et al. (2009) analyzed gasoline and bioethanol sprays for swirl-type GDI injectors using the LISA model, used extensively for hollow cone spray simulations (Lee et al. 2004; Lucchini et al. 2005). The numerical studies predicted spray development, spray morphology and spray breakup approximately. Similar observations were also reported by Gao et al. (2005). Liu et al. (2016) simulated spray atomization with a pressure swirl injector using LISA with TAB as a secondary breakup model. They concluded that these models could predict the physical phenomenon of atomization quite well. In addition, results showed the ability of the model to capture local SMD and global spray shapes well. Wang et al. (2017) evaluated the effect of deposit on spray behaviour in the Euler Lagrangian framework. They performed simulation using CONVERGE. The blob method was implemented to inject spray into a constant volume chamber. At the same time, spray breakup was predicted by the KH-ACT-RT model. Spray simulation results showed longer penetration, smaller spray cone angle, and larger droplet size on the formation of deposits. Also, simulation results were in good agreement with the experiments.

Generally, the TAB model is applied for the low-velocity sprays. It is mainly grounded on the physical mechanisms that occur during the spray droplet breakup. However, it has few shortcomings, which many researchers attempted to overcome. Beatrice et al. (1995) concluded that TAB underpredicted the droplet lifetime, reducing its accuracy for high-pressure sprays. However, Bianchi and Pelloni (1999) optimized the model constant and validated it with a correlation of Pilch and Erdman (1987). There were also disagreements in the prediction of liquid penetration length. Tanner (1997) and Park et al. (2002) tried to improve this discrepancy in their study. Matysiak (2007) concluded that the TAB model underestimates the disruptive drag force effect. After taking motivation from Matysiak (2007), Sula and Papalexandris (Jaynes 1957) proposed a modified TAB model, including the influence of drag forces. The model showed the potential to predict global macroscopic spray parameters such as liquid and vapour penetrations better. However, vapour penetration was overpredicted compared to the experimental values.

Pischke et al. (2010) modelled gasoline direct injection using a combined high Eulerian–Lagrangian spray model. Primary and secondary atomization was modelled with a combined LISA-KH-TAB approach. This combined model was validated with the experimental data, and it showed good agreement with the macroscopic and microscopic spray characteristics. Zamani et al. (2016) studied the simulation of a spray of GDI multi-hole injector. They concluded that model coefficient validated for single plume could not be used for multi plume spray simulation without modification due to plume-to-plume interactions and pressure drag. They also presented new coefficients for multi-hole injectors. Huh and Gosman (1991) developed a model to consider internal turbulent stress on the primary breakup. As per this model, initial disturbance on the liquid–gas interface arises from turbulent flow inside the liquid jet, further growing as per the KH instability. It was found that the Huh Gosman-KH-RT model predicted the spray morphology of multi-hole GDI injectors in a better way than the KH-RT model under various ambient and fuel injection pressures (Li et al. 2014).

Li et al. (2021) coupled Wu-Faeth (WF) turbulent primary breakup theory with KH-RT instability theory to investigate the effect of nozzle-produced turbulence and aerodynamic force on GDI spray atomization. This study provided insights into the relationship between various mechanisms influencing spray atomization and the method of calibrating the proposed atomization model for the GDI engines. Figure 4.13 shows simulation results of WF-KH-RT and KH-RT with two different sets of model coefficients. It was found that WF KH-RT showed highly resembling

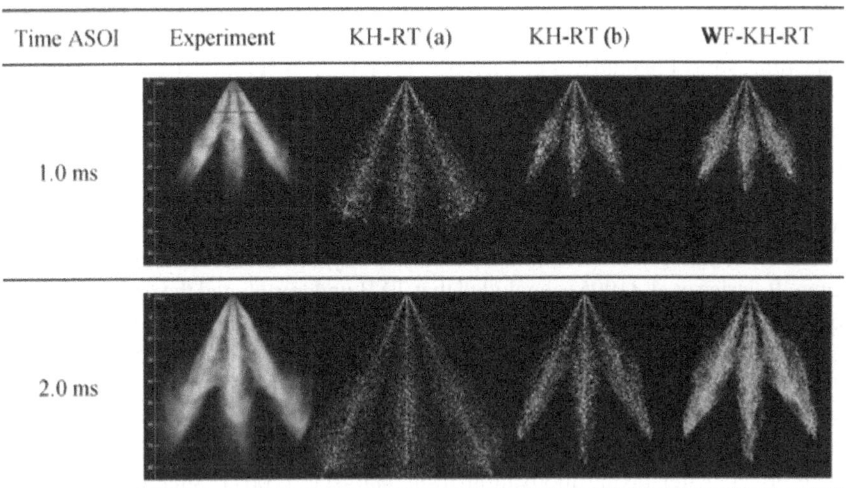

Fig. 4.13 Comparison of spray morphology for experimental, KHRT, and WF-KH RT simulations (Li et al. 2021)

Table 4.2 Hybrid atomization model (Bella et al. 2002)

Hybrid atomization model	
Primary breakup	
Low–medium injection pressure	WAVE
High injection pressure	Turbulence-cavitation-aerodynamic induced model
Secondary breakup	
12 < We < 16 (vibrational)	TAB
16 < We < 45 (bag)	DDB
45 < We < 100 (chaotic)	DDB + WAVE
100 < We < 1000 (stripping)	WAVE
We > 1000 (catastrophic)	WAVE + RT

morphology against experimental, where both shape and penetration are predicted well.

Table 4.2 represents a hybrid atomization model for different *We* (Bella et al. 2002). The secondary breakup model selection criteria for the vibrational regime was TAB (Tanner 1997). For bag breakup, the deformation and breakup (DDB) approach was applied (Ibrahim et al. 1993). For chaotic breakup, DDB was used with the Wave breakup model. The competition between these two models determines the breakup model that could be used. At a higher *We*, the RT model was implemented in competition with the KH model (Patterson and Reitz 1998).

4.5 Summary

Researchers have done substantial work to model the spray breakup to understand the spray breakup phenomenon in IC engine combustion chambers. There are numerous publications and articles on spray modelling, which is evidence of the success of this approach for many applications. Modelling and simulations have opened doors for developing innovative technologies and engine optimization. However, literature shows that model calibration/validation with the experimental data by tuning model constants and parameters is essential. The validated models capture the actual spray breakup process quite well. This chapter covers important research studies available in the field related to IC engine sprays. Advanced spray development may result in a superior and improved understanding of the in-cylinder combustion process, which would help reduce the brake-specific fuel consumption and tailpipe emissions.

4.6 Future Directions

Despite advancements, spray breakup models have provided opportunities to improve modelling aspects to address its drawbacks. There are grey areas of spray breakup, which require attention. A few of them are: (i) the capability to capture the breakup of the entire spray area from dense to dilute, (ii) the ability to predict nozzle turbulence and cavitation and their interdependent effect, (iii) modelling of needle deformation and its impact. Spray modelling for applications such as electrostatic sprays, acoustic induced spray breakup could also be explored. Another limitation is difficulty in the inclusion of internal surface roughness of the nozzle in numerical modelling. The capacity of spray modelling to resolve flow field with LES/DNS models and capturing the changes in fuel properties without any a priori requirement of recalibration. Development of cavitation models, which considers multiple phases, nucleation, and condensation, is required. The sole purpose of spray modelling is to predict the combustion phenomena accurately. Therefore, more research effort is required to evaluate the spray model for combustion studies. The existing spray models often over-predict the effect of cross-flow under swirl supported diesel engine simulations. Furthermore, the impact of large-scale convection on spray dispersion has been scarcely studied. As mentioned in this section, combined experimental and simulation efforts are required to investigate many of these challenging problems.

References

Arcoumanis C, Gavaises M, French B (1997) Effect of fuel injection processes on the structure of diesel sprays. SAE Trans 106:1025–1064. Retrieved June 25, 2021 from http://www.jstor.org/stable/44730737

Asgarian A, Heinrich M, Schwarze R, Bussmann M, Chattopadhyay K (2020) Experiments and modelling of the breakup mechanisms of an attenuating liquid sheet. Int J Multiph Flow 130:103347. https://doi.org/10.1016/j.ijmultiphaseflow.2020.103347

Badra JA, Sim J, Elwardany A, Jaasim M, Viollet Y, Chang J, Amer A, Im HG (2016) Numerical simulations of hollow-cone injection and gasoline compression ignition combustion with naphtha fuels. J Energy Resour Technol 138(5). https://doi.org/10.1115/1.4032622

Bardi M, Payri R, Malbec LMC, Bruneaux G, Pickett LM, Manin J, Bazyn T, Genzale CL (2012) Engine combustion network: comparison of spray development, vaporization, and combustion in different combustion vessels. Atom Sprays 22(10). https://doi.org/10.1615/AtomizSpr.2013005837

Baumgarten C (2006) Mixture formation in internal combustion engines. Springer, Berlin Heidelberg

Beale JC (1999) Modelling fuel injection using the Kelvin–Helmholtz/Rayleigh–Taylor hybrid atomization model in KIVA-3V, MS thesis in mechanical engineering, University of Wisconsin, Madison

Beale JC, Reitz RD (1999) Modelling spray atomization with the Kelvin–Helmholtz/Rayleigh–Taylor hybrid model. Atom Sprays 9(6). https://doi.org/10.1615/AtomizSpr.v9.i6.40

Beatrice C, Belardini P, Berteli C, Camerotti M, Cirillo N (1995) Fuel jet models for multidimensional diesel combustion calculation: an update. SAE Trans 104:194–204. Retrieved June 25, 2021 from http://www.jstor.org/stable/44633211

Bekdemir C, Somers LMT, De Goey LPH (2008) Numerical modelling of diesel spray formation and combustion. MS thesis, Eindhoven University of Technology

Bella G, Rocco V, Ubertini S (2002) Combustion and spray simulation of a DI turbocharged diesel engine. SAE Trans 111:2549–2565. Retrieved June 25, 2021 from http://www.jstor.org/stable/44743269

Bianchi GM, Pelloni P (1999) Modelling the diesel fuel spray breakup by using a hybrid model. SAE technical paper 1999-01-0226. https://doi.org/10.4271/1999-01-0226

Bravo L, Kweon CB (2014) A review on liquid spray models for diesel engine computational analysis. Army Research Lab Aberdeen Proving Ground MD. Accession number: ADA603658

Chen S, Xing Y, Liu X, Zhao L (2020) Numerical investigation of the effect of the injection angle on the spray structures of an air-blast atomizer. Eng Comput. https://doi.org/10.1108/EC-03-2020-0175

Davanlou A, Lee JD, Basu S, Kumar R (2015) Effect of viscosity and surface tension on breakup and coalescence of bicomponent sprays. Chem Eng Sci 131:243–255. https://doi.org/10.1016/j.ces.2015.03.057

Deshpande SS, Gurjar SR, Trujillo MF (2015) A computational study of an atomizing liquid sheet. Phys Fluids 27(8):082108. https://doi.org/10.1063/1.4929393

Dombrowski N, Johns WR (1963) The aerodynamic instability and disintegration of viscous liquid sheets. Chem Eng Sci 18(3):203–214. https://doi.org/10.1016/0009-2509(63)85005-8

Duret B, Reveillon J, Menard T, Demoulin FX (2013) Improving primary atomization modelling through DNS of two-phase flows. Int J Multiph Flow 55:130–137. https://doi.org/10.1016/j.ijmultiphaseflow.2013.05.004

Duronio F, De Vita A, Allocca L, Anatone M (2020) Gasoline direct injection engines–a review of latest technologies and trends. Part 1: spray breakup process. Fuel 265:116948. https://doi.org/10.1016/j.fuel.2019.116948

Gao J, Jiang D, Huang Z, Wang X (2005) Experimental and numerical study of high-pressure-swirl injector sprays in a direct injection gasoline engine. Proc Inst Mech Eng Part A J Power Energy 219(8):617–629. https://doi.org/10.1243/095765005X31333

Hossainpour S, Binesh AR (2009) Investigation of fuel spray atomization in a DI heavy-duty diesel engine and comparison of various spray breakup models. Fuel 88(5):799–805. https://doi.org/10.1016/j.fuel.2008.10.036

Huh KY (1991) A phenomenological model of diesel spray atomization. In: Proceedings of the international conference on multiphase flows' 91-Tsukuba

Hwang SS, Liu Z, Reitz RD (1996) Breakup mechanisms and drag coefficients of high-speed vaporizing liquid drops. Atom Sprays 6(3). https://doi.org/10.1615/AtomizSpr.v6.i3.60

Ibrahim EA, Yang HQ, Przekwas AJ (1993) modelling of spray droplets deformation and breakup. J Propul Power 9(4):651–654. https://doi.org/10.2514/3.23672

Jaynes ET (1957) Information theory and statistical mechanics. Phys Rev 106(4):620. https://doi.org/10.1103/PhysRev.106.620

Kawaharada N, Thimm L, Dageförde T, Gröger K, Hansen H, Dinkelacker F (2020) Approaches for detailed investigations on transient flow and spray characteristics during high-pressure fuel injection. Appl Sci 10(12):4410. https://doi.org/10.3390/app10124410

Khaleghi H, Farani Sani H, Ahmadi M, Mohammadzadeh F (2021) Effects of turbulence on the secondary breakup of droplets in diesel fuel sprays. Proc Inst Mech Eng Part D J Autom Eng 235(2–3):387–399. https://doi.org/10.1177/0954407020958581

Lee CS, Kim HJ, Park SW (2004) Atomization characteristics and prediction accuracies of hybrid breakup models for a gasoline direct injection spray. Proc Inst Mech Eng Part D J Autom Eng 218(9):1041–1053. https://doi.org/10.1243/0954407041856746

Li ZH, He BQ, Zhao H (2014) Application of a hybrid breakup model for the spray simulation of a multi-hole injector used for a DISI gasoline engine. Appl Therm Eng 65(1–2):282–292. https://doi.org/10.1016/j.applthermaleng.2013.12.063

Li Y, Huang Y, Luo K, Liang M, Lei B (2021) Development and validation of an improved atomization model for GDI spray simulations: coupling effects of nozzle-generated turbulence and aerodynamic force. Fuel 299:120871. https://doi.org/10.1016/j.fuel.2021.120871

Liu Y, Xiang Q, Li Z, Yao S, Liang X, Wang F (2018) Experiment and simulation investigation on the characteristics of diesel spray impingement based on droplet impact phenomenon. Appl Sci 8(3):384. https://doi.org/10.3390/app8030384

Liu RC, Le JL, Song WY, Yang SH (2016) LISA model for simulation of liquid sheet breakup in swirl injection. In: Material science and environmental engineering: the proceedings of 2016 international workshop on material science and environmental engineering (IWMSEE2016), pp 653–662. https://doi.org/10.1142/9789813143401_0071

Lucchini T, D'Errico G, Nordin N (2005) CFD modelling of gasoline sprays. SAE technical paper 2005-24-086. https://doi.org/10.4271/2005-24-086

Magnotti GM, Genzale CL (2017) Exploration of turbulent atomization mechanisms for diesel spray simulations. SAE technical paper no. 2017-01-0829. https://doi.org/10.4271/2017-01-0829

Matysiak A (2007) EV zur S tropfenbeladener S in einem V. Euler-Lagrange Verfahren zur Simulation tropfenbeladener Strömung in einem Verdichtergitte Ph.D. Thesis, Department of Mechanical Engineering, Helmut Schmidt Universität - Universität der Bundeswehr Hamburg

Niu YY, Wu CH, Huang YH, Chou YJ, Kong SC (2020) Evaluation of breakup models for liquid side jets in supersonic cross flows. Numer Heat Transf Part A Appl 79(5):353–369. https://doi.org/10.1080/10407782.2020.1847513

Nowruzi H (2019) Review on simulation of non-reacting fuel spray. Am J Mech Eng 7(1):1–8. https://doi.org/10.12691/ajme-7-1-1

Nsikane D, Mustafa K, Ward A, Morgan R, Mason D, Heikal M (2017) Statistical approach on visualizing multi-variable interactions in a hybrid breakup model under ECN spray conditions. SAE Int J Engines 10(5):2461–2477. Retrieved June 25, 2021 from https://www.jstor.org/stable/26422627

O'Rourke PJ, Amsden AA (1987) The TAB method for numerical calculation of spray droplet breakup. SAE technical paper 872089.https://doi.org/10.4271/872089

Park SH, Kim HJ, Suh HK, Lee CS (2009) Atomization and spray characteristics of bioethanol and bioethanol blended gasoline fuel injected through a direct injection gasoline injector. Int J Heat Fluid Flow 30(6):1183–1192. https://doi.org/10.1016/j.ijheatfluidflow.2009.07.002

Park JH, Yoon Y, Hwang SS (2002) Improved TAB model for prediction of spray droplet deformation and breakup. Atom Sprays 12(4). https://doi.org/10.1615/AtomizSpr.v12.i4.20

Patterson M, Reitz R (1998) Modelling the effects of fuel spray characteristics on diesel engine combustion and emission. SAE Trans 107:27–43. Retrieved June 25, 2021 from http://www.jstor.org/stable/44736506

Pickett L (2008) Engine modeling user's group meeting at the SAE congress, Detroit, MI

Pilch M, Erdman CA (1987) Use of breakup time data and velocity history data to predict the maximum size of stable fragments for acceleration-induced breakup of a liquid drop. Int J Multiph Flow 13(6):741–757. https://doi.org/10.1016/0301-9322(87)90063-2

Pischke P, Martin D, Kneer R (2010) Combined spray model for gasoline direct injection hollow-cone sprays. Atom Sprays 20(4). https://doi.org/10.1615/AtomizSpr.v20.i4.60

Rao DCK, Karmakar S, Basu S (2017) Atomization characteristics and instabilities in the combustion of multi-component fuel droplets with high volatility differential. Sci Rep 7(1):1–15. https://doi.org/10.1038/s41598-017-09663-7

Reitz RD, Bracco FV (1982) Mechanism of atomization of a liquid jet. Phys Fluids 25(10):1730–1742. https://doi.org/10.1063/1.863650

Rostami E, Moghaddam MH (2020) The velocity and viscosity impact on the annular spray atomization of different fuels. Combust Theor Model 1–35. https://doi.org/10.1080/13647830.2020.1845399

Schmidt D, Nouar I, Senecal P, Rutland J, Martin J, Reitz R, Hoffman J (1999) Pressure-swirl atomization in the near field. SAE Trans 108:471–484. Retrieved June 25, 2021 from http://www.jstor.org/stable/44743386

Senecal PK, Richards KJ, Pomraning E, Yang T, Dai MZ, McDavid RM, Cartesian ANPCC (2007) A new parallel cut-cell cartesian CFD code for rapid grid generation applied to in-cylinder diesel engine simulations. SAE technical papers 2007-01-0159. https://doi.org/10.4271/2007-01-0159

Shao C, Luo K, Chai M, Fan J (2018a) Sheet, ligament, and droplet formation in swirling primary atomization. AIP Adv 8(4):045211. https://doi.org/10.1063/1.5017162

Shao C, Luo K, Chai M, Fan J (2018b) Sheet, ligament and droplet formation in swirling primary atomization. AIP Adv 8(4):045211. https://doi.org/10.1063/1.5017162

Shimasaki SI, Taniguchi S (2011) Formation of uniformly sized metal droplets from a capillary jet by electromagnetic force. Appl Math Model 35(4):1571–1580. https://doi.org/10.1016/j.apm.2010.09.033

Shrestha K, Van Strien J, Singh N, Inthavong K (2020) Primary breakup and atomization characteristics of a nasal spray. PLoS ONE 15(8):e0236063. https://doi.org/10.1371/journal.pone.0236063

Smirnov NN, Nikitin VF, Tyurenkova VV (2012) Nonequilibrium diffusion combustion of liquid fuel droplets and sprays modelling. Heat Transf Res 43(1). https://doi.org/10.1615/HeatTransRes.v43.i1.10

Som S, Aggarwal SK (2010) Effects of primary breakup modelling on spray and combustion characteristics of compression ignition engines. Combust Flame 157(6):1179–1193. https://doi.org/10.1016/j.combustflame.2010.02.018

Sonawane U, Kalwar A, Agarwal AK (2020) Microscopic and macroscopic spray characteristics of gasohols using a port fuel injection system. SAE technical paper no. 2020-01-0324. https://doi.org/10.4271/2020-01-0324

Sovani SD, Sojka PE, Sivathanu YR (1999) Prediction of drop size distributions from first principles: the influence of fluctuations in relative velocity and liquid physical properties. Atom Sprays 9(2). https://doi.org/10.1615/AtomizSpr.v9.i2.20

Stiesch G (2003) Modelling engine spray and combustion processes. Springer, Berlin Heidelberg

Sula C, Grosshans H, Papalexandris MV (2020) Assessment of droplet breakup models for spray flow simulations. Flow Turbul Combust 105:889–914. https://doi.org/10.1007/s10494-020-00139-9

Tanner F (1997) Liquid jet atomization and droplet breakup modeling of non-evaporating diesel fuel sprays. SAE Trans 106:127–140. Retrieved June 25, 2021 from http://www.jstor.org/stable/44730666

Tanner FX, Weisser G (1998) Simulation of liquid jet atomization for fuel sprays by means of a cascade drop breakup model. SAE technical paper no. 980808.https://doi.org/10.4271/980808

Von Helldorff H, Micklow GJ (2019) Primary and secondary spray breakup modelling for internal combustion engine applications. J Multidiscip Eng Sci Technol (JMEST) 6(4). ISSN: 2458-9403

Wang B, Jiang Y, Hutchins P, Badawy T, Xu H, Zhang X, Rack A, Tafforeau P (2017) Numerical analysis of deposit effect on nozzle flow and spray characteristics of GDI injectors. Appl Energy 204:1215–1224. https://doi.org/10.1016/j.apenergy.2017.03.094

Web Source (2021). https://cfdflowengineering.com/turbulent-multiphase-combustion/. Accessed on 25th May 2021

Yang W, Jia M, Sun K, Wang T (2016) Influence of density ratio on the secondary atomization of liquid droplets under highly unstable conditions. Fuel 174:25–35. https://doi.org/10.1016/j.fuel.2016.01.078

Zamani H, Hosseini V, Afshin H, Allocca L, Baloo M (2016) Large eddy simulation of GDI single-hole and multi-hole injector sprays with comparison of numerical break-up models and coefficients. J Applied Fluid Mech 9(2)

Zhao J, Ren Y, Tong Y, Lin W, Nie W (2021) Atomization of a liquid jet in supersonic cross-flow in a combustion chamber with an expanded section. Acta Astronaut 180:35–45. https://doi.org/10.1016/j.actaastro.2020.11.051

Part III
CI Engine Modeling

Chapter 5
Modeling and Simulation of Diesel Engines Using CFD and Its Applications in Optimizing Various In-Cylinder Techniques

Pradeep Menon and Mayank Mittal

Abstract A four-stroke IC engine cycle undergoes the following processes—intake, compression, combustion, expansion and exhaust, which are governed by complex fluid dynamics, chemical kinetics and heat transfer phenomena and occur in milli/nanosecond timescales inside the engine cylinder making it difficult to visualize experimentally. A better understanding of the in-cylinder phenomena helps design better engines in terms of performance, fuel consumption and emissions. Optimization of the engine processes through physical testing can be time and cost intensive. CFD analysis helps visualize the various in-cylinder processes, analyze them better and evaluate various engine optimization parameters without having to test on an actual engine, hence it is comparatively effortless to do back-to-back comparisons to identify the effect of a single parameter, which reduces the engine development time and cost drastically. CFD analysis of Diesel engines is particularly useful since there are many in-cylinder optimization parameters in a diesel engine in comparison with a conventional gasoline engine, due to the direct injection and heterogeneous nature of combustion. This chapter aims to give an overview of the general methodology followed in performing CFD simulations in diesel engines towards the objective of enhancing the performance and minimizing the fuel consumption and emissions. Here a comprehensive coverage of the complete process of modeling and simulation is attempted, starting with the basic input data required, typical assumptions used, fundamental governing equations and discretization, model formulation, initial conditions, boundary conditions, model validation, various sub-models used and the analysis of results. The effect of in-cylinder techniques like combustion chamber geometry, injection timing, multiple injections and exhaust gas recirculation on performance and emissions is discussed through a brief review of published literature. It is hoped that the reader will get an insight into the procedure followed in diesel engine CFD analysis and some of its applications in the optimization through various in-cylinder techniques.

Keywords Diesel engine · CFD · Emission · In-cylinder technique

P. Menon · M. Mittal (✉)
Indian Institute of Technology Madras, Chennai, India
e-mail: mmittal@iitm.ac.in

A. K. Agarwal et al. (eds.), *Engine Modeling and Simulation*, Energy, Environment, and Sustainability, https://doi.org/10.1007/978-981-16-8618-4_5

5.1 Introduction

The internal combustion (IC) engine has been the most popular prime mover for more than a century. The most significant application of course is in automobiles ranging from small 50 cc gasoline engines in mopeds to large diesel engines in heavy duty trucks. IC engines are also used in off-highway applications like agricultural equipment, pumps, industrial and domestic power generators, construction equipment, aeroplanes, ships and many more. The adaptability and versatility of an IC engine is hard to match. Even in today's world when there is so much flak for the petroleum fuelled IC engine being a major source of air pollution, it is very difficult to find a like-to-like replacement for it meeting the cost, space, performance targets as well as the practicality and versatility.

Conventional IC engines run on either gasoline or diesel. Diesel engines are popular due to their higher thermal efficiency and higher peak torque for the same size, especially in heavy duty automobile and power generation applications. The major drawback of diesel engines is the higher amount of pollutants in the form of oxides of nitrogen (NOx) and particulate matter (PM) in comparison with gasoline engines. With tighter emission regulations coming up, reducing these pollutants while retaining the advantages of diesel engines become even more challenging. From years of engine development, researchers have identified certain methods by which NOx and PM could be reduced (Heywood 2018). Usually a combination of these methods and its optimization are used for emission control. The optimum combination however could vary from engine to engine and for the same engine from one operating point to another. In-cylinder emissions or engine out emissions refer to the emissions coming out from the engine cylinder, which could be controlled using in-cylinder techniques. After-treatment systems are used to reduce the pollutants between the engine cylinder and the vehicle tail pipe. These have become very expensive and bulky nowadays, and vehicle manufacturers try to minimize the investment on those systems, making in-cylinder techniques more important.

In-cylinder techniques deal with the optimization of in-cylinder variables to achieve the design targets of performance, fuel economy and emissions. These variables include the combustion chamber shape, fuel injection parameters like injection pressure, number of injections, start of injection and injection duration, and exhaust gas recirculation (EGR). A four-stroke IC engine cycle starts with the intake stroke where fresh air is inducted into the cylinder through the intake valves, followed by its compression. In diesel engines, the fuel injection occurs somewhere before the top dead center (TDC). The injected liquid fuel must vaporize and mix with the surrounding air to ignitable proportions. Once the temperature exceeds the auto-ignition temperature, the mixture gets ignited marking the start of combustion. Combustion results in the expansion of gases with the rise in cylinder pressure and temperature. Towards the bottom dead center in the expansion stroke, the exhaust valve opens, and the combustion products flow out to the exhaust port. All these processes occur in a very short time frame typically in milli/nanosecond timescales and are governed by complex fluid dynamics, chemical kinetics, and heat transfer phenomena. It is very difficult to capture or visualize these processes on a physically running engine (Stiesch 2003).

Optimization of these in-cylinder engine parameters through physical testing can be time and cost intensive. A combination of these parameters which is optimum for one operating point may not be the best for other points. Also, what is best for performance could be the worst for emissions or vice versa. Engine manufacturers spend a lot of efforts in finding the optimum configuration for all operating points. This is an iterative process till one arrives at the best combination, and to do this on physical prototypes is expensive and time consuming. This is where Computational Fluid Dynamics (CFD) tools pitch in and give tremendous flexibility to the manufacturer to try out different combinations virtually and arrive at the right design for the right operating point quickly.

It is evident that the in-cylinder phenomena which occur during the engine operation are quite complex. Even though CFD sounds like a magical tool, it is not as simple as it seems to prepare an engine model that could accurately predict the engine behaviour. Therefore, researchers apply suitable assumptions and prepare simplified models which could relate to the physical processes. Commercial CFD software applications have in-built modules tailor-made for internal combustion engine simulations (Richards et al. 2017). These applications help in simulating the charge air motion within the cylinder, fuel–air interaction, chemical kinetics of the combustion process and emission formation, without having to build an engine and test it. Hence, a CFD study helps visualize and analyze the various in-cylinder processes. It is relatively effortless to perform a parametric analysis isolating the effect of each parameter one by one, which reduces the engine development time and cost drastically.

This chapter is organized with following sections. First the typical objectives of a diesel engine CFD simulation are identified, then the broad classification of engine simulation is discussed. This is followed by the outline of the general methodology followed in performing CFD simulations in diesel engines. Then the mathematical modeling of the physical processes through governing equations is discussed. Subsequently the different input parameters required to perform the analysis are discussed which form the initial and boundary conditions for the governing equations. The importance of minimizing the computational time and cost is discussed through grid independence study. Then the typical methods used for model validation are discussed followed by various post-processing or analysis which could be performed using the data generated through CFD. The chapter is wrapped up with a brief discussion on the application of CFD in diesel engine optimization with in-cylinder techniques through a review of published literature. It is hoped that the reader will get an insight into the procedure followed in diesel engine CFD analysis and some of its applications in the optimization through various in-cylinder techniques.

5.2 Objectives of Diesel Engine Simulation

The main objectives of an engine manufacturer are always higher performance (higher power/torque), lower fuel consumption and less emissions at all operating points. CFD helps achieving these objectives at various levels. Specific to diesel engines, the objectives could be put into the following two broad categories.

- Study charge air motion within the engine cylinder
  - Effect of intake port geometry
  - Effect of valve geometry
  - Effect of valve lift and timing
  - Study internal EGR and residual gas fraction in one cycle
- Study combustion and its effects on performance and emissions
  - Air–fuel spray interaction
  - Rate of pressure rise and its effects on performance and noise
  - Rate of emission formation.

5.3 Types of Engine CFD Simulation

Based on the objectives of the analysis, diesel engine CFD analysis could be performed in two broad types—open cycle simulation and closed cycle simulation.

5.3.1 *Open Cycle Simulation*

This type of simulation considers the full four strokes of the engine cycle. The CFD model could include the intake manifold, intake port, intake valves, intake valve seats, the engine cylinder, piston, exhaust valves, exhaust valve seats, exhaust port and exhaust manifold. This is usually used to simulate the effect of the port geometry, valve lift and timing on charge air motion. The turbulence levels within the cylinder could be simulated. The backflow which could occur during the valve open period and the mass of residual gases from the previous cycle trapped could also be predicted with reasonable accuracy through this. Simulation of charge air motion with flow geometries are bound to be more accurate than performing simulations with fixed conditions of charge air motion at the start of simulation based on assumptions or experience as accurately measured information is hard to acquire. Usually flow boundary conditions are required for such models. This type of simulation starts with many assumptions in initial conditions at the start of the cycle, hence it is recommended to perform three to four cycles to negate the effect of initial conditions on the results and usually the fifth cycle is bound to give realistic results. One must perform full 720 crank angle degree (CAD) simulation four to five times to get satisfactory results. Therefore, this type of simulation even though gives better results, is computation intensive.

5.3.2 Closed Cycle Simulation

When the objective of the simulation is focused only on the combustion and its effect on performance and emissions, one could go for closed cycle simulation. It performs the simulation from intake valve close (IVC) time to exhaust valve open (EVO) time, that is, the part of the cycle when both valves are closed, hence the name. Since the total crank angles covered from IVC to EVO will typically be in the range of around 200 CAD, this type of simulation is faster and computationally less intensive. But the advantage of performing multiple cycles to negate the effect of initial conditions is not possible in this. Hence, the model must be defined sufficiently well to give satisfactory results, which again depends on the objectives of the simulation. If one wishes to give the prediction of the performance or emissions in absolute terms, for example in terms of gross IMEP or NOx in g/kg of fuel, then the model must be well detailed and must take care of initial conditions and boundary conditions especially well. If the objective is to do a back-to-back comparison between changes in one or more parameters, the absolute value is of less significance as the relative change in performance or emission between the two simulations is good enough to predict the effect of the design change on an actual engine.

Closed cycle simulations could be further simplified in engines with a symmetric combustion chamber design with vertical centrally mounted fuel injector, by performing sector simulations, that is, simulating only one sector containing one plume of injector fuel spray. This helps in achieving even faster results at less computational cost. Table 5.1 gives an overall comparison between open cycle and closed cycle simulations.

One could also do a full cycle simulation at the required operating condition, obtain the in-cylinder conditions at IVC, and map those conditions in subsequent closed cycle simulations. This improves the results considerably in comparison with a standalone closed cycle simulation.

5.4 Outline of the Methodology

This section aims to give an outline of the methodology followed in performing diesel engine CFD simulation. The flow of processes could be depicted as shown in Fig. 5.1. The objectives decide whether to follow an open cycle approach or a closed cycle approach. According to the approach, the input parameters of initial conditions and boundary conditions vary. Appropriate data must be identified to perform model validation, more details of which is discussed in Sect. 5.11. A well detailed CAD model helps in achieving better results. Also, the CFD software, one uses, puts some

Table 5.1 Comparison of open cycle and closed cycle simulations

Parameter	Open cycle simulation	Closed cycle simulation
Simulation objectives	Effect of port geometry Effect of valve timing and lift Effect of backflow and internal EGR Effect of combustion chamber shape Effect of fuel injection parameters Effect of EGR	Effect of combustion chamber shape Effect of fuel injection parameters Effect of EGR
Geometry required	Intake and exhaust ports, intake and exhaust valves, seats and guides, parts of cylinder head, liner and piston which form the combustion chamber	Parts of cylinder head, liner and piston which form the combustion chamber
Computation cost/time	High	Low
Advantage	More accurate results as the flow variable calculations are more realistic rather than based on assumptions	Much faster and hence cheaper computation with relatively good accuracy Symmetric combustion chambers can use sector simulations for further speed
Disadvantage	High computation time Complex geometry required, leading to difficulties in meshing	Comparatively less accurate results due to the assumptions especially related to initial conditions of flow variables

constraints in geometries, for example closed valves must be within 0.2 mm gap with valve seat (Richards et al. 2017). Such care as needed must be taken while preparing the 3D model. Many commercial softwares these days give the choice of automatic grid or mesh generation, which could be further refined manually if required (Richards et al. 2017). The mesh near the walls, valve head and seat, and at the regions where the geometry varies considerably, the gradients of velocity and temperature could be high, and a fine grid would be needed to give good results. Commercial softwares take care of this in automatic meshing; however, one can verify and adjust the local mesh refinement manually too.

CFD model and sub-models are built keeping in mind the objectives. The input parameters required to prepare these models include the initial conditions, boundary conditions, operating condition details like fuel injection timing, duration, rate of injection, etc. It also includes basic engine geometrical parameters like compression ratio, connecting rod length, crank offset, etc. which help simulate the piston motion during the engine cycle. There are various other inputs too which are required to solve the governing equations and are usually used in fine tuning the model and to ensure convergence.

Grid independence study is discussed in more detail in Sect. 5.10. With grid independence study, one is equipped with the optimum model to proceed with the

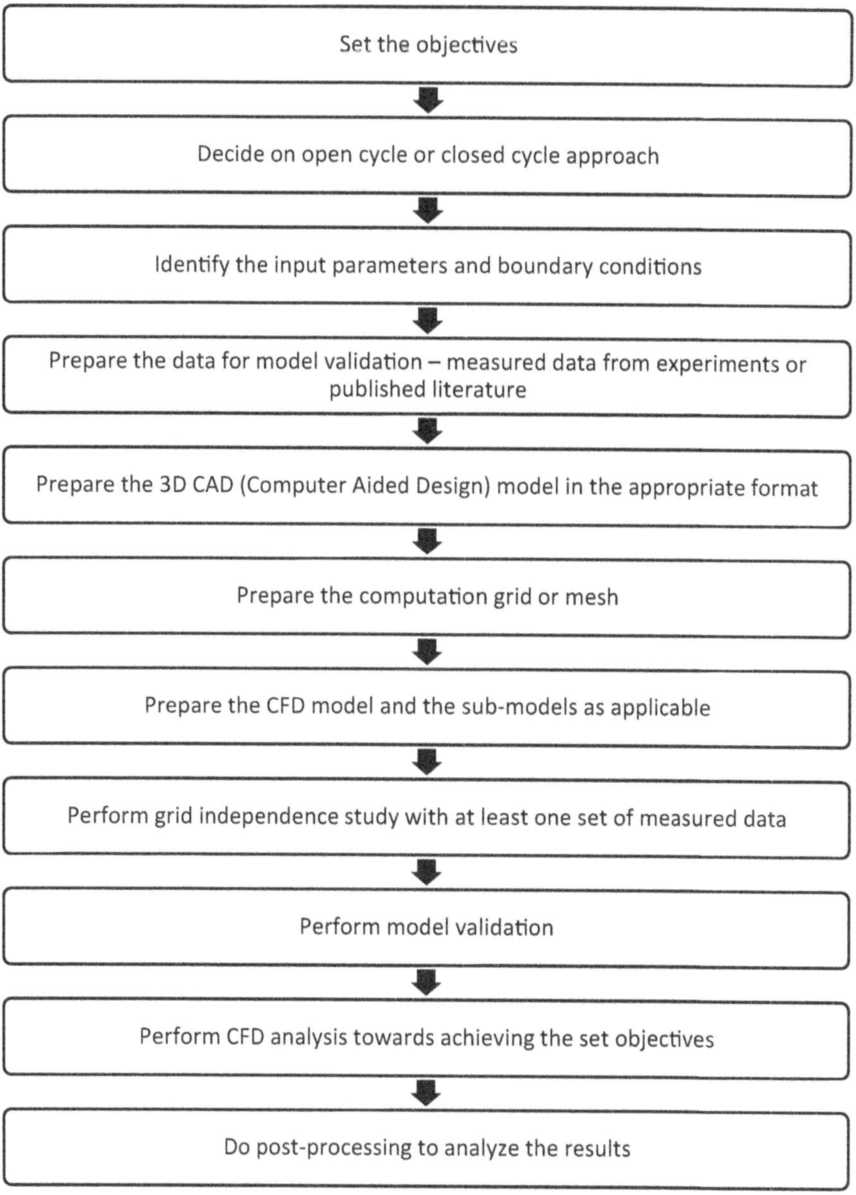

Fig. 5.1 Outline of the methodology

CFD analysis. Appropriate model validation is to be performed to prove the capability of the model to predict the required results by comparing with measured or known data. Once a well validated model is available, there is complete flexibility to do CFD analysis in a parametric fashion and do post-processing to analyze the results.

5.5 Governing Equations—Mathematical Model Formulation

In diesel engine simulation, be it a 0D, 1D or 3D CFD analysis, the thermodynamic variables of interest are pressure and temperature inside the cylinder, and the composition of products of combustion particularly the pollutants like NOx and PM with respect to crank angle. All of these are calculated by solving the relevant governing equations by applying necessary assumptions and suitable initial and boundary conditions. In computation fluid dynamics, the dynamics of fluid flow are governed by the equation of state and the conservation equations, that is, the conservation of mass, momentum, and energy. This section discusses the details of the governing equations.

Compressible form of the mass conservation equation is given by the well-known continuity equation (Ferziger and Peric 2002; White 1999)

$$\frac{\partial \rho}{\partial t} + \frac{\partial \rho u_i}{\partial x_i} = S$$

where, u is the velocity, ρ is the density and S is the source term.

Compressible form of the momentum conservation equation is given by the famous Navier Stokes equation (Ferziger and Peric 2002; White 1999)

$$\frac{\partial \rho u_i}{\partial t} + \frac{\partial \rho u_i u_j}{\partial x_j} = -\frac{\partial P}{\partial x_i} + \frac{\partial \sigma_{ij}}{\partial x_j} + S_i$$

where, P is the pressure, and σ_{ij} is the viscous stress tensor given by (Richards et al. 2017)

$$\sigma_{ij} = \mu_t\left(\frac{\partial u_i}{\partial x_j} + \frac{\partial u_j}{\partial x_i}\right) + \left(\mu' - \frac{2}{3}\mu_t\right)\left(\frac{\partial u_k}{\partial x_k}\delta_{ij}\right)$$

where, μ' is the dilatational viscosity (usually set to zero), δ_{ij} is the Kronecker delta and μ_t is the turbulent viscosity given by

$$\mu_t = \mu + C_\mu \rho \frac{k^2}{\varepsilon}$$

where, μ is the molecular viscosity, C_μ is a turbulence model constant, k is the turbulent kinetic energy, and ε is the turbulent dissipation.

The conditions inside the cylinder will always be turbulent, hence we consider the turbulence model always. If turbulence is not being considered, $\mu_t = \mu$ in the above equation.

Gas flow could be considered as compressible flow, and liquid flow could be considered as incompressible flow. An appropriate equation of state must be chosen for the compressible flow. Ideal gas laws are "idealistic" and the conditions inside the cylinder many a time deviate from ideal gas assumptions. Therefore, a more appropriate equation of state such as Redlich-Kwong equation of state, given below, could be used (Richards et al. 2017).

$$P = \frac{RT}{v - \beta_{rk} v_c} - \frac{\left(\alpha_{rk} \frac{p_c v_c^2}{\sqrt{T_r}}\right)}{v^2 + \beta_{rk} v_c v}$$

where, $v_c = \frac{RT_c}{p_c}$ is the critical volume, T_c is the critical temperature, p_c is the critical pressure, α_{rk} represents the attractive forces between molecules, and β_{rk} represents the volume of the molecules. Most CFD softwares give the option to choose the thermodynamic properties as functions of temperature alone (ideal gas) or as functions of temperature and pressure (real gas). Critical temperature and pressure of air could be used here since the main constituent is air, and the values are 133 K and 37.7 MPa, respectively.

Redlich-Kwong equation is found to give relatively poor results in the liquid phase, while Peng-Robinson equation is found to give better results. Redlich-Kwong-Soave equation is found to give good results for hydrocarbons (Richards et al. 2017).

Energy equation with convection and diffusion terms, given below, is solved for specific internal energy conservation. Flow velocities inside the cylinder are not as high as to generate discontinuities like shock waves, hence solving for specific internal energy alone instead of total energy seems appropriate (Richards et al. 2017).

$$\frac{\partial \rho e}{\partial t} + \frac{\partial u_j \rho e}{\partial x_j} = -P\frac{\partial u_j}{\partial x_j} + \sigma_{ij}\frac{\partial u_i}{\partial x_j} + \frac{\partial}{\partial x_j}\left(K_t \frac{\partial T}{\partial x_j}\right) + \frac{\partial}{\partial x_j}\left(\rho D \sum_m h_m \frac{\partial Y_m}{\partial x_j}\right) + S$$

where, ρ is the density, e is the specific internal energy, P is the pressure, σ_{ij} is the stress tensor, T is the temperature, D is the mass diffusion coefficient, h_m is the species enthalpy, Y_m is the mass fraction of species m, and S is the source term.

K_t is the turbulent conductivity given by

$$K_t = K + c_p \frac{\mu_t}{Pr_t}$$

where, K is the conductivity, $Pr_t = \frac{c_p \mu_t}{K_t}$ is the turbulent Prandtl number, c_p is the specific heat at constant pressure, and μ_t is the turbulent viscosity. Turbulent Prandtl number can be given as an input in the model.

In the energy equation, the source term S accounts for turbulent dissipation. The pressure work term $-P\frac{\partial u_j}{\partial x_j}$ accounts for compression and expansion, the viscous dissipation term $\sigma_{ij}\frac{\partial u_i}{\partial x_j}$ accounts for kinetic energy viscously dissipating into heat and the species diffusion term $\frac{\partial}{\partial x_j}\left(\rho D \sum_m h_m \frac{\partial Y_m}{\partial x_j}\right)$ accounts for energy transport due to species diffusion.

Species conservation equation, given below, is solved to estimate the mass fraction of all species in the cylinder (Richards et al. 2017; Turns 2000)

$$\frac{\partial \rho_m}{\partial t} + \frac{\partial \rho_m u_j}{\partial x_j} = \frac{\partial}{\partial x_j}\left(\rho D_t \frac{\partial Y_m}{\partial x_j}\right) + S_m$$

where, ρ_m is the density of species m, u is the velocity, Y_m is the mass fraction of species m in a cell and is calculated as:

$$Y_m = \frac{M_m}{M_{tot}} = \frac{\rho_m}{\rho_{tot}}$$

where, M_m is the mass of species m, M_{tot} is the total mass of the cell, ρ_{tot} is the total density of the cell, $\rho_m = Y_m \rho$, and D_t is the turbulent mass diffusion coefficient given by

$$D_t = \frac{\nu_t}{Sc_t}$$

where, Sc_t is the turbulent Schmidt number, whose value can be given as an input in the model.

The source term S_m accounts for evaporation, combustion chemical reactions and other sub-models. The diffusion coefficients are calculated independently of the species.

Pollutants like NOx and soot could be modelled as passives. Passives could also be used to track charge air motion similar to what one does in particle image velocimetry (PIV) technique. Passive transport is governed by the passive transport equation given below

$$\frac{\partial \rho \phi}{\partial t} + \frac{\partial \rho u_i \phi}{\partial x_i} = \frac{\partial}{\partial x_i}\left(\rho D \frac{\partial \phi}{\partial x_i}\right) + S$$

where, ϕ is the passive scalar, ρ is the density, u is the velocity, S is the source term and D is the diffusion coefficient given by

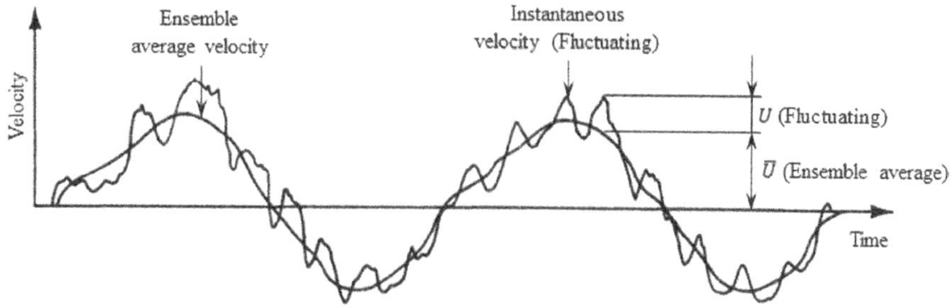

Fig. 5.2 Velocity components of in-cylinder charge motion

$$D = \frac{\nu}{Sc}$$

where, ν is the kinematic viscosity and Sc is the Schmidt number, which can be given as input during passives modelling.

Turbulence is an integral part of diesel engine combustion. It enhances the rate of mixing of air and fuel and in terms of the transport equations, it enhances the rate of mixing of momentum, energy, and species. Typically, two types of turbulence models are used—Reynolds Averaged Navier Stokes (RANS) and Large Eddy Simulation (LES).

RANS models are two equation models including the standard $k - \varepsilon$ model, RNG (Renormalization Group) $k - \varepsilon$ model, Standard $k - \omega$ model, etc. In these models, the flow variables like velocity are split into an ensemble mean and a fluctuating term. Figure 5.2 shows the ensemble mean and fluctuating component in in-cylinder charge motion. From Fig. 5.2, the instantaneous velocity at any time is given by

$$u = \overline{U} + U$$

where, $\overline{U}$ is the ensemble mean and U is the fluctuating component.

This RANS decomposed term is substituted into the continuity equation and Navier Stokes equation to get the turbulence transport equations. These also require two additional equations—one for turbulent kinetic energy and one for dissipation, given by

$$\frac{\partial \rho k}{\partial t} + \frac{\partial \rho u_i k}{\partial x_i} = \tau_{ij}\frac{\partial u_i}{\partial x_j} + \frac{\partial}{\partial x_j}\left(\frac{\mu}{Pr_k}\frac{\partial k}{\partial x_j}\right) - \rho\varepsilon + \frac{C_s}{c}S_s$$

$$\frac{\partial \rho \varepsilon}{\partial t} + \frac{\partial \rho u_i \varepsilon}{\partial x_i} = \frac{\partial}{\partial x_j}\left(\frac{\mu}{Pr_k}\frac{\partial \varepsilon}{\partial x_j}\right) + c_{\varepsilon 3}\rho\varepsilon\frac{\partial u_i}{\partial x_i} + \left(c_{\varepsilon 1}\tau_{ij}\frac{\partial u_i}{\partial x_j} - c_{\varepsilon 2}\rho\varepsilon + C_s S_s\right)\frac{\varepsilon}{k} + S - \rho R$$

where, u_i is the instantaneous velocity (sum of ensemble mean and the fluctuating component), c is an empirical constant (usually taken as 1.5), S is the user defined source term, S_s is the source term arising from the interaction with spray, $c_{\varepsilon 1}, c_{\varepsilon 2}$ and $c_{\varepsilon 3}$ are the model constants for compression and expansion periods, R is given by

$$R = \frac{C_\mu \eta^3 \left(1 - \frac{\eta}{\eta_0}\right)}{1 + \beta \eta^3} \frac{\varepsilon^2}{k}$$

where, η is given by

$$\eta = \frac{k}{\varepsilon} \left| S_{ij} \right| = \frac{k}{\varepsilon} \sqrt{2 S_{ij} S_{ij}}$$

$R = 0$ for standard $k - \varepsilon$ model.

In LES models, the difference is in the splitting of flow variable into a resolved field and a sub-grid field. For IC engine applications, RANS models like RNG $k - \varepsilon$ model is proven and widely used.

These governing equations discussed above must be solved to obtain the pressure, temperature, and composition of gases inside the cylinder at each crank angle. Commercial CFD softwares generally use the finite volume approach due to their ability to solve complex geometries with simplicity, where the integral form of the governing equations is solved. Finite volume approach solves equations as fluxes on the cell faces.

In finite volume approach, the variables could be stored in a collocated grid or a staggered grid arrangement (Tannehill et al. 1997). Figure 5.3 shows a collocated grid arrangement where all variables are stored at cell center (Richards et al. 2017). If using collocated grid arrangement, adequate care must be taken to avoid undesired checkerboarding of pressure or velocity which leads to spurious results. Schemes such as Rhie-Chow interpolation method could be used to avoid this checkerboarding.

To calculate the flux at the cell face, the velocity value has to be interpolated. This could be done by averaging the velocity of adjacent cells (central difference method) or by using upwind techniques. Central difference method gives second

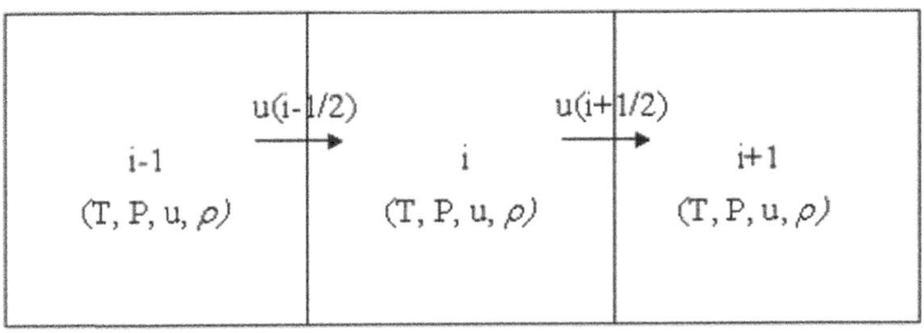

Fig. 5.3 Collocated grid arrangement (Richards et al. 2017). © Convergent Science, Inc. All rights reserved. Used with permission

order accurate results, but the prediction of transport of convective property (velocity here) may not be correct. Upwind techniques are more physically correct as they can predict the transport of property always downstream of flow and avoid spurious values which could result by the averaging technique. Upwind schemes have order of accuracy as one, even though it is less than central difference technique, physically correct result is ensured. The order of accuracy could be enhanced by modifying the upwind schemes. One will have the option to select the method and control the convergence, correctness, and order of accuracy of results accordingly.

If we look at the governing equations, there is no dedicated equation to solve for pressure, it is not calculated directly from the equation of state. There should be some sort of pressure–velocity coupling to solve for pressure. There are various pressure–velocity coupling techniques used conventionally like Marker and Cell method (MAC), Simplified Marker and Cell Method (SMAC), Semi Implicit Method for Pressure Linked Equations (SIMPLE), SIMPLE Revised (SIMPLER), SIMPLE Corrected/Consistent (SIMPLEC), Pressure Implicit Splitting of Operators (PISO), etc. (Tannehill et al. 1997). PISO is also an extended version of the SIMPLE family of methods. SIMPLE family and PISO are the most popular pressure–velocity coupling techniques available.

A typical solution procedure used in CFD simulations is shown in Fig. 5.4 (Richards et al. 2017). The simulation starts with the initial conditions. These values are used to calculate various parameters associated with the different sub-models including source terms as applicable. Now, the estimation of thermodynamic parameters and species mass begin.

First the momentum equation is solved. Then using appropriate pressure–velocity coupling, equation of state and density-pressure relation through pressure ratio, pressure, velocity, temperature, and species mass are estimated by solving the relevant governing equations. Based on the acceptable tolerance on error, this pressure–velocity coupling loop is iterated. Once the pressure estimate of acceptable error is available, this is used to estimate density. Subsequently, species mass transport equation is solved to get the mass fraction of different species. Then energy equation is solved to estimate the temperature. Using the updated values, pressure ratio is updated, and pressure is recalculated. Using these updated values, velocity is calculated again.

For convergence and optimum computational time, there are certain numbers termed CFL (Courant-Friedrichs-Lewis) numbers which relate the time step size with grid size. The larger the number, lesser the computation time, but it could also result in the solution diverging when run in explicit mode. Over-relaxation or under-relaxation could be used to accelerate convergence.

5.6 Input Data—The Link Between Experiments and CFD

In the previous section, we have seen how the physical processes occurring inside the engine cylinder are expressed as mathematical models through the governing

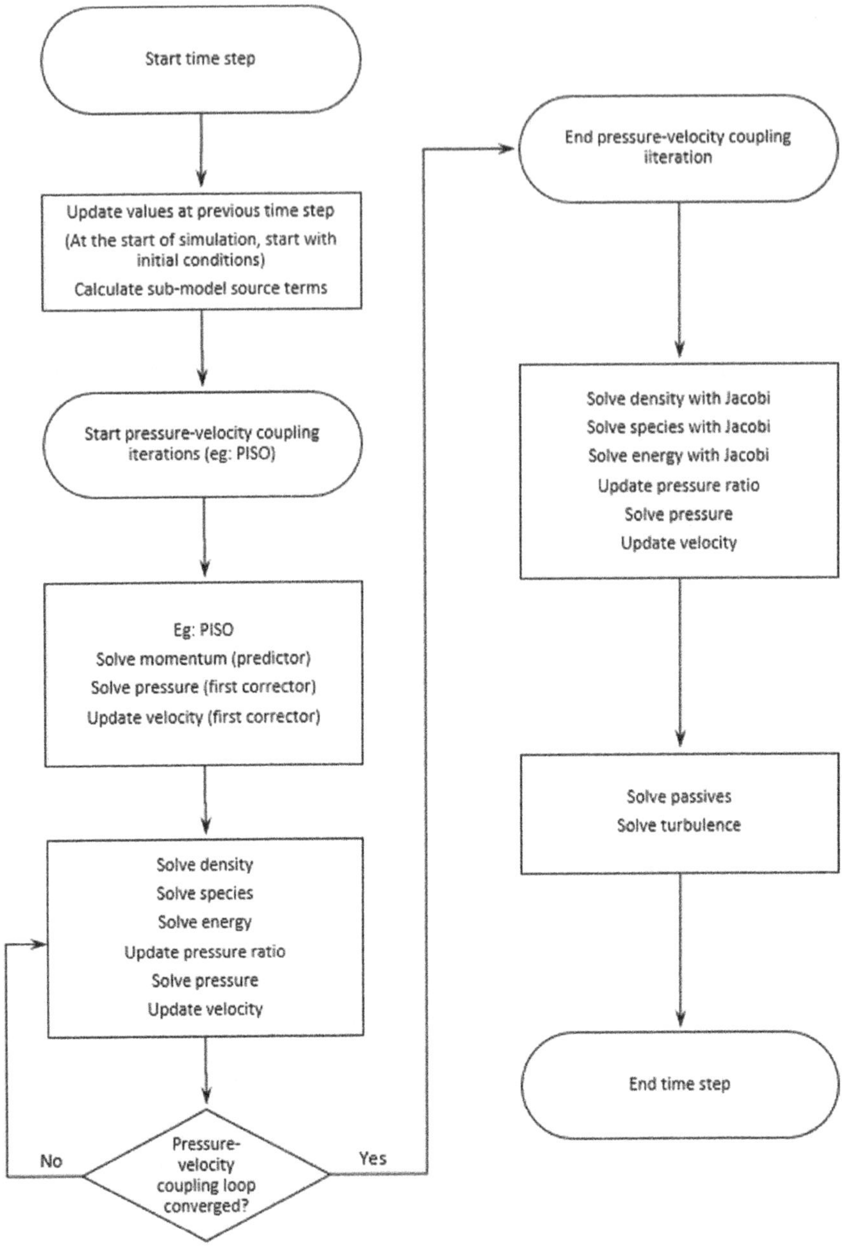

Fig. 5.4 Typical solution procedure in CFD simulation

equations. All these governing equations are partial differential equations in time and space, which require initial conditions and boundary conditions to solve. These conditions are obtained from the physical testing of the engine. All CFD models represent to an actual system, an IC engine in present context, and the objectives are meant to be achieved on the actual engine. CFD is a tool to predict the effect a design would have on an actual engine. To make a good CFD model, it should correlate well to the physical engine to the maximum extent possible. This is achieved through proper data acquisition by conducting experiments on the actual engine. The more data that we have, the better is the chance to make a good CFD model.

The input data required for diesel engine simulation irrespective of open cycle or closed cycle are the basic geometric specifications of the engine like bore, stroke, connecting rod length, compression ratio, valve timings and lift, etc.; engine speed, mass flow rates of fuel and air; fuel injection pressure; fuel injection timing, duration and quantity of fuel injected; fuel system specifications like number of orifices in the injector, orifice diameter, nozzle tip protrusion, nozzle spray angle, spray solid cone angle, etc. and composition of the intake charge. Specific to open cycle simulation, the intake manifold and exhaust manifold pressure and temperature on a crank angle resolved basis if available is ideal. For closed cycle simulation, crank angle resolved in-cylinder pressure and temperature are needed. Closed cycle simulations also need the initial conditions for turbulence level within the cylinder. The above set of data contribute to the initial conditions and boundary conditions necessary to solve the governing equations. For model validation (discussed in more detail in Sect. 5.11), crank angle resolved in-cylinder pressure is necessary for both types of simulation. Emission measurements if available could also be used in model validation as applicable.

We will now see the thermodynamic measurements usually done on a diesel engine and how they are measured briefly. Measurement of geometric specifications are not discussed here. Engines are tested on test rigs with dynamometers for loading the engine. The test set-up should also have the necessary arrangement to avoid pressure fluctuations in the intake flow.

Intake manifold and exhaust manifold pressures are usually measured using piezoresistive sensors which give the absolute pressure measurement. In-cylinder pressure is measured using piezoelectric pressure sensors which detect the relative pressure changes inside the cylinder. However, for combustion analysis, we need absolute pressure which is obtained through pegging. Pegging is a process where the relative pressure is referenced to some known absolute pressure, for example, the in-cylinder pressure at bottom dead center (BDC) is referenced to intake manifold pressure whose absolute value is known (Burnt and Pond 1997). Proper care must be taken in pegging and TDC estimation which could lead to error in combustion phasing with respect to crank angle. The crank angle resolution should be as low as possible (for example 0.1 CAD) to generate better data for analysis. It is also recommended to record the in-cylinder pressure data for 100–120 cycles at least and then take the average of that to eliminate the cycle-to-cycle variations and get smoothened data.

Modern engines are equipped with crank speed sensor and cam phase sensor which together detect engine speed and compression TDC. These sensors are usually Hall effect type sensors which detect changes in magnetic flux in combination with a toothed wheel mounted on the crankshaft or camshaft. Optical encoders could also be used for this purpose. This will require a separate encoder disc with equally spaced alternate transparent and opaque strips or holes which interpret engine speed through the frequency at which an external light source passes through the disc.

Fuel mass flow rate could be measured through gravimetric measurement using a weighing balance which is probably the simplest yet accurate method. This will require some bulky installation in the test set-up though. Other methods include installing a small turbine type electronic flow meter in the fuel line, but the disadvantage is that it measures the volume flow rate whose calibration also depends on fuel viscosity and pressure (Kirkpatrick 2021). While installation is compact, calibration is tedious and measurement may not be as reliable as weighing balance method. In all methods, there will be some fuel return from the injector or common rail which also needs to be accounted for by connecting the return downstream of the measuring point.

Air mass flow rate could be measured using a constant temperature hot film anemometer. This instrument can give the flow rate within 1%. The flow measured by the flow meter may not get trapped inside the cylinder and participate in combustion due to leakage losses in the intake system and blow-by losses, so it is recommended to verify the air mass by measuring the exhaust flow rate as well (Kirkpatrick 2021).

Manifold gas temperature could be measured using thermocouples of appropriate range. In-cylinder temperature is usually not measured due to the difficulties associated with it. It is very difficult to locate a thermocouple and pressure sensor simultaneously in the cramped region between the valves, injector, and coolant jacket. In addition to this, the response time needed for real time temperature data acquisition is very high which is not practically possible. The temperature stratification that exists within the engine cylinder also makes real time combustion temperature data acquisition very difficult.

If the CFD objectives include emission reduction, it is advisable to acquire emission measurement data as well, which could be used in model validation too. Chief regulated pollutants from diesel engines are oxides of nitrogen (NOx), particulate matter (PM)—mainly soot, unburned hydrocarbons (HC), and carbon monoxide (CO). The measurement technique for each pollutant is different, for example, NOx could be measured using chemiluminescence analyzer, HC could be estimated using flame ionization detection (FID) method, CO could be measured using non-dispersive infra-red radiation (NDIR) method while soot could be estimated using a smoke meter like Hartridge smoke meter (Kirkpatrick 2021). There are other techniques also to measure these pollutants and the one available at the test rig could be used, but the results also must be treated with appropriate caution.

It is not always necessary to test the engine to generate the data. Sometimes, published literature might be available with detailed test data which could be used for the CFD study. However, having one's own test rig is always useful if one wishes to generate more data or wants to recreate a test condition later. Moreover, it is desirable to have a good control over the test conditions and experimental data to prepare an efficient CFD model as the initial conditions and boundary conditions form the key input data to solve the governing equations of the mathematical CFD model.

5.7 Typical Initial Conditions

Solving unsteady state partial differential equations require initial conditions for the dependent variables. These initial conditions are the values of the dependent variables like pressure, temperature, turbulent kinetic energy, etc. at the crank angle corresponding to the start of simulation. As discussed in Sect. 5.6, these are obtained from the measured data generated through the experiments.

In closed cycle simulation, the initial conditions needed are the composition of the charge at IVC, the pressure and temperature inside the engine cylinder at IVC, and the turbulent kinetic energy and dissipation rate of the charge motion at IVC. Out of these, usually only the in-cylinder pressure would be available from the measured data. Temperature could be estimated based on the trapped mass calculated from the measured air flow.

Turbulence related parameters are estimated based on parameters like swirl number, and it is also a usual practice to use the values recommended by the commercial software developer based on the combustion model. In open cycle simulation, these are calculated through simulations and as discussed in Sect. 5.3.1, after four cycles the turbulence levels stabilize inside the cylinder. The data thus generated from open cycle simulation could be used for further closed cycle simulation to improve the model accuracy. Swirl number is a typical parameter used for turbulence inside the cylinder. It is estimated by rig level measurements of intake port, valves, and engine cylinder. The same set-up could also measure the flow discharge coefficients of the valves, which are used in open cycle simulation. There is limited literature available with published data on the swirl or turbulence related initial conditions used in CFD. One such work (Menon et al. 2020) on a DI diesel engine with a swirl number 2 used the values of turbulent kinetic energy as 62.0271 m^2/s^2 and turbulent dissipation as 17,183.4 m^2/s^3 at a very low load and a high load condition at 1800 rpm.

Composition of charge is difficult to accurately estimate especially if there is some internal EGR or backflow during the valve open period. In engines equipped with external EGR, the external EGR quantity would be known accurately as a percentage of fresh air. The number of species in air is also a variable and must be fixed while specifying the charge composition.

EGR or residual gases comprise of the combustion products which has carbon dioxide (CO_2) and water vapour (H_2O) as the major constituents along with various hydrocarbons, oxides of nitrogen, oxygen, nitrogen, and other minor species. The effect of EGR or residual gas on combustion is to act as a diluent, i.e., to displace and reduce the concentration of oxygen in the intake charge. Reducing the oxygen concentration results in less oxygen available for combustion, hence decreases the overall temperature. Another significant effect is the thermal effect, that is, increasing the specific heat at constant pressure causing the charge to absorb more heat reducing the overall temperature. There is also some dissociation (endothermic reactions) of CO_2 and H_2O which causes drop in temperature, which is called chemical effect of EGR. There is also a density effect associated with EGR which is at higher temperature (combustion products) which has an effect of reducing the fresh charge density. Previous studies (Ladommatos et al. 1996a, b, 1997) had established that the main effect of CO_2 or H_2O as EGR constituent is to act as diluent, and its thermal and chemical effects, independently are much less when compared to the dilution effect. Therefore, it is safe to take either CO_2 alone or a combination of CO_2 and H_2O to simulate residual gas fraction or EGR.

While specifying the residual mass fraction in closed cycle simulation, one could use the thermal effect of EGR due to the change in C_p value of the charge to fine-tune the polytropic index during compression. Polytropic index could be estimated as the slope of log plot of pressure–volume for the cycle. The measured air mass flow rate will not directly reflect the mass of air inside the cylinder during IVC due to backflow, residual gas fraction and changes in temperature. So, the initial condition of temperature is fixed using these constraints to match the measured air mass flow rate and polytropic index of compression to the maximum extent possible.

In open cycle simulation, the amount of residual gas trapped, and its temperature are automatically calculated during the full cycle. The other initial conditions like the mass flow rate of air, its temperature and pressure at the crank angle corresponding to the start of simulation are also available from measurements. The exhaust mass flow rate and exhaust gas temperature in the manifold are also available from measurements. If there is external EGR, its quantity and temperature could also be measured.

5.8 Typical Boundary Conditions

In the case of a partial differential equation in three dimensions, one has to specify three spatial boundary conditions for the dependent variable to solve the equation. In the case of unsteady state equation, time based initial condition is also needed. In CFD, there are broadly four types of boundary conditions based on the manner in which the dependent variable appears in the relation, which are discussed below.

5.8.1 Dirichlet Boundary Condition

In this type of boundary condition, the absolute value of the dependent variable is directly specified at the boundary. For example, the x component of velocity at a wall is taken as zero (no slip boundary condition).

5.8.2 Neumann Boundary Condition

Here, the boundary condition is in the form of the gradient of the dependent variable. For example, adiabatic condition at a wall implies that there is no heat transfer or temperature gradient in the normal direction.

5.8.3 Robbins Boundary Condition

In this type of boundary condition, the gradient of the dependent variable is a function of the dependent variable itself. For example, convective heat transfer occurring over a boundary will be represented in terms of temperature gradient as a function of the temperature difference between the wall and the surrounding fluid.

5.8.4 Mixed Boundary Condition

This type of boundary condition as the name signifies is a mixture of two or more of the above three types applied at certain regions over a boundary.

There is one more classification based on how or where the boundary condition is applied. This classification is as follows.

5.8.5 Flow Boundary Condition

As discussed in Sect. 5.3.1, this type of boundary condition is required in open cycle simulation, where there is flow of mass across the boundary. One could specify a velocity boundary condition and a pressure boundary condition to fully define the flow conditions to solve the momentum equation. The velocity boundary condition could be in terms of mass flow rate too. These values are derived from the measurement data. The boundary condition must also take care of backflow conditions which could occur in some cases. To solve the energy equation, temperature boundary conditions also have to be defined. Turbulence equations need boundary conditions for turbulent kinetic energy and dissipation rate. To solve for the species mass fractions, there must be boundary conditions for each of the species. Similarly, if passives are part of the simulation, boundary conditions must be specified for that as well. All these boundary conditions could be Dirichlet or Neumann or even Robbins type of boundary condition which depends on the physics of the problem at the boundary.

5.8.6 Wall Boundary Condition

This type of boundary condition as the name implies, applies to fixed or moving walls of the model like the surface of piston, liner wall, valve head, valve seats, etc. which form the geometric boundary of the computational domain. These boundary conditions must be specified for both open cycle and closed cycle simulations. In modern direct injection diesel engines, the turbulence levels inside the engine are high, so it may not be possible to resolve the viscous sub-layer (Warsi 1993). Hence the velocity boundary conditions specified for the walls must take care of this effect, therefore it is recommended to use law-of-the-wall type boundary conditions for velocity and temperature, which is a logarithmic curve fit of the turbulent boundary layer.

Temperature must also be specified for all boundaries. In-cylinder temperatures are rarely measured due to the difficulties associated with it, while the temperatures inside the intake manifold and exhaust manifold could be measured easily. So, the wall temperatures of the manifolds could be acquired directly from the experiments whereas the temperatures of the piston, cylinder head, liner and the ports have to be estimated. From measurements and simulations done on modern diesel engines, the surface temperature on the liner and cylinder head is around 200 °C at the rated torque and power conditions. The temperature on the piston surface is usually around 250–300 °C at the rated condition. Typical coolant temperatures are also in the range of 85–110 °C in the rated steady state conditions. The data acquired from the measurement on DI diesel engine piston is shown in Fig. 5.5, measured liner temperature on DI diesel engine at 11 bar bmep is shown in Fig. 5.6. Cylinder head temperatures measured at different locations of an SI engine is also shown for reference in Fig. 5.7 (Heywood 2018).

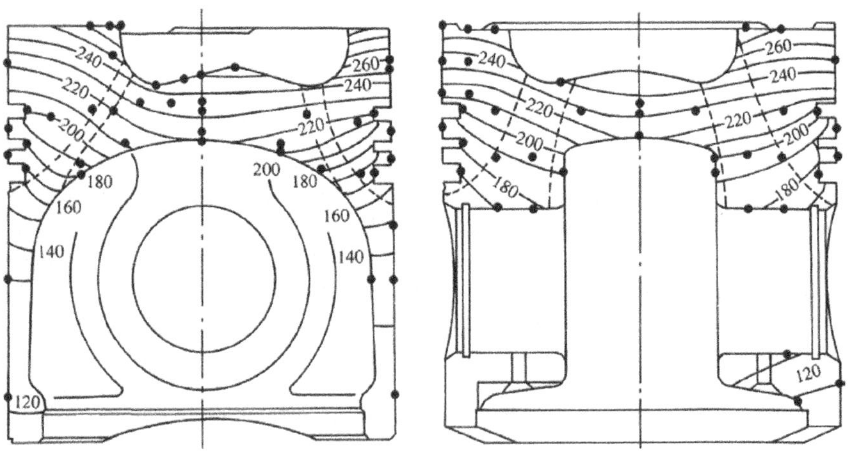

Fig. 5.5 Temperature profile on a DI Diesel engine piston (Furuhama and Suzuki 1979)

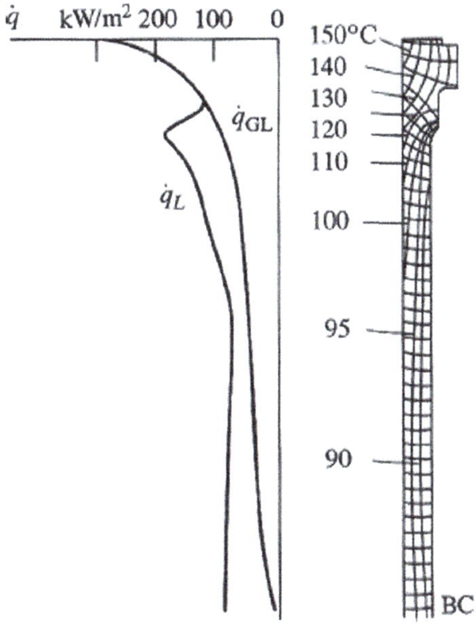

Fig. 5.6 Liner wall temperatures in a Diesel engine (Woschni 1979)

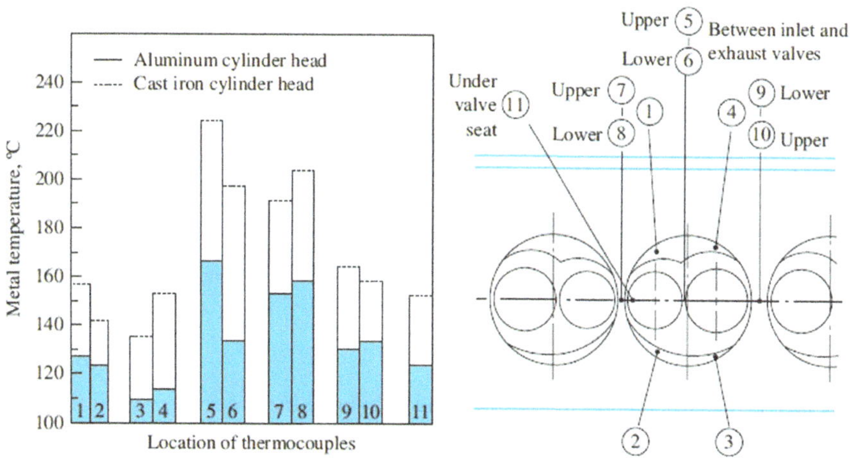

Fig. 5.7 Measured temperature in an SI engine cylinder head (Heywood 2018)

Figure 5.8 shows the temperature gradients that is typically found in IC engines between burned gases and coolant (Heywood 2018).

Figure 5.8 shows the instantaneous gas temperature T_g, average gas temperature $\overline{T}_g$, convection heat transfer $\dot{q}_{CV}$, radiation heat transfer $\dot{q}_R$, conduction through the liner $\dot{q}_{CN}$, convection heat transfer to coolant $\dot{q}_{CV}$ at temperature T_c.

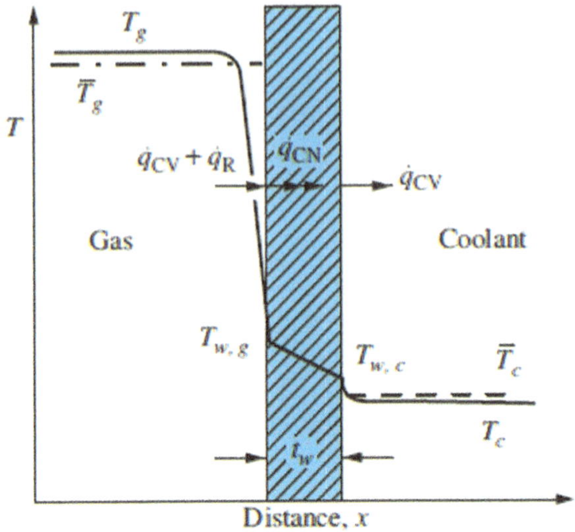

Fig. 5.8 Temperature gradient between burned gas and coolant (Heywood 2018)

With this experience and the measured temperature of the coolant, one could safely guess the starting value for the boundary wall temperatures which could be fine-tuned later if required during model validation.

There are various wall heat transfer models also available in turbulence modeling like O'Rourke and Amsden (1997), Han and Reitz (1997), Angelberger et al. (1997), etc. which could be selected.

Turbulence boundary conditions must also be specified for the walls. Neumann boundary condition of zero normal gradient could be given for turbulent kinetic energy for all walls, and Dirichlet boundary conditions for the turbulent dissipation could be specified which would usually be calculated near the wall by a relation (Richards et al. 2017) of the form

$$\varepsilon = \frac{C_\mu^{0.75} k^{1.5}}{\kappa y}$$

where, ε is the turbulent dissipation, k is the turbulent kinetic energy, y is the distance from the wall to the middle of the cell, C_μ is a turbulence model constant, and κ is von Karman's constant of 0.4187.

5.8.7 Periodic Boundary Condition

As discussed in Sect. 5.3.2, symmetric combustion chambers could be simplified using sector simulations. Such simulations use periodic boundary conditions.

Most of the commercial CFD software also give the option to specify temporally and spatially varying boundary conditions which would increase the accuracy of the results at a higher computational cost. It is up to the user to select the level of detail in selecting the boundary condition depending on the objectives set.

5.9 Different Sub-models

This section discusses the different sub-models used in diesel engine CFD simulations. A brief overview of different sub-models is given with specific details and references of some of the common sub-models used.

5.9.1 *Piston Motion*

In IC engine, during the 720 CAD rotation of the crankshaft, the piston will execute four strokes governed by the geometry of the slider crank mechanism. Piston motion is governed by the equation

$$z = 2a - l\left[1 - \cos\left\{\sin^{-1}\left(\frac{a\sin\theta + x}{l}\right)\right\}\right] - a(1 - \cos\theta)$$

where, z is the position of piston along piston motion direction, a is half stroke, l is the connecting rod length, x is the crank pin offset, and θ is the crank angle. These parameters are given as inputs in the engine specification details.

5.9.2 *Reaction Mechanism*

Diesel is a complex multicomponent hydrocarbon fuel. Combustion reaction occurs when there is appropriate conditions of equivalence ratio and temperature inside the engine cylinder with energy more than the activation energy required to break the chemical bonds. A global one step reaction of a hydrocarbon is represented as follows (Olikara and Borman 1975)

$$x_{13}\left\{C_nH_mO_lN_k + \left(\frac{\left(n + \frac{m}{4}\right) - \left(\frac{l}{2}\right)}{\phi}\right)[O_2 + 3.76N_2 + 0.0444Ar]\right\} \rightarrow$$
$$x_1H + x_2O + x_3N + x_4H_2 + x_5OH + x_6CO + x_7NO + x_8O_2 + x_9H_2O + x_{10}CO_2 + x_{11}N_2 + x_{12}Ar$$

where, ϕ is the equivalence ratio.

In reality, this global single step reaction never happens in a single step, rather the reaction will proceed through hundreds of elementary reactions through the formation of numerous radicals or active species. The number of elements involved in all the reactions are a maximum of four—C, H, O and N. It is impossible to include all elementary reactions and study their chemical kinetics to arrive at the combustion rate and products. Researchers have come up with reduced species reaction mechanism with a smaller number of species and elementary reactions which approximately give similar results. Most of the commercial CFD Softwares give the options to choose the reaction mechanism and many a time, it is linked with the combustion model selected. For example, commercial software CONVERGE gives the option of a mechanism with about 42 species through 168 reactions. These reactions are defined in the CONVERGE library and are activated based on the combustion model selection. The reaction data also has the pre-exponential factor, the temperature exponent and the activation energy required for the calculation of the forward reaction rate from the Arrhenius equation. The reverse reaction rate is calculated from the forward reaction rate and the equilibrium constant determined from the thermodynamic properties (Richards et al. 2017).

5.9.3 Spray Modelling

Liquid diesel after injection undergoes break-up, air entrainment due to drop drag force, inter-drop collision, turbulent dispersion, wall impingement and finally evaporation. These processes could be modelled in many ways. This section discusses some of the popular models in each process. As seen in the previous section, diesel is a complex multi-component hydrocarbon, so it is difficult to model diesel as it is and most of the commercial softwares use surrogates to model diesel in the liquid phase and gaseous phase. It is important to select the appropriate surrogate in each phase by looking into the properties listed in the software database.

Diesel fuel injectors used in modern direct injection engines are electronically operated mechanical injectors. There will be a lag between the electric signal from the engine management system to the actual start of injection, termed as the hydraulic delay. This delay could vary anywhere between 2 and 7 CAD depending on the injector and engine management system. Similarly, the injectors will not be able to deliver fuel at a uniform injection rate. Also, depending on the injection duration (period for which the injector nozzles remain open), it might be possible that the injector is not fully open, especially during pilot injections. So, the rate of injection varies from one operating point to another and from one injector to another. This rate of injection is also a user input for CFD simulation. One has the option of measuring the approximate rate of injection through experiments based on the variation in fuel injection pressure and the actual fuel mass delivered, but it is a tedious process. Most CFD studies arrive at an ideal rate of injection through trial and error method. Depending on the objectives, for example if the objective is only a back-to-back comparison of designs, one could also go for a simple square profile for the injection rate.

The process of fuel delivery into the cylinder depends on the following parameters—injection quantity, duration, rate of injection, injection pressure and coefficient of discharge. Most softwares use three of these parameters to fix the amount of fuel delivered in a crank angle. So, one must play around with these parameters to fine tune the model to match the experiment data. Usually the injection quantity and injection pressure are the most accurate information available from the measured data. Actual injection duration could differ from the duration taken from the electric signal due to the hydraulic lag and inertia of the injector.

Many standard models are available to model the liquid fuel injection droplet size distribution. In blob injection model the drop size is equal to the effective nozzle diameter calculated from the discharge coefficient. There are probabilistic distributions models like Chi square distribution, Rosin Rammler distribution, etc. which are based on Sauter Mean Diameter.

After fuel injection, the next physical process that occurs is the break-up of the liquid jet into droplets which happens due to the surrounding air drag which causes distortion and breakup of the drop. Drop drag and distortion is usually modelled with the classic Taylor Analogy Breakup (TAB) based model (O'Rourke and Amsden 1987) which is based on the analogy between an oscillating drop and a spring-mass-damper system. It is also recommended to use a dynamic drop drag model if available to account for the departure from the spherical shape of the drop to the extreme case of a disk.

There are many spray breakup mechanisms available in most of the commercial softwares to estimate the atomization and breakup. The most popular models are discussed here. Kelvin–Helmholtz (KH) Rayleigh–Taylor (RT) models, Taylor Analogy Breakup (TAB), etc. are the most popular ones.

Kelvin–Helmholtz (KH) model is based on the liquid jet stability analysis (Reitz and Bracco 1986). KH-ACT model (Som and Aggarwal 2010) is another available model which is a modified KH model which also considers aerodynamics, cavitation, and turbulence. Rayleigh–Taylor (RT) model is another common model used to predict breakup. While some models which use this instability mechanism neglect gas and liquid phase viscosity (Ricart et al. 1997), there are other models which also consider the effect of viscosity (Senecal et al. 2007). There have been models developed by combing the KH and RT models which have been found to give better initial conditions for the evaporation and subsequent combustion and emissions calculations (Patterson and Reitz 1998). KH and RT based models are the most popularly used (Xin et al. 1998; Hwang et al. 1996) and most commercial softwares give options to tune the breakup time through constants where one can get a faster breakup if required.

The droplets formed after breakup will undergo collision and coalescence. Various models are available to predict droplet collision and subsequent coalescence. The widely used ones for collision are O'Rourke scheme (O'Rourke 1981), No time counter (NTC) scheme (Schmidt and Rutland 2000) which are all probability-based models working on random number generation to decide the probability of collision. NTC scheme has been reported to give faster and more accurate results under certain conditions (Richards et al. 2017). Collision outcomes are modelled by schemes like

O'Rourke (1981), Post and Abraham (2002), etc. Grid size also plays and important role in calculating the collisions as most models consider collisions happening within one grid only. It is recommended to refine the grid size for the spray region to get better resolution of results.

During fuel injection, depending on the injection pressure, in-cylinder pressure and turbulence and the injection timing, there are chances of the spray impinging on the wall. Wall being cooler usually lead to the formation of a thin fuel film. It is important to model the wall impingement phenomenon to accurately predict droplet evaporation and mixing. Rebound/slide model (Naber and Reitz 1988; Manuel et al. 1991) is one of the most widely used models. In this model, the incoming jet droplet might either rebound or slide on the wall depending on its Weber number. Wall film/splash model is another model that is used (Stanton et al. 1998).

Popular droplet evaporation models are Frossling model (Amsden et al. 1989) and Chiang model (Chiang et al. 1992), which calculate how the droplet radius change over time. These models also use heat transfer correlations to the drop to evaluate the droplet temperature and the rate of evaporation. Wall film evaporation models can also be separately included which could incorporate the conduction, convection, boiling and evaporation of the film.

5.9.4 Combustion Modelling

Once the droplet has evaporated and mixed with the surrounding air, the next phase is the combustion. This section discusses some of the popular combustion models used commercially. SAGE chemistry solver model is one of the most popular general combustion model which uses the Arrhenius equation to calculate the reaction rate for the elementary reactions based on the local conditions in a cell (Senecal et al. 2003). Extended coherent flamelet model (ECFM) (Colin et al. 2003) is usually used to simulate premixed combustion hence more useful in spark ignition engines. For diesel applications, this needs to be applied to three zones—fuel only, air or air plus residuals, mixing zone—thus making it ECFM3Z model. Some studies which used ECFM3Z model used a different model to predict the auto-ignition (Moiz et al. 2015).

Shell CTC model is another commonly used (Kong and Reitz 1993; Tan and Reitz 2004) diesel combustion model which separately models the ignition delay period using the Shell model which was originally designed to predict knock in gasoline engine (Halstead et al. 1977; Schapertons and Lee 1985). The CTC (characteristic time combustion) model (Abraham et al. 1985; Xin et al. 1997) as the name indicates is based on a characteristic time parameter which is a function of chemical kinetics time and turbulent mixing time, to arrive at equilibrium conditions. It considers seven species fuel, O_2, N_2, CO_2, H_2O, CO, and H_2 and the mass fractions are solved using atom balance, relations of equilibrium constant and equilibrium water–gas shift reaction with Newton–Raphson method.

A study (Moiz et al. 2015) done on a diesel engine by comparing different combustion models performance with experimental data in terms of ignition delay, flame lift-off length, flame penetration length and global flame structures yielded the following results. ECFM3Z gave relatively decent estimates in general in comparison with the SHELL CTC model. SAGE model was found to be computation intensive with longer run-time, but gave good correlation for ignition delay and flame lift-off length.

5.9.5 Turbulence Modelling

In-cylinder turbulence is best modelled with the RNG $k - \varepsilon$ model as it accounts for the flow compressibility which has been found to have significant influence in combustion, especially soot emissions (Han and Reitz 1995). RNG $k - \varepsilon$ model has also been found to give better results when compared with the standard $k - \varepsilon$ model due to its better capability of predicting small eddies (Abay et al. 2018). The other popular models were discussed in Sect. 5.5.

5.9.6 Wall Heat Transfer Model

There are various wall heat transfer models as discussed in Sect. 5.8.6. O'Rourke and Amsden wall heat transfer model (Amsden 1997) is given by

$$k\frac{\partial T}{\partial x_i} = \frac{\mu_m c_p F\left(T_f - T_w\right)}{Pr_m y} n_i$$

where,

$$F = \begin{cases} 1.0, & y^+ < 11.05 \\ \dfrac{\frac{y^+ Pr_m}{Pr_t}}{\frac{1}{\kappa} ln(y^+) + B + 11.05\left(\frac{Pr_m}{Pr_t} - 1\right)}, & y^+ < 11.05 \end{cases}$$

$$y^+ = \frac{\rho \mu_\tau y}{\mu_m}$$

k is the thermal conductivity, κ is von Karman constant, B is the law of the wall parameter, Pr_m is the molecular Prandtl number, Pr_t is the turbulent Prandtl number, T_f is the fluid temperature, T_w is the wall temperature, and μ_τ is the shear speed. This model has been found to give good results with $\kappa = 0.42$ and $B = 5.5$.

Han and Reitz model (1997) takes into account the unsteady compressible effects. It is given by

$$k\frac{\partial T}{\partial x_i} = \begin{cases} \frac{\mu_m c_p (T_f - T_w)}{Pr_m y} n_i, \ y^+ < 11.05 \\ \frac{\rho c_p \mu_\tau T_f ln\left(\frac{T_f}{T_w}\right) n_i}{2.1 ln(y^+) + 2.513}, \ y^+ > 11.05 \end{cases}$$

Angelberger model (1997) is given by

$$k\frac{\partial T}{\partial x_i} = \frac{\rho c_p \mu_\tau T_w \ln\left(\frac{T_f}{T_w}\right) n_i}{\theta^+}$$

where,

$$\theta^+ = \begin{cases} Pr y^+, \ y^+ \leq 13.2 \\ 2.075 \ln(y^+) + 3.9, \ y^+ > 13.2 \end{cases}$$

5.9.7 Emissions Modelling

In most CFD applications, there would be separate sub-models used for NOx and PM emissions calculations. The other species would be estimated directly from the combustion model itself. Here the common emission sub-models are discussed briefly.

NOx emission modelling. The universally accepted extended Zel'dovich mechanism is used in most applications and is given by the following reactions

$$O + N_2 \rightleftharpoons NO + N$$
$$N + O_2 \rightleftharpoons NO + O$$
$$N + OH \rightleftharpoons NO + H$$

The first two reactions were given initially (Bowman 1975), and the third reaction was added later (Lavoie et al. 2007). The direction of these reactions is governed by the rate constants and the reaction rates depend on the temperature.

NO could further be converted to NO_2 through reactions such as (Merryman and Levy 1073)

$$NO + HO_2 \rightarrow NO_2 + OH$$

The reverse reaction is also possible if conditions are favourable.

$$NO_2 + O \rightarrow NO + O_2$$

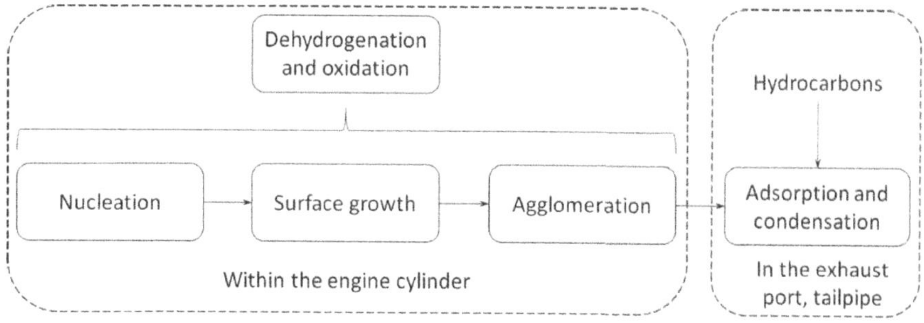

Fig. 5.9 Stages of formation of particulate matter

CONVERGE uses a factor of 1.533, i.e., the ratio of molecular weights of NO_2 to NO, to convert the calculated NO into NOx (Richards et al. 2017).

PM emission modelling. PM is the term given to plain carbon soot, its agglomerates and particles which have unburned hydrocarbons deposited or condensed on them. The basic structure forming PM is soot. Figure 5.9 shows the stages of formation of particulate matter in a diesel engine (Amann and Siegla 1981).

At high temperatures during combustion, the fuel hydrocarbons undergo pyrolysis and/or oxidation and form many unsaturated hydrocarbons and polycyclic aromatic hydrocarbons. These are thought to be the precursors for the soot formation, from which the initial soot particles are formed (nucleation) via the condensation of gas phase species. These small particles then undergo surface growth, where the small particles attach to each other form bigger surfaces and the process is called agglomeration. Other hydrocarbons also tend to deposit or condense on these surfaces. Simultaneously, some of the soot particles formed undergo oxidation to form CO or CO_2. This formation and oxidation reactions happen in parallel, and the balance of these two decide the net soot coming out of the engine. Formation and oxidation are also governed by temperature and hence freeze after a certain temperature.

Soot concentration, as discussed above, is determined by the competing reactions of soot formation and soot oxidation. One of the most popular soot formation models (Han et al. 1996) is the Hiroyasu model (Hiroyasu and Kadota 1976), and the soot oxidation model (Kim and Huh 1998) which has been used with it in most studies is the NSC (Nagle Strickland-Constable) model (Nagle and Strickland-Constable 1962). Total hydrocarbons are considered as precursors of soot formation. The agglomeration of soot particles and the condensation of hydrocarbons on the soot surface are not considered in this model. This is an empirical model which has tuning parameters which can be adjusted to match the experimental data.

Some commercial softwares also provide more detailed phenomenological models which are based on simplified physical processes which also include the soot coagulation as well. Models like Gokul (Vishwanathan and Reitz 2010), Dalian (Jia et al. 2009), Waseda (Kaminaga et al. 2008) all use similar Arrhenius relation for soot formation and NSC based oxidation models with additional models for soot coagulation. Gokul model has been found to be more suitable for low temperature

Case	1	2	3	4	5
Elements average	104,230	249,530	300,130	360,143	446,765
Cylinder pressure	1.987×10^6	3.004×10^6	3.213×10^6	3.213×10^6	3.213×10^6

Fig. 5.10 Summary of grid independence study based on peak firing pressure (Hamid et al. 2020)

diesel application, Dalian model for diesel HCCI application and Waseda for medium duty diesel application (Richards et al. 2017). The soot precursors in each case also might be different.

5.10 The Importance of Grid Independence Study

CFD analysis of engine phenomena are indeed time saving as well as cost saving, but one must keep in mind the computation cost as well as time. As computation time increases, computation cost also increases, and it affects other simulations which might be in queue. It is also ideal to consume as much less storage and memory re-sources of the computer while simulating. Hence, it is desirable to optimize the computation effort required to achieve the required results. This is done by a grid independence study.

The main factor which affect computation time is the number of grids in the computation domain. Without doubt, the finer the grid, the more accurate or resolved the results. It is up to the user to determine what level of accuracy or resolution is required. Also, depending on the nature of the problem and the solver used, it is a common observation that beyond a certain grid size, further refinement in grid size will not alter the results much.

Specific to IC engine simulations, grid independence study could be performed by comparing the in-cylinder pressure. One could compare the peak pressure with various base grid sizes or could compare the overall trend of the pressure trace. Figure 5.10 shows a comparison which used peak pressure as the reference in a study on a diesel engine (Hamid et al. 2020).

Figure 5.11 shows a comparison which used the overall pressure trace as the reference in a study on a diesel engine (Mauro et al. 2018). The comparison could be done in motoring condition or firing condition too.

5.11 Model Validation

Before one can rely on the results estimated by a CFD model, it is essential to prove that the model is capable of predicting the required results with the required level of accuracy. The process by which this is established is called model validation. Here, the results predicted by the model are compared with the data from measurements to

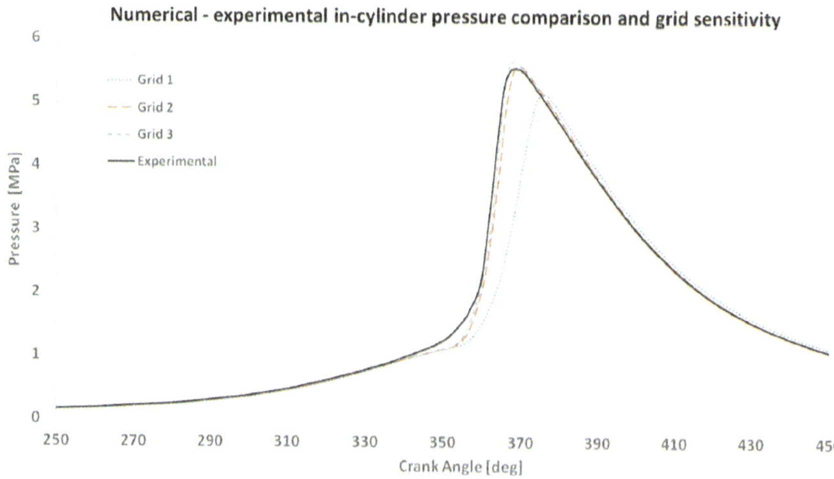

Fig. 5.11 Grid independence study comparing pressure trace (Mauro et al. 2018)

see the correlation. Higher the correlation, better the model. This is one of the very important phases of the CFD study which determines the usefulness of the results generated by the CFD model.

To prove that the data calculated by the model, ideally all the result parameters have to be compared with experimental data, which depends on the set objectives. For instance, if the objective of the simulation is to predict the performance and emissions of a diesel engine, then it is desirable to have a comparison of the performance and emissions predicted by the CFD model with the measured data.

Diesel engine performance is estimated using the in-cylinder pressure vs. crank angle trace and the instantaneous heat release rate vs. crank angle trace. These are the two most popular parameters which are used to validate the CFD model. Figure 5.12 shows a model validation where pressure trace and heat release rate trace are compared with experimental data (Brijesh et al. 2014).

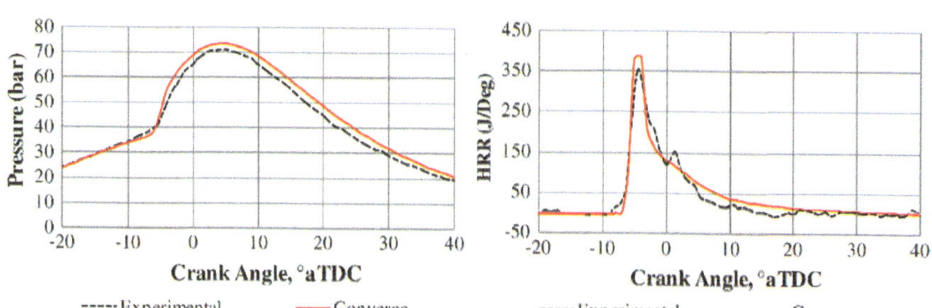

Fig. 5.12 Model validation with pressure trace and rate of heat release curve (Brijesh et al. 2014)

CFD models can predict the pollutant formation on a crank angle basis, but experi-mentally it is difficult to get that information. So, the emission values are usually compared between the CFD estimated value at EVO and the measured value. Figure 5.13 shows a model validation where the emission values are compared in a natural gas—diesel dual fuel engine (Shu et al. 2019). Here, the authors have validated the model at four operating points and the emissions predictions were satisfactory. The trends of variation of NOx, HC and CO between the four operating points observed in experiments were captured in simulation.

Figure 5.14 shows another model where NOx and soot emission values have been compared (Mobasheri and Peng 2012a). Here, the crank-angle based variation of NOx and soot are plotted, however the stabilized value at the end of simulation is compared with the measured value.

There are some studies done on intake flow where the model validation is done based on the velocity vector field of the intake flow by comparing the results with observations from optical engines (Perini et al. 2018). Figure 5.15 shows one such model validation where the tangential velocity in a horizontal plane at height d_z is predicted and compared with measured data obtained through PIV experiments for two swirl ratios (R_s) at different crank angles.

There are studies done where model is validated against data generated from optical engines. One such study compared the soot emissions from experiments

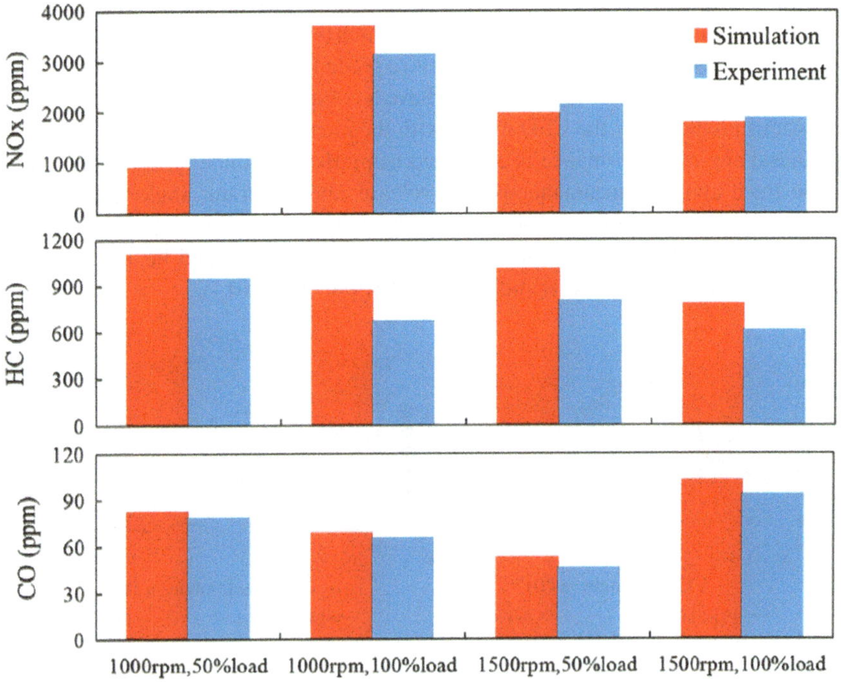

Fig. 5.13 Model validation comparing emissions (Shu et al. 2019)

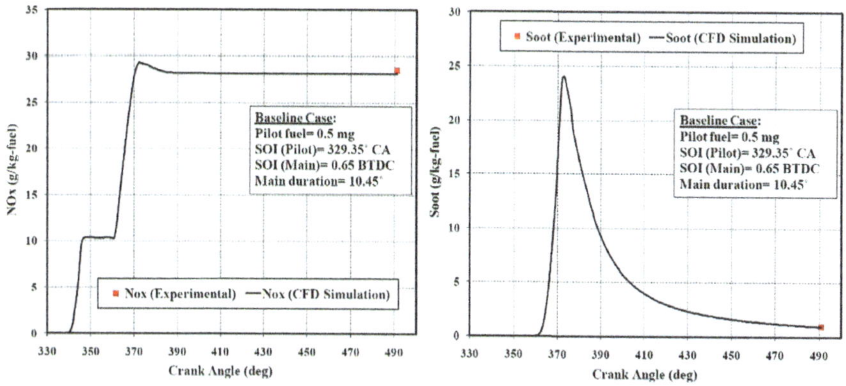

Fig. 5.14 Model validation plotting emission formation, but comparing the stabilized value (Mobasheri and Peng 2012a)

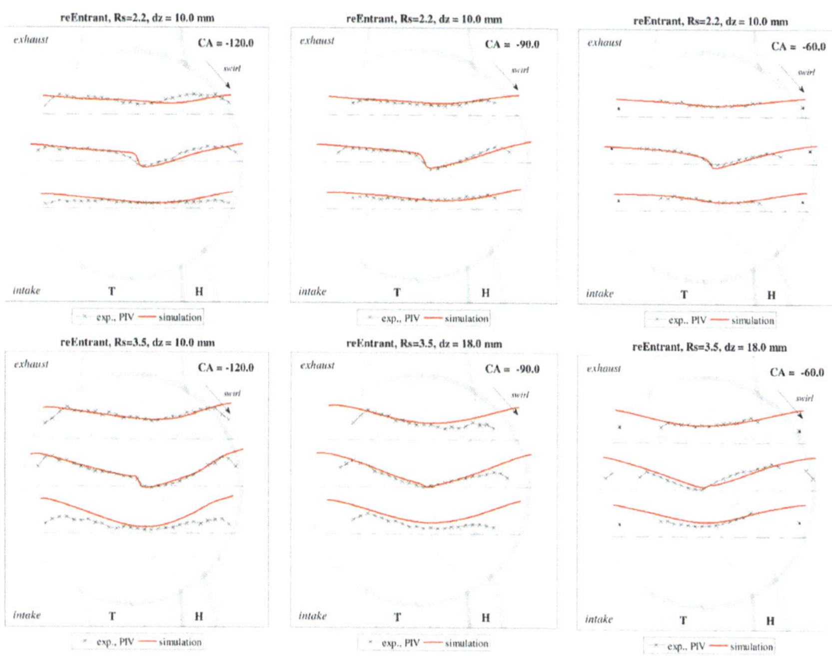

Fig. 5.15 Model validation with measured velocity vector field (Perini et al. 2018)

done using planar laser induced incandescence (PLII), Fig. 5.16 shows the model validation for soot mass fraction contours (Kong et al. 2017). Here the soot mass fraction contours at varying distance from the injector (x axis) at time after start of

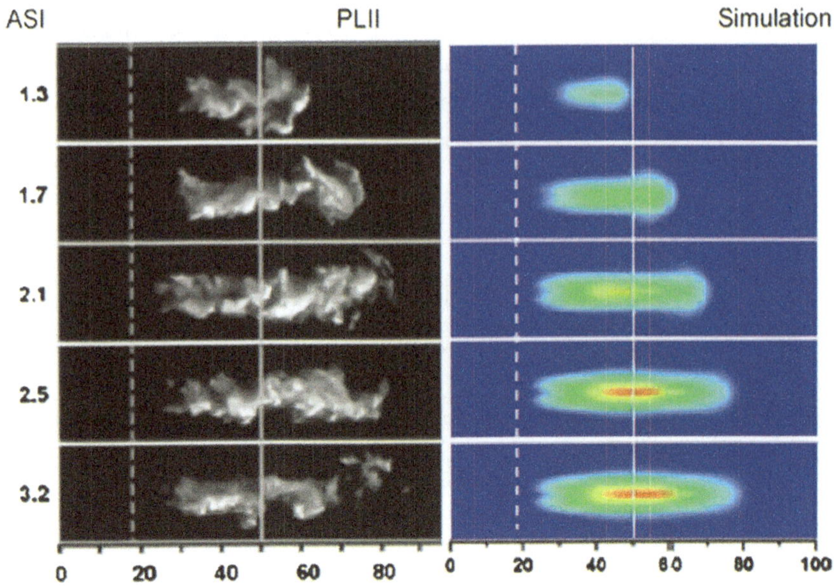

Fig. 5.16 Model validation of soot mass fraction with optical engine experiment (Kong et al. 2017)

injection (ASI) is compared between simulation and planar laser-induced incandescence (PLII) experiments. Simulation is found to predict well especially after 2 ms after start of injection.

However, it may not be practical always to get all data measured, especially emissions, as the measurement methods available in the test rig might not give the complete or accurate information. This is particularly true in soot as the popular soot measuring methods use dilution tunnel or smoke meters which give a qualitative result rather than a quantitative result. Soot is commonly measured in units of Filter Smoke Number (FSN) and it is difficult to accurately convert it to units of g/kW-h or g/kg fuel. Even though correlations exist for approximate conversion to quantitative data, the values may not be very accurate. If one has access to accurate measured data of emissions, it is desirable to add that in the model validation, as it helps in making the model more robust.

It is worth noting that there are many studies done previously which compare only the pressure trace or both pressure and heat release rate curves for model validation and discuss on the emissions comparison of the models (Puri et al. 2017; Sharma et al. 2010; Shuai et al. 2009; Lee et al. 2015; Zhu et al. 2004). Moreover as mentioned earlier, the level of model validation depends on the objectives, for instance if the objective is mainly to compare the effects of various in-cylinder strategies, rather than accurately predicting the absolute values of performance and emission parameters in each case, one could do model validation without emission data as well.

5.12 Analysis of Results

The advantage of CFD simulation is that you can generate a lot of data in a short time. It is up to the user to decide what all data is to be generated based on the objectives. One must select only the bare minimum data required for one's analysis to keep the computation cost and storage space requirement less. For instance, CFD simulation is capable of generating spatially resolved data including velocity and temperature, but one may not need this data for analysis if the objective is solely to estimate say emissions or fuel consumption. In that case, one could select only relevant data to generate 3D output and also at the relevant crank angles to minimize cost and storage space.

The way we use the data generated by CFD simulation is called post-processing which in itself is limitless. It is up to the user to select the data and decide on how to present it in the best possible manner. Most common data presented in diesel engine CFD simulation are the in-cylinder pressure profile and rate of heat release rate as in any thermodynamic simulation, which were already discussed in Sect. 5.11. Accumulated heat release or cumulative heat release is another parameter which is used frequently to analyze combustion, Fig. 5.17 shows one such comparison from a study done on a diesel engine comparing different turbulence models (Jagus and Jiang 2011).

If one is interested in studying the charge flow during the intake and compression stroke till IVC, velocity magnitudes could be plotted with the directions indicated

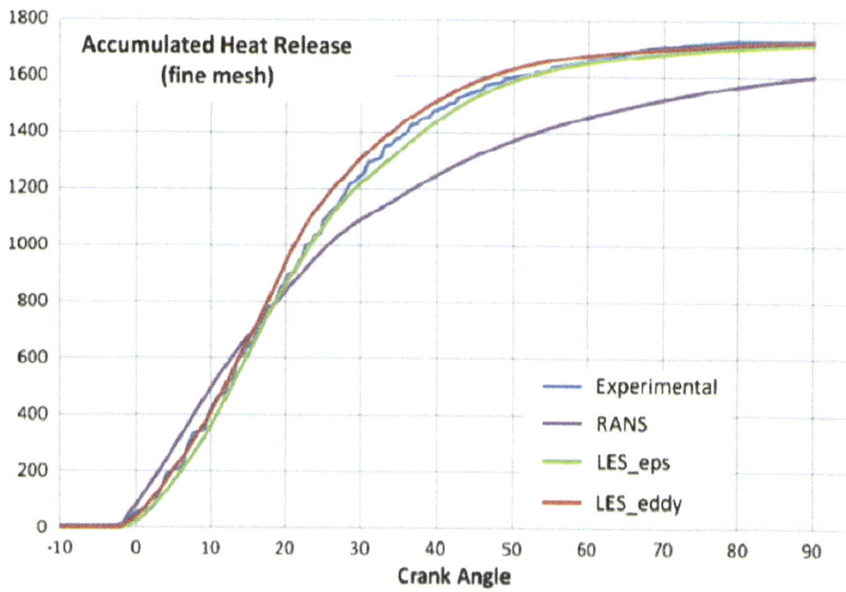

Fig. 5.17 Accumulated/cumulative heat release (Jagus and Jiang 2011)

by the arrows. Such an example is shown in Fig. 5.18 which compares flow field between two piston bowl designs (Li et al. 2010). One could also plot the pressure, temperature and turbulent kinetic energy at different sections and crank angles of interest as shown in Fig. 5.19 (Sharma et al. 2010).

In studies focusing on soot emissions, one could go for plots of temperature and soot as shown in Fig. 5.20 (Shi and Reitz 2008a). Temperature plot is shown in Fig. 5.20a and soot concentration in Fig. 5.20b for a study done by varying the injector spray angle, swirl ratio and injection pressure.

Local equivalence ratio is another factor which influences soot and NOx formation which can also be plotted as has been done in some studies as shown in Fig. 5.21 (Khan et al. 2018). Regions of high equivalence ratio result in incomplete combustion and consequently higher soot formation, whereas equivalence ratio closer to

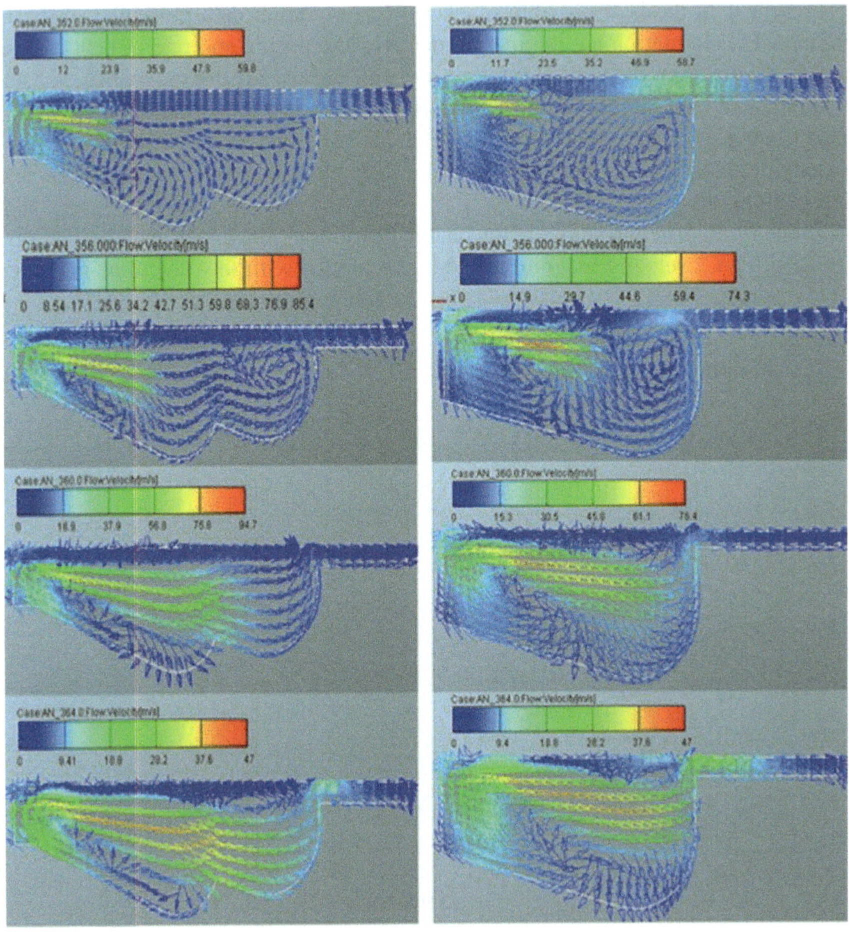

Fig. 5.18 Plot of velocity vector field (Li et al. 2010)

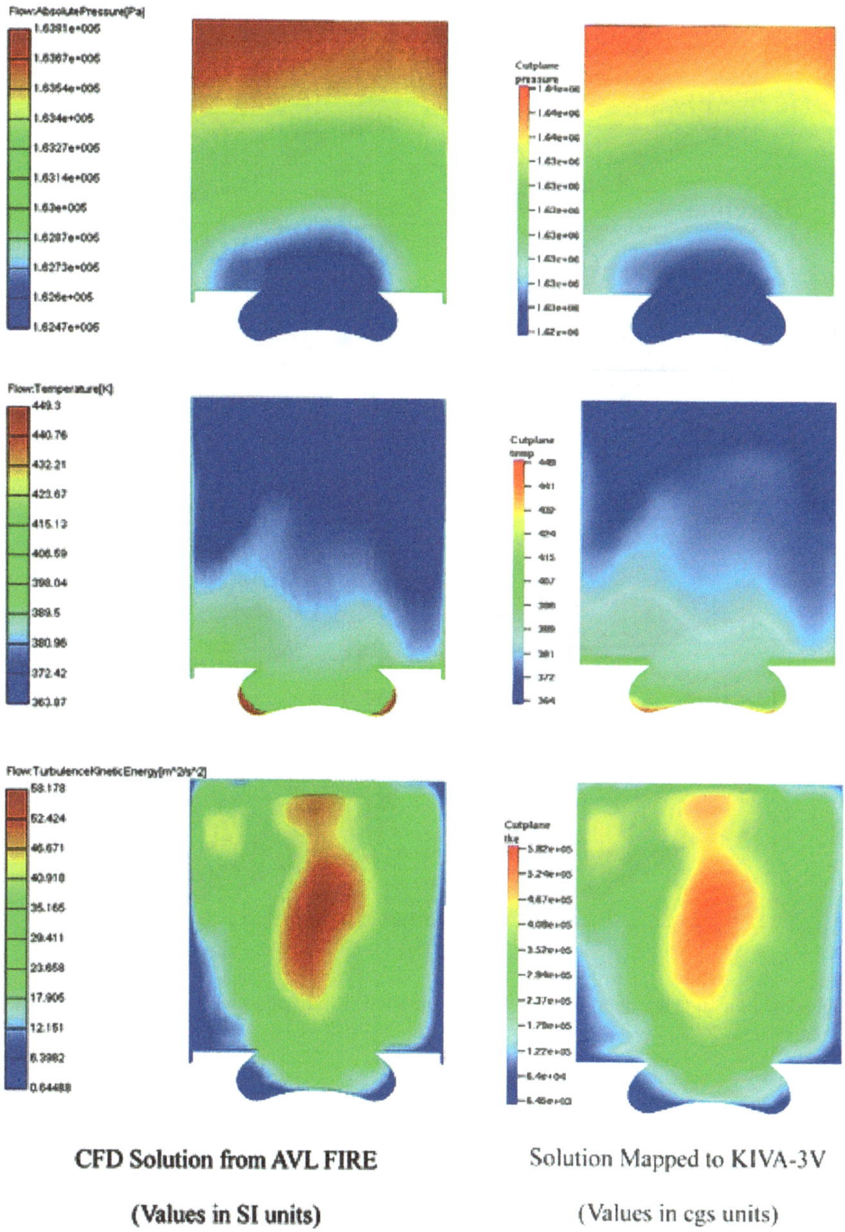

Fig. 5.19 Temperature, pressure, and turbulent kinetic energy contours (Sharma et al. 2010)

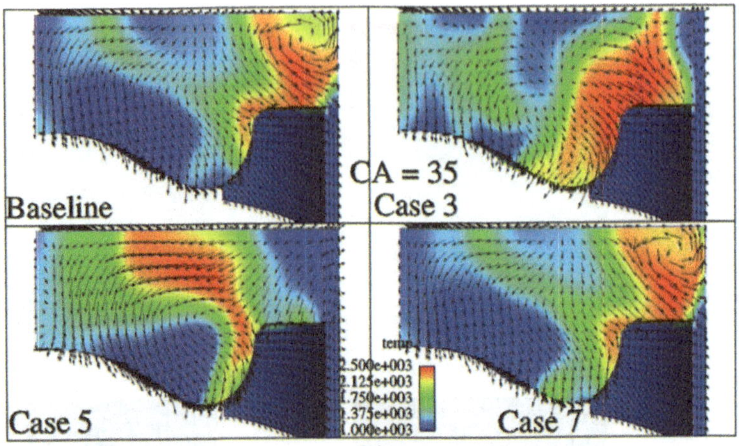

(a)

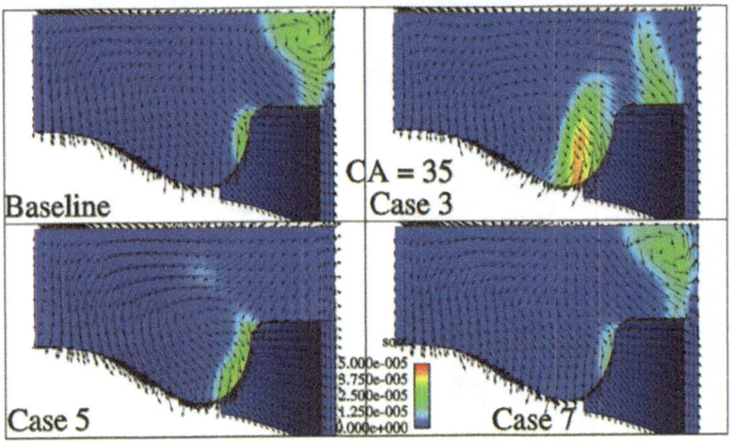

(b)

Fig. 5.20 Plot showing temperature and concentration of soot and oxygen during fuel injection (Shi and Reitz 2008a)

stoichiometric ratio could be a result of better mixing and hence produce less soot and more NOx.

One could simulate the behaviour of liquid spray impingement on piston bowl and see how the jet splits, how the film forms, etc. Figure 5.22 shows one such application where the droplet behaviour and the temperature distribution after combustion are shown (Li et al. 2010).

In models estimating soot emissions, it might be worthwhile to consider the OH radical concentration as well since OH radical presence is also an important factor in soot oxidation apart from O and O_2. Figure 5.23 shows one such plot of soot, OH density and soot particle size (Dp) at different crank angles (Vishwanathan and Reitz 2015).

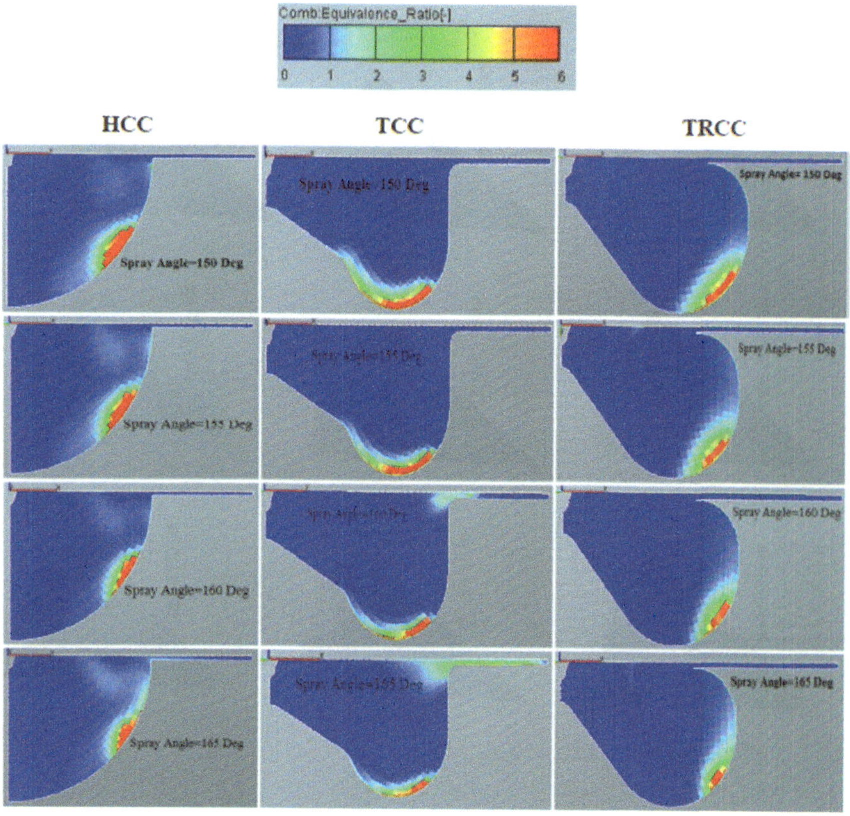

Fig. 5.21 Plot showing local equivalence ratio (Khan et al. 2018)

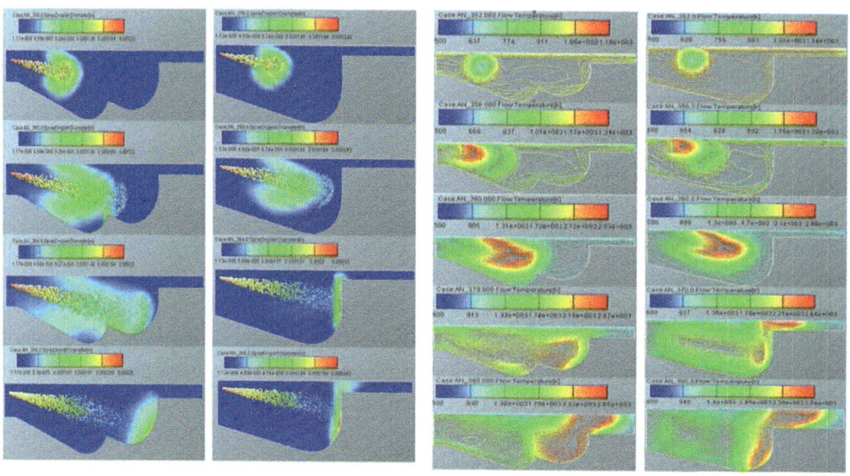

Fig. 5.22 Fuel impingement on piston bowl and corresponding temperature profile (Li et al. 2010)

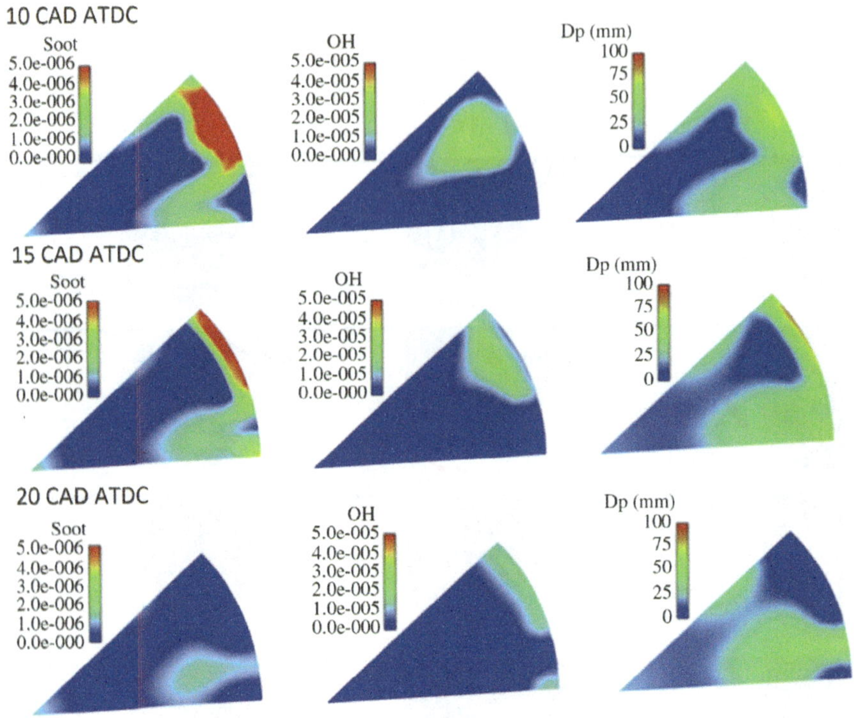

Fig. 5.23 Plot showing soot and OH density (Vishwanathan and Reitz 2015)

When comparing the effect of two or more parameters on emissions or performance, it would be worthwhile to use surface plots as shown in Fig. 5.24 (Shi and Reitz 2008a).

5.13 Application of CFD in Engine Optimization with In-Cylinder Techniques

One of the key applications of CFD in diesel engines is in the optimization of engine combustion chamber and fuel system to achieve better performance, lower fuel consumption and emissions. These methods are commonly known as in-cylinder techniques as the changes are done on in-cylinder parameters like the combustion chamber shape, fuel injection timing, duration, multiple injections and EGR. This section discusses briefly about how CFD analysis has been used in engine optimization with in-cylinder techniques particularly combustion chamber shape change, main injection timing, use of a pilot injection and EGR, by various researchers.

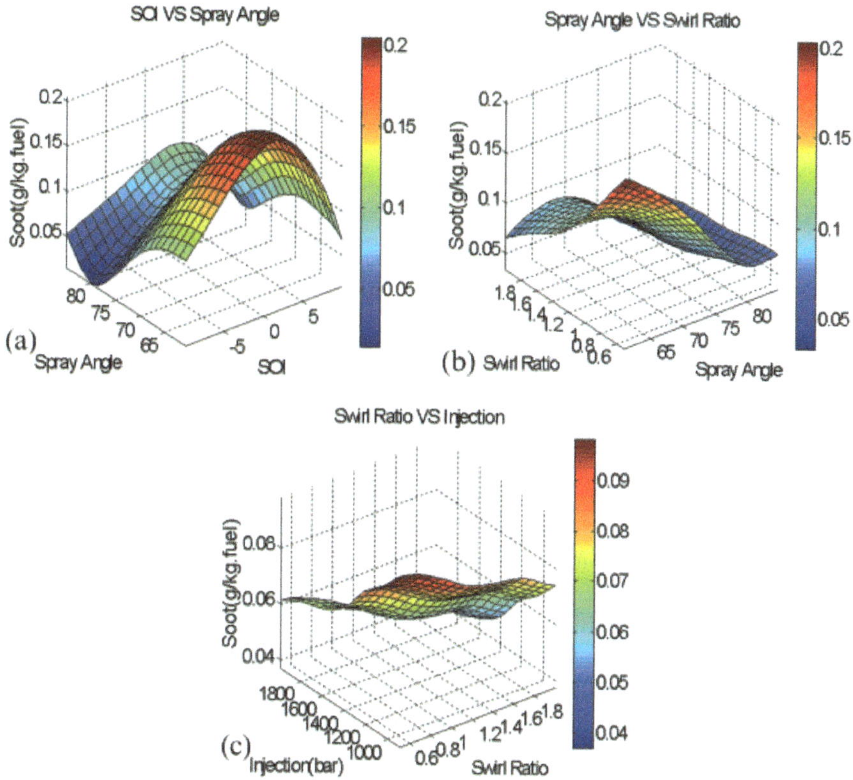

Fig. 5.24 Surface plot to illustrate the inter-dependence of multiple variables (Shi and Reitz 2008a)

5.13.1 Combustion Chamber Geometry

In direct injection diesel engines, the combustion chamber is formed by the walls of the cylinder head, liner, and the piston bowl. The charge air motion, its flow patterns, and velocities at the time of injection govern the rate of mixing of the fuel and air. The shape of the combustion chamber and the injection timing affect how the liquid jet impinges on the wall. Hence, the shape of the combustion chamber plays a significant role in determining the mixing of fuel and air, as the geometry will determine the turbulence levels of the charge at the time of fuel injection for the pre-mixed combustion (Rao et al. 1993).

Figure 5.25 shows the basic geometrical parameters of a typical direct injection diesel piston bowl (Menon et al. 2020). The variation of these parameters influences the charge motion within the combustion chamber and hence the fuel–air mixing. Here, D_{max} is the bowl diameter, D_t is the throat diameter, h is the bowl depth, p is the pip height, l is the lip height, R is the toroidal arc radius and α is the pip cone angle.

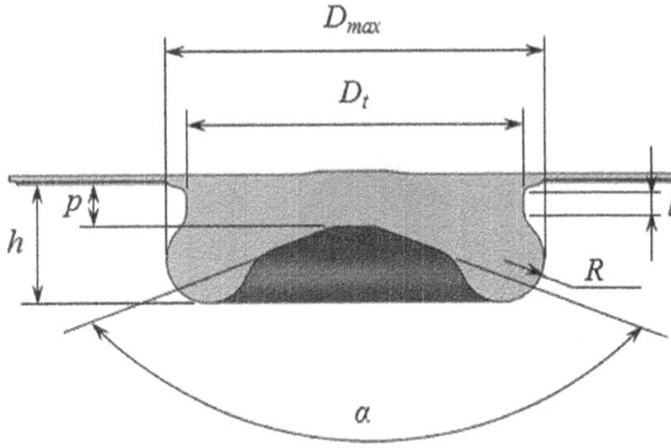

Fig. 5.25 Geometrical parameters of a typical diesel engine piston bowl (Menon et al. 2020)

As mentioned earlier, the combustion chamber geometry influences how the fuel jet strikes the wall and behaves afterwards, so the injection timing and spray angle at one operating point for one bowl geometry may not be the ideal for another. This is one of the main objectives of combustion optimization, to find the right combination for the required application.

Studies suggest that the fuel jet striking well within the bowl towards the center, neither towards the squish nor to the bottom regions, result in the best utilization of air (Dolak et al. 2010). This is especially suitable for multiple injections as the interval between the two injections could be controlled in such a way as to maximize the air utilization. A stepped piston bowl geometry helps achieve this by intentionally making the two injections to strike at two different locations, second below the first which will ensure more air above and below the spray. In conventional piston bowl geometries, the burned gases from the first injection combustion will displace some of the air available for the second injection spray which would result in higher soot. To reduce NOx, the injection timings of the two injections have to be optimized.

Studies have shown that bowl width and depth have direct implication on soot and NOx respectively, and it has been observed that the narrower the bowl, higher the soot, and NOx is high if the bowl is too deep or too shallow (Mobasheri and Peng 2012b). The same study also found out that if the spray strikes the wall early, it will result in less soot and more NOx, and this can be achieved by having a longer lip. A parametric CFD study on piston bowl geometry and injection timing noted that soot and fuel consumption was less with higher bowl diameters, even though this resulted in higher NOx, and this behaviour did not change with any modification of the arc radius (Rajamani et al. 2012). The same study found that the arc radius did not have any impact on NOx. The study proceeded to find out the effect of a higher cone angle was to give lower soot, BSFC and higher NOx. The same effect was observed by advancing the injection timing. This study also noted the significance of the spray impingement point on the piston bowl on emissions and BSFC—lower soot and BSFC can be achieved by targeting higher impingement point, with the compromise of higher NOx. The study also remarked that with wider bowls, the chances of wall impingement are less so the spray targeting becomes less critical, however wider bowls might affect part load trade-off between NOx and soot, low load HC and CO especially in low compression ratio applications.

Re-entrant bowls have been found to give higher mixing rates resulting in less soot, and permits retarding the main injection timing and hence reduce NOx without affecting BSFC (Middlemiss 1978; Saito et al. 1986). Simultaneous reduction of NOx and soot has been found possible through optimum injection timing in combination with a highly re-entrant bowl with lower pip height which generates high levels of turbulence (Prasad et al. 2011). Upcoming generations of modern diesel engines seem to be towards wider, shallower re-entrant designs (Bauder et al. 2005; Matsui et al. 2008; Crabb et al. 2013).

Combustion chamber optimization studies done on a DI engine found that the width of the throat had direct effect on NOx and soot where deep bowls with narrow throat and low pip height producing lower NOx, while wider throat bowls with bigger toroidal radius producing less soot. In this study, the lip length was found to have less influence on NOx and soot. Design changes which reduce soot were found to reduce hydrocarbon emissions too. Here too, wall impingement was found to have an effect on soot and the authors noted that wall wetting could be eliminated or reduced by advancing the injection timing and adopting smaller spray cone angles to reduce soot (Lim and Min 2005). There are some studies which had contradictory opinion about wall wetting and have reported that higher wall impingement or wall wetting has resulted in less soot and they have attributed this to be a result of better mixing attained through the jet striking the wall (Okajima and Kumagai 1990).

There have been studies which observed that soot emission is more sensitive to changes in combustion chamber design, while NOx is not as sensitive (Risi et al. 2003). Therefore, it is recommended to modify the combustion chamber design in such a way as to reduce soot and use other in-cylinder techniques to reduce NOx. Studies have reported that the conditions which result in less soot at high loads produce higher soot at low loads, especially at very low loads (Miles and Andersson 2016; Ge et al. 2010). As far as the combustion chamber design is concerned for higher loads, studies recommend wider bowls to reduce soot, while to reduce soot at lower loads they recommend narrow bowls (Shi and Reitz 2008b; Ge et al. 2009a; Fasolo et al. 2005). Most of the studies recommend optimizing the combustion chamber design for high load and use other techniques to reduce emissions at lower loads (Shi and Reitz 2008a; Ge et al. 2009a, b). Most of the studies favoured a bigger bowl diameter and shallow depth for the higher load case to reduce soot (Mobasheri and Peng 2012b; Rajamani et al. 2012; Miles and Andersson 2016; Shi and Reitz 2008b).

5.13.2 Main Injection Timing

The start of fuel injection (SOI) or the injection timing is a very significant operating parameter which affects the performance and emissions of a diesel engine. The timing has a bearing on the air–fuel mixing rate which is crucial in deciding the rate of pressure rise as well as NOx and soot. Retarded SOI has been found to reduce NOx and is a well-known and widely used measure to control NOx emissions. Retarding the main injection timing reduces the ignition delay as the cylinder contents are already at a higher temperature and conditions are favourable for auto-ignition, hence there is effectively less pre-mixed combustion which results in less NOx. The effect of late injection is in limiting the spatial development of local zones of high temperature thereby lowering the average temperature within the combustion chamber (Aoyama et al. 1990). Earlier studies show that retarding the main injection timing reduces the ignition delay which in turn reduces the amount of fuel prepared for the premixed phase of combustion leading to less NOx and more soot (He et al. 2015; Tennison and Reitz 2001). An associated phenomenon is the lower peak firing pressure and temperature and hence lower gross IMEP.

Since there is always a known NOx—soot trade-off in diesel engines, it is known from many studies that retarding the fuel injection timings reduce NOx but with an associated increase in soot and BSFC as per many studies (Kouremenos et al. 2001a; Zeng et al. 2006). One such study was done on a six-cylinder turbocharged diesel engine at several engine speeds all at full load, by advancing and retarding the injection timing about the timing for maximum torque at each speed (Fulton and Leviticus 1993). The results showed while NOx was reduced by retarding the timing, soot was higher and there was a drop in power and efficiency. 2-D image studies showed that the flame temperatures are lower with slower rate of burning when late injection timings are used, which result in less NOx and more soot (Bakenhus and Reitz 1999). CFD studies on a four-cylinder diesel engine at full load found that NOx could be reduced with retarded injection timing and if higher injection pressure is used even soot also could be reduced (Puri et al. 2017). Another CFD study done on an engine at 1000 rpm noted that retarded injection timings result in lower cylinder pressure, temperature, cumulative heat release and NOx, while advanced injection timings result in the increase of all three parameters (Jayashankara and Ganesan 2010).

There have also been studies done on re-entrant bowl which found out that there is significant swirl amplification and better squish flow which aid better mixing and hence permit to retard the fuel injection with minimum impact on soot and BSFC (Middlemiss 1978; Saito et al. 1986; Ikegami et al. 1990; Kidoguchi et al. 1999). So, it is essential to have the right combination of combustion chamber and fuel injection parameters to get the full benefit at all operating points. Studies have highlighted the fact that the optimum injection timing could be different to achieve the best performance, lowest emissions, and lowest fuel consumption, and again the optimum timing could vary from one operating point to the other (Shi and Reitz 2008a; Risi et al. 2003; Ge et al. 2009a). So, it is up to the combustion developers to adjust the SOI based on the specific requirement.

5.13.3 Pilot Injection

Pilot injection refers to a short fuel injection process where around 5–20% of the total fuel quantity per cycle could be injected around 5–20 CAD before the start of the main injection. Pilot injection fuel quantity is less, thus less amount of fuel gets prepared for the premixed combustion and the main injection fuel will be injected into the hot gases of pilot combustion, which reduces the ignition delay of the main injection and hence result in lower NOx and lower rate of pressure rise which also reduces the combustion noise.

Studies performed at light loads have found that up to 25% reduction in NOx could be attained through pilot injection (Minami et al. 1995). Researchers have also established that the quantity of fuel in pilot injection and the separation between pilot and main injection could be optimized for each operating point for emissions as well as performance (Sharma et al. 2010). An experimental study at low speed low load observed that the pilot separation had a significant role in optimization for emissions (Beatrice et al. 2001). The opportunity of simultaneous reduction of NOx and soot by optimizing the pilot injection fuel quantity and timing was highlighted by a computational study (Mobasheri et al. 2011). Such an optimization study performed in a mechanical fuel injection system found that NOx could be reduced by up to 30% without significant increase in soot at different loads (Dürnholz et al. 1994). Another experimental study on the effect of pilot quantity and timing found that the increase in soot associated with the introduction of a pilot injection can be offset by optimizing the pilot fuel quantity and timing (Nishimura et al. 1998).

A study done by varying the injection quantity in a double injection scheme concluded that the rate of premixed combustion was slower when a smaller quantity was injected in the pilot injection and resulted in less NO, however when the pilot quantity was increased, both NO and soot were found to be less (Sindhu et al. 2017). This study also noted that the split injection strategy could reduce NO with minimum impact on performance when compared to retarded injection timing plus EGR combination which hampers the engine performance considerably. Another experimental study also noted that the pilot injection quantity and timing affected NOx and soot irrespective of speed and load (Badami et al. 2001), but had a slightly different observation than stated earlier—higher pilot quantity increased both NOx and soot and the NOx increase was attributed to the increase in the mean cylinder temperature even though the premixed combustion was less and the increase in soot was attributed to the reduction in pre-mixed combustion. This study also stated when the pilot injection separation is increased, there is a reduction in the ignition delay of the main injection fuel till a certain limit, and this in turn reduces NOx and increases soot.

A study which compared a single injection scheme with a double injection scheme found that with pilot injection, the peak rate of heat release is lower, but the peak value remains for longer duration when compared with a single injection case, hence giving a more even heat release which could result in lower soot (Ehleskog et al. 2007). Another study highlighted that one could retard the main injection when using a pilot injection as the pilot combustion ensures shorter delay of the main injection fuel, and thereby reduce NOx without any BSFC penalty (Aoyama et al. 1990). One study saw that NOx was found to be increasing till a certain advancement of pilot injection timing and after a certain CAD, further advancement reduced NOx, which highlights the importance of separation between pilot injection and main injection (Sharma et al. 2010).

With pilot injection, the ignition sites of the second injection was found closer to the nozzle (Beatrice et al. 2001; Amagai et al. 1999) which reduces the chances of wall impingement, and in one of these studies (Amagai et al. 1999), it was also found that pilot injection results in faster combustion than single injection. The effect of pilot quantity on soot was seen in a study on a transparent engine, where low pilot quantity was found to reduce soot (Koyanagi et al. 1999). This study also recommended to resort to EGR to reduce NOx further. The significance of pilot quantity on soot was found in another experimental study as well which also remarked that pilot injection could result in an increase or decrease of NOx based on the changes in ignition delay and the level of premixed combustion (Lee and Reitz 2003). While adding a pilot injection presents the opportunity to simultaneously reduce NOx and soot by optimizing the injection timings, a study found that if the soot emissions can increase drastically when the main injection is retarded beyond a certain limit, since it increases the ignition delay and the burned gas temperature falls drastically in the expansion stroke which reduces soot oxidation (Bakenhus and Reitz 1999).

The advantage of using retarded injection timing with a pilot injection in reducing emissions was established in various studies (Aoyama et al. 1990; Bakenhus and Reitz 1999; Shundoh et al. 1992). The importance of pilot injection quantity, SOI of pilot and the separation between the pilot injection and the main injection on emissions has also been highlighted in several research works (Sharma et al. 2010; Mobasheri et al. 2011; Dürnholz et al. 1994; Nishimura et al. 1998; Sindhu et al. 2017). Therefore, there is immense scope to study the pilot injection quantity, SOI and separation as well as main injection timing in engine optimization and CFD is a very useful tool in this aspect to simplify the optimization process.

5.13.4 Exhaust Gas Recirculation (EGR)

Introducing a regulated quantity of exhaust gases into the intake port or intake manifold is known as exhaust gas recirculation (EGR). There also exists a certain level of internal EGR and residuals from previous cycle which is typical during the valve overlap period and exhaust backflow during the gas exchange process. EGR has mainly three effects on the intake charge—dilution, thermal and chemical. EGR displaces some oxygen from the intake charge which reduces the mass fraction of oxygen available for combustion—this is the dilution effect. EGR increases the heat capacity of the charge due to the changes in gas composition—this is the thermal effect. Exhaust gases are composed mostly of CO_2 and H_2O which tend to dissociate (endothermic reaction)—this is the chemical effect. The net effect of these three effects is to reduce the in-cylinder temperature, and hence reduce NOx. Studies done on EGR and the dilution, chemical and thermal effects of CO_2 and H_2O independently found that the dilution effect reduces the cylinder temperature and pressure which results in significantly lower NOx, but the lack of oxygen increases soot (Ladommatos et al. 1996a, b, 1997). The studies also found out that the thermal and chemical effects were much less when compared to the dilution effects.

An early study found that there is significant increase in soot when using EGR and retarded timing to reduce NOx (Yu and Shahed 1981). A study done on the optimization of diesel engine found out that EGR results in higher soot and BSFC (Bazari and French 1993). In some later studies, it was found that it is possible to reduce NOx and soot simultaneously with retarded injection and EGR (Kimura et al. 2002). A numerical study showed that when EGR is used in combination with retarded main injection and pilot injection, it results in further reduction in NOx (Mobasheri et al. 2011). A comparative study on various in-cylinder techniques at a high load and a very low load showed that the addition of EGR significantly reduced NOx and increased soot in both cases, and when a combination of pilot injection with retarded main injection and EGR is applied, the effect of EGR seems to be the most dominant (Menon et al. 2020).

One study on a four cylinder naturally aspirated diesel engine operating at part load recommended that adding EGR to the intake charge without reducing the original intake mass will reduce the penalty on soot emissions (Ladommatos et al. 1998). Similar result was obtained through a computational study as well where it was found that EGR reduces NOx with a sharp rise in soot, but when the boost pressure is increased (which effectively increases the intake air mass), there is less penalty on soot (Kouremenos et al. 2001b). Tests done on a heavy-duty diesel engine also had a similar observation that EGR with higher boost pressure reduced soot considerably without much change in NOx (Desantes et al. 2000). This study also suggested that EGR affects soot oxidation rather than soot formation by observing that soot emissions could be correlated to the gas temperature at 75% heat release, partial pressure of oxygen at the end of combustion, and the in-cylinder pressure.

Experimental study done on a turbocharged diesel engine showed that as the EGR rate is increased, the rate at which NOx reduced was found higher and this effect was seen much higher at lower air–fuel ratios (Jacobs et al. 2003). Experiments done on a two-cylinder diesel engine showed that with EGR, the ignition delay was increased due to a reduction in the rates of auto-ignition reactions, while the activation energy and order of the reactions were found unaffected (Nitu et al. 2002). Another study which conducted experiments found that when more EGR is applied, there is higher heat loss to the wall because of increased flame radiation, which result in lower in-cylinder temperatures and hence less NOx (Egnell 2000). Another study done at 25% load with high EGR found that simultaneous reduction of NOx and soot could be achieved with a penalty of higher BSFC, CO and HC, which could be overcome by using an advanced injection timing (Alriksson et al. 2005).

From the above discussion, it is evident that the combustion chamber design governs the charge air motion, velocity field and turbulence levels of the intake charge at the time of fuel injection. The orientation of the fuel injector, number of holes, injection pressure, injection timing and injection duration affect how the fuel–air mixing happens in combination with the air flow generated by the combustion chamber design. Retarded injection timing can reduce NOx, but at the cost of increased soot and drop in power and fuel economy. However, with modern engines equipped with electronic fuel injection systems, one can administer an appropriate pilot injection and these drawbacks could be eliminated. With the introduction of EGR, drastic reduction of NOx can be achieved.

In the previous sections, we could see that each in-cylinder technique has its own effect on performance and emission parameters, and the effect is not the same at all load conditions. It is therefore extremely important to select the right combination of strategies for each operating condition. It is the goal of an engine developer to find the right combination of these parameters to get the desired output. This is where CFD becomes a very important tool in cutting the development cost as well as time. The optimum combination of these parameters could be different for best power, lowest emissions, and lowest fuel consumption. Again, the combination could be different at different speed-load conditions. Therefore, the development engineers always look for the best trade-off based on the requirement and once the best combination is selected, the engine is said to have been optimized at that point. There are certain constraints, for example, combustion chamber geometry cannot be changed from high load to low load in an engine. Hence, it is worthwhile to follow previous studies and optimize the combustion chamber geometry for the high load operating points. The other load conditions must be optimized for performance or emissions or a trade-off between the two using other techniques. With the help of CFD, one could even devise a design of experiments to do back to back comparisons with one change at a time which helps arrive at the best combination for the given engine.

5.14 Summary

In this chapter, the focus is on the significance of CFD simulation in engine development and how it is applied in engine research. This would serve as a reference point for researchers starting their journey in diesel engine CFD simulation. A brief introduction to different simulation categories and a description of the usually followed methodology are given. In Sect. 5.5 the fundamental governing equations underlying the various processes and the different variables therein are introduced. Discussions in Sects. 5.7 and 5.8 talk about the initial conditions and boundary conditions with references to the values used in literature based on actual measurements would serve as a good starting point for beginners. Section 5.9 describes the different sub-models and wherever possible comparisons between different models are provided with supporting literature which would enable the reader to select the appropriate model.

Sections 5.10 and 5.11 introduce the reader to the concepts of grid independence and model validation which are two vital aspects in getting fast and reliable results. Commonly followed methods are discussed with reference to published literature. This would give the reader full understanding and flexibility to choose the parameters according to the model under study. It is worth noting that there may not be a single correct or wrong method in terms of grid independence and model validation, it is all about how relevant it is to the study and how reliable the model outcome is.

The same is applicable in terms of post-processing which is discussed in Sect. 5.12 by providing some of the results sought through CFD simulations. It is important to note that the parameters or plots discussed here are not exhaustive. CFD opens a wide window of opportunity for the user to visualize and interpret different parameters and variables in engines. It is up to the user to understand the capability of the model and harness its powers by interpreting the data and presenting it in the most appropriate manner.

The chapter is closed with a discussion on one of the most important applications of CFD simulations in engines—optimization of engine performance and emissions through in-cylinder techniques. Section 5.13 discusses this in detail in terms of the common modifications done in combustion chamber shape, fuel injection parameters and EGR, and how they affect the performance and emissions through published literature. This section would serve as a concise review of published data on in-cylinder methods applied in diesel engines, and hopefully inspire new researchers to identify new problems and seek their solutions.

References

Abay K, Colak U, Yuksek L (2018) Computational fluid dynamics analysis of flow and combustion of a diesel engine. J Therm Eng 4(2, Special Issue 7):1878–1895

Abraham J, Bracco FV, Reitz RD (1985) Comparisons of computed and measured premixed charge engine combustion. Combust Flame 60(3):309–322. https://doi.org/10.1016/0010-2180(85)90036-7

Alriksson M, Rente T, Denbratt I (2005) Low soot, low NOx in a heavy-duty diesel engine using high levels of EGR. SAE Technical Paper, 2005-01-3836. https://doi.org/10.4271/2005-01-3836

Amagai K, Hashimoto Y, Arai M (1999) Ignition and combustion characteristics of two-stage injection diesel spray. JSAE Rev 20:407–411. https://doi.org/10.1016/S0389-4304(99)00023-5

Amann CA, Siegla DC (1981) Diesel particulates—what they are and why. Aerosol Sci Technol 1:73–101. https://doi.org/10.1080/02786828208958580

Amsden AA (1997) KIVA-3V: a block structured KIVA program for engines with vertical or canted valves. Los Alamos National Laboratory Technical Report

Amsden AA, O'Rourke PJ, Butler TD (1989) KIVA-II: a computer program for chemically reactive flows with sprays. Los Alamos National Laboratory Technical Report

Angelberger C, Poinsot T, Delhaye B (1997) Improving near-wall combustion and wall heat transfer modeling in SI engines computations. SAE Technical Paper, 972881. https://doi.org/10.4271/972881

Aoyama T, Mizuta J, Oshima Y (1990) NOx reduction by injection control. SAE Technical Paper, 900637. https://doi.org/10.4271/900637

Badami M, Millo F, D'amato D (2001) Experimental investigation on soot and NOx formation in a DI common rail diesel engine with pilot injection. SAE Technical Paper, 2001-01-0657. https://doi.org/10.4271/2001-01-0657

Bakenhus M, Reitz RD (1999) Two-color combustion visualization of single and split injections in a single-cylinder heavy-duty DI diesel engine using an endoscope-based imaging system. SAE Technical Paper, 1999-01-1112. https://doi.org/10.4271/1999-01-1112

Bauder R, Gruber M, Michels E, Pamio Z, Schiffgens H, Wimmer W (2005) The new Audi 4.2 L V8 TDI-engine—part 2: thermodynamics, application and exhaust after-treatment. MTZ Worldwide 66:29–32. https://doi.org/10.1007/BF03227801

Bazari Z, French B (1993) Performance and emissions trade-offs for a HSDI diesel engine—an optimization study. SAE Technical Paper, 930592. https://doi.org/10.4271/930592

Beatrice C, Belardini P, Bertoli C, Lisbona MG, Migliaccio M (2001) Combustion patterns in common rail DI engines inferred by experiments and CRD computations. Combust Sci Technol 162:235–261. https://doi.org/10.1080/00102200108952143

Bowman CT (1975) Kinetics of pollutant formation and destruction in combustion. Prog Energy Combust Sci 1:33–45. https://doi.org/10.1016/0360-1285(75)90005-2

Brijesh P, Harshvardhan A, Sreedhara S (2014) A study of combustion and emissions characteristics of a compression ignition engine processes using a numerical tool. Int J Adv Eng Sci Appl Math 6(1–2):17–30. https://doi.org/10.1007/s12572-014-0107-9

Burnt M, Pond C (1997) Evaluation of techniques for absolute cylinder pressure correction. SAE Technical Paper, 970036 (1997). https://doi.org/10.4271/970036

Chiang CH, Raju MS, Sirignano WA (1992) Numerical analysis of a convecting, vaporizing fuel droplet with variable properties. Int J Heat Mass Transf 35(5):1307–1324. https://doi.org/10.1016/0017-9310(92)90186-V

Colin O, Benkenida A, Angelberger C (2003) 3D modeling of mixing, ignition and combustion phenomena in highly stratified gasoline engines. Oil Gas Sci Technol Revue d'IFP Energies Nouvelles 58(1):47–62. https://doi.org/10.2516/ogst:2003004

Crabb D, Fleiss M, Larsson J, Somhorst J (2013) New modular engine platform from Volvo. MTZ Worldwide 74:4–11. https://doi.org/10.1007/s38313-013-0081-9

de Risi A, Donateo T, Laforgia D (2003) Optimization of the combustion chamber of direct injection diesel engines. SAE Technical Paper, 2003-01-1064. https://doi.org/10.4271/2003-01-1064

Desantes JM, Arregle J, Molina SA, Lejeune M (2000) Influence of the EGR rate, oxygen concentration and equivalent fuel/air ratio on the combustion behavior and pollutant emissions of a heavy-duty diesel engine. SAE Technical Paper, 2000-01-1813. https://doi.org/10.4271/2000-01-1813

Dolak J, Shi Y, Reitz RD (2010) A computational investigation of stepped-bowl piston geometry for a light duty engine operating at low load. SAE Technical Paper, 2010-01-1263. https://doi.org/10.4271/2010-01-1263

Dürnholz M, Endres H, Frisse P (1994) Preinjection a measure to optimize the emission behavior of DI diesel engine. SAE Technical Paper, 940674. https://doi.org/10.4271/940674

Egnell R (2000) The influence of EGR on heat release rate and NO Formation in a DI diesel engine. SAE Technical Paper, 2000-01-1807. https://doi.org/10.4271/2000-01-1807

Ehleskog R, Ochoterena RL, Andersson S (2007) Effects of multiple injections on engine-out emission levels including particulate mass from an HSDI diesel engine. SAE Technical Paper, 2007-01-0910. https://doi.org/10.4271/2007-01-0910

Fasolo B, Doisy A-M, Dupont A, Lavoisier F (2005) Combustion system optimization of a new 2 liter diesel engine for EURO IV. SAE Technical Paper, 2005-01-0652. https://doi.org/10.4271/2005-01-0652

Ferziger JH, Peric M (2002) Computational methods for fluid dynamics, 3rd edn. Springer, Berlin

Fulton B, Leviticus L (1993) Variable injection timing effects on the performance and emissions of a direct injection diesel engine. SAE Technical Paper, 932385. https://doi.org/10.4271/932385

Furuhama S, Suzuki H (1979) Temperature distribution of piston rings and piston in high speed diesel engine. Bull JSME 22(174):1788–1795

Ge H-W, Shi Y, Reitz RD, Wickman DD, Zhu G, Zhang H, Kalish Y (2009a) Heavy-duty diesel combustion optimization using multi-objective genetic algorithm and multi-dimensional modelling. SAE Technical Paper, 2009-01-0716. https://doi.org/10.4271/2009-01-0716

Ge H-W, Shi Y, Reitz RD, Wickman DD, Willems W (2009b) Optimization of a HSDI diesel engine for passenger cars using a multi-objective genetic algorithm and multi-dimensional modelling. SAE Int J Engines 2(1):691–713. https://doi.org/10.4271/2009-01-0715

Ge H-W, Shi Y, Reitz RD, Wickman DD, Willems W (2010) Engine development using multi-dimensional CFD and computer optimization. SAE Technical Paper, 2010-01-0360. https://doi.org/10.4271/2010-01-0360

Halstead MP, Kirsh LJ, Quinn CP (1977) The autoignition of hydrocarbon fuels at high temperatures and pressures—fitting of a mathematical model. Combust Flame 30:45–60. https://doi.org/10.1016/0010-2180(77)90050-5

Hamid MF, Idroas MY, Mazlan M, Shukriwani S, Teoh YH, Sharzali CM, Miskam MA, Abdullah MK (2020) Numerical investigation of the characteristics of the in-cylinder air flow in a compression-ignition engine for the application of emulsified biofuels. Processes 8:1517 MDPI. https://doi.org/10.3390/pr8111517

Han Z, Uludogan A, Hampson G, Reitz R (1996) Mechanism of soot and NOx emission reduction using multiple-injection in a diesel engine. SAE Technical Paper, 960633. https://doi.org/10.4271/960633

Han Z, Reitz RD (1995) Turbulence modeling of internal combustion engines using RNG k-ε models. Combust Sci Technol 106(4–6):267–295. https://doi.org/10.1080/00102209508907782

Han Z, Reitz RD (1997) A temperature wall function formulation for variable density turbulence flow with application to engine convective heat transfer modeling. Int J Heat Mass Transf 40(3):613–625. https://doi.org/10.1016/0017-9310(96)00117-2

He S, Du B-G, Feng L-Y, Fu Y, Cui J-C, Long W-Q (2015) A numerical study on combustion and emission characteristics of a medium-speed diesel engine using in-cylinder cleaning technologies. Energies 8(5):4118–4137. https://doi.org/10.3390/en8054118

Heywood JB (2018) Internal combustion engine fundamentals. McGraw-Hill Education

Hiroyasu H, Kadota T (1976) Models for combustion and formation of nitric oxide and soot in direct injection diesel engines. SAE Technical Paper, 760129. https://doi.org/10.4271/760129

Hwang SS, Liu Z, Reitz RD (1996) Breakup mechanisms and drag coefficients of high-speed vaporizing liquid drops. Atom Sprays 6:353–376

Ikegami M, Fukuda M, Yoshihara Y, Kaneko J (1990) Combustion chamber shape and pressurized injection in high-speed direct-injection diesel engines. SAE Technical Paper, 900440. https://doi.org/10.4271/900440

Jacobs T, Assanis D, Filipi Z (2003) The impact of exhaust gas recirculation on performance and emissions of a heavy-duty diesel engine. SAE Technical Paper, 2003-01-1068. https://doi.org/10.4271/2003-01-1068

Jagus K, Jiang X (2011) Large eddy simulation of diesel fuel injection and mixing in a HSDI engine. Flow Turbul Combust 87:473–491. https://doi.org/10.1007/s10494-011-9338-z

Jayashankara B, Ganesan V (2010) Effect of fuel injection timing and intake pressure on the performance of a DI diesel engine—a parametric study using CFD. Energy Convers Manage 51:1835–1848. https://doi.org/10.1016/j.enconman.2009.11.006

Jia M, Peng Z, Xie M (2009) Numerical investigation of soot reduction potentials with diesel homogeneous charge compression ignition combustion by an improved phenomenological soot model. J Automob Eng 223(3):395–412. https://doi.org/10.1243/09544070JAUTO993

Kaminaga T, Kusaka J, Ishii Y (2008) A three-dimensional numerical study on exhaust gas emissions from a medium-duty diesel engine using a phenomenological soot particle formation model combined with detailed chemistry. Int J Engine Res 9(4):283–295. https://doi.org/10.1243/14680874JER00908

Khan S, Panuab R, Bose PK (2018) Combined effects of piston bowl geometry and spray pattern on mixing, combustion and emissions of a diesel engine: a numerical approach. Fuel 225:203–217. https://doi.org/10.1016/j.fuel.2018.03.139

Kidoguchi Y, Yang C, Miwa K (1999) Effect of high squish combustion chamber on simultaneous reduction of NOx and particulate from a direct-injection diesel engine. SAE Technical Paper, 1999-01-1502. https://doi.org/10.4271/1999-01-1502

Kim H, Huh H (1998) Collapse analysis of space-framed vehicle models by finite element limit analysis. SAE Technical Paper, 980549. https://doi.org/10.4271/980549

Kimura S, Ogawa H, Matsui Y, Enomoto Y (2002) An experimental analysis of low temperature and premixed combustion for simultaneous reduction of NOx and particulate emissions in direct injection diesel engines. Int J Engine Research 3(4):249–259. https://doi.org/10.1243/146808702762230932

Kirkpatrick AT (2021) Internal combustion engines: applied thermosciences, 4th edn. Wiley, Hoboken

Kong SC, Sun Y, Rietz RD (2017) Modeling diesel spray flame liftoff, sooting tendency, and NOx emissions using detailed chemistry with phenomenological soot model. J Eng Gas Turb Power 129

Kong S-C, Reitz RD (1993) Multidimensional modeling of diesel ignition and combustion using a multistep kinetics model. J Eng Gas Turbines Power 115(4):781–789. https://doi.org/10.1115/1.2906775

Kouremenos D, Hountalas D, Binder K, Raab A, Schnabel M (2001a) Using advanced injection timing and EGR to improve DI diesel engine efficiency at acceptable NO and soot levels. SAE Technical Paper, 2001-01-0199. https://doi.org/10.4271/2001-01-0199

Kouremenos DA, Hountalas DT, Binder KB (2001b) The effect of EGR on performance and pollutant emissions of heavy-duty diesel engines using constant and variable AFR. SAE Technical Paper, 2001-01-0198. https://doi.org/10.4271/2001-01-0198

Koyanagi K, Öing H, Maly R (1999) Optimizing common rail-injection by optical diagnostics in a transparent type diesel engine. SAE Technical Paper, 1999-01-3646. https://doi.org/10.4271/1999-01-3646

Ladommatos N, Abdelhalim S, Zhao H, Hu Z (1996a) The dilution, chemical and thermal effects of exhaust gas recirculation on diesel engine emissions—part 1: effect of reducing inlet charge oxygen. SAE Technical Paper, 961165. https://doi.org/10.4271/961165

Ladommatos N, Abdelhalim SM, Zhao H, Hu Z (1996b) The dilution, chemical, and thermal effects of exhaust gas recirculation on diesel engine emissions—part 2: effects of carbon dioxide. SAE Technical Paper, 961167. https://doi.org/10.4271/961167

Ladommatos N, Abdelhalim SM, Zhao H, Hu Z (1997) The dilution, chemical, and thermal effects of exhaust gas recirculation on diesel engine emissions—part 3: effects of water vapour. SAE Technical Paper, 971659. https://doi.org/10.4271/971659

Ladommatos N, Abdelhalim SM, Zhao H, Hu Z (1998) Control of oxides of nitrogen from diesel engines using diluents while minimising the impact on particulate pollutants. Appl Therm Eng 18:936–980. https://doi.org/10.1016/S1359-4311(98)00031-3

Lavoie GA, Heywood JB, Keck JC (2007) Experimental and theoretical investigation of nitric oxide formation in internal combustion engines. Combust Sci Technol 1:313–326. https://doi.org/10.1080/00102206908952211

Lee T, Reitz RD (2003) The effects of split injection and swirl on a HSDI diesel engine equipped with a common-rail injection system. SAE Technical Paper, 2003-01-0349. https://doi.org/10.4271/2003-01-0349

Lee J, Lee S, Kim J, Kim D (2015) Bowl shape design optimization for engine-out PM reduction in heavy duty diesel engine. SAE Technical Paper, 2015-01-0789. https://doi.org/10.4271/2015-01-0789

Li X, Sun Z, Du W, Wei R (2010) Research and development of double swirl combustion system for a DI diesel engine. Combust Sci Technol 182(8):1029–1049. https://doi.org/10.1080/00102200903544271

Lim J, Min K (2005) The effects of spray angle and piston bowl shape on diesel engine soot emissions using 3-D CFD simulation. SAE Technical Paper, 2005-01-2117. https://doi.org/10.4271/2005-01-2117

Manuel A, Gonzalez D, Borman G, Reitz R (1991) A study of diesel cold starting using both cycle analysis and multidimensional calculations. SAE Technical Paper, 910180. https://doi.org/10.4271/910180

Matsui R, Shimoyama K, Nonaka S, Chiba I, Hidaka S (2008) Development of high-performance diesel engine compliant with Euro-V. SAE Technical Paper, 2008-01-1198. https://doi.org/10.4271/2008-01-1198

Mauro S, Şener R, Gül MZ, Lanzafame R, Messina M, Brusca S (2018) Internal combustion engine heat release calculation using single-zone and CFD 3D numerical models. Int J Energy Environ Eng 9:215–226. https://doi.org/10.1007/s40095-018-0265-9

Menon P, Kamble T, Mittal M (2020) A computational study and experiments to investigate the combustion and emission characteristics of a small naturally aspirated diesel engine through changes in combustion chamber geometry, injection parameters and EGR. IOP Conf Ser: Mater Sci Eng 912:042031

Merryman EL, Levy A (1975) Nitrogen oxide formation in flames: the roles of NO_2 and fuel nitrogen. In: Proceedings of fifteenth international symposium on combustion, p. 1073 (The combustion institute). https://doi.org/10.1016/S0082-0784(75)80372-9

Middlemiss I (1978) Characteristics of the Perkins 'squish lip' direct injection combustion system. SAE Technical Paper, 780113. https://doi.org/10.4271/780113

Miles PC, Andersson O (2016) A review of design considerations for light-duty diesel combustion systems. Int J Engine Res 17(1):6–15. https://doi.org/10.1177/1468087415604754

Minami T, Takeuchi K, Shimazaki N (1995) Reduction of diesel engine NOx using pilot injection. SAE Technical Paper, 950611. https://doi.org/10.4271/950611

Mobasheri R, Peng Z (2012) CFD modelling of the effects of injection timing on the combustion process and emissions in an HSDI diesel engine. In: Proceedings of the ASME 2012 internal combustion engine division spring technical conference ICES2012, Torino, Piemonte, Italy. 6–9 May 2012

Mobasheri R, Peng Z (2012) Analysis of the effect of re-entrant combustion chamber geometry on combustion process and emission formation in a HSDI diesel engine. SAE Technical Paper, 2012-01-0144. https://doi.org/10.4271/2012-01-0144

Mobasheri R, Peng Z, Mirsalim SM (2011) CFD evaluation of effects of split injection on combustion and emissions in a DI diesel engine. SAE Technical Paper, 2011-01-0822. https://doi.org/10.4271/2011-01-0822

Moiz A, Som S, Bravo L, Lee S (2015) Experimental and numerical studies on combustion model selection for split injection spray combustion. SAE Technical Paper, 2015-01-0374. https://doi.org/10.4271/2015-01-0374

Naber J, Reitz R (1988) Modeling engine spray/wall impingement. SAE Technical Paper, 880107. https://doi.org/10.4271/880107.SAE

Nagle J, Strickland-Constable RF (1962) Oxidation of carbon between 1000–2000 C. Proc Fifth Carbon Conf 1:154. https://doi.org/10.1016/B978-0-08-009707-7.50026-1

Nishimura T, Satoh K, Takahashi S, Yokota K (1998) Effects of fuel injection rate on combustion and emission in a DI diesel engine. SAE Technical Paper, 981929. https://www.jstor.org/stable/44736668

Nitu B, Singh I, Zhong L, Badreshany K, Henein N, Bryzik W (2002) Effect of EGR on autoignition, combustion, regulated emissions and aldehydes in DI diesel engines. SAE Technical Paper, 2002-01-1153. https://doi.org/10.4271/2002-01-1153

O'Rourke PJ (1981) Collective drop effects on vaporizing liquid sprays. Ph.D. Thesis, Princeton University, Princeton

O'Rourke PJ, Amsden AA (1987) The TAB method for numerical calculation of spray droplet breakup. SAE Technical Paper, 872089. https://doi.org/10.4271/872089

Okajima S, Kumagai S (1990) Experimental investigation of soot formation and NOx reduction by impinging spray in a closed vessel. In: 23rd symposium (international) on combustion (The Combustion Institute), vol 2, no 1, pp 275–279. https://doi.org/10.1016/S0082-0784(06)80270-5

Olikara C, Borman G (1975) A computer program for calculating properties of equilibrium combustion products with some applications to I.C. engines. SAE Technical Paper, 750468. https://doi.org/10.4271/750468

Patterson M, Reitz R (1998) Modeling the effects of fuel spray characteristics on diesel engine combustion and emission. SAE Technical Paper, 980131. https://doi.org/10.4271/980131.CST

Perini F, Zha K, Busch S, Kurtz E, Peterson RC, Warey A, Reitz RD (2018) Piston geometry effects in a light-duty, swirl-supported diesel engine: Flow structure characterization. Int J Engine Res 19(10):1079–1098

Post SL, Abraham J (2002) Modeling the outcome of drop-drop collisions in diesel sprays. Int J Multiph Flow 28(6):997–1019. https://doi.org/10.1016/S0301-9322(02)00007-1

Prasad BVVSU, Sharma CS, Anand TNC, Ravikrishna RV (2011) High swirl-inducing piston bowls in small diesel engines for emission reduction. Appl Energy 88(7):2355–2367. https://doi.org/10.1016/j.apenergy.2010.12.068

Puri T, Soni L, Deshpande S (2017) Combined effects of injection timing and fuel injection pressure on performance, combustion and emission characteristics of a direct injection diesel engine numerically using CONVERGE CFD tool. SAE Technical Paper, 2017-28-1953. https://doi.org/10.4271/2017-28-1953

Rajamani V, Schoenfeld S, Dhongde A (2012) Parametric analysis of piston bowl geometry and injection nozzle configuration using 3D CFD and DoE. SAE Technical Paper, 2012-01-0700. https://doi.org/10.4271/2012-01-0700

Rao K, Winterbone D, Clough E (1993) Influence of swirl on high pressure injection in hydra diesel engine. SAE Technical Paper, 930978. https://doi.org/10.4271/930978

Reitz RD, Bracco FV (1986) Mechanisms of breakup of round liquid jets. Encyclopedia of Fluid Mechanics Gulf Publishing Company

Ricart LM, Xin J, Bower GR, Reitz RD (1997) In-cylinder measurement and modeling of liquid fuel spray penetration in a heavy-duty diesel engine. SAE Technical Paper, 971591. https://doi.org/10.4271/971591

Richards KJ, Senecal PK, Pomraning E (2017) CONVERGE v2.3 manual. Convergent Science, Inc., Madison

Saito T, Daisho Y, Uchida N, Ikeya N (1986) Effects of combustion chamber geometry on diesel combustion. SAE Technical Paper, 861186. https://doi.org/10.4271/861186

Schapertons H, Lee W (1985) Multidimensional modeling of knocking combustion in SI engines. SAE Technical Paper, 850502. https://doi.org/10.4271/850502

Schmidt DP, Rutland CJ (2000) A new droplet collision algorithm. J Comput Phys 164(1):62–80. https://doi.org/10.1006/jcph.2000.6568

Senecal PK, Pomraning E, Richards KJ (2003) Multi-dimensional modeling of direct-injection diesel spray liquid length and flame lift-off length using CFD and parallel detailed chemistry. SAE Technical Paper, 2003-01-1043. https://doi.org/10.4271/2003-01-1043

Senecal PK, Richards KJ, Pomraning E, Yang T, Dai MZ, McDavid RM, Patterson MA, Hou S, Shethaji T (2007) A new parallel cut-cell cartesian CFD code for rapid grid generation applied to in-cylinder diesel engine simulations. SAE Technical Paper, 2007-01-0159. https://doi.org/10.4271/2007-01-0159

Sharma CS, Anand TNC, Ravikrishna RV (2010) A methodology for analysis of diesel engine in-cylinder flow and combustion. Progr Comput Fluid Dyn 10(3):157–167. https://doi.org/10.1504/PCFD.2010.033327

Shi Y, Reitz RD (2008) Assessment of optimization methodologies to study the effects of bowl geometry, spray targeting and swirl ratio for a heavy-duty diesel engine operated at high-load. SAE Int J Eng 1(1):537–557 https://www.jstor.org/stable/26308302

Shi Y, Reitz RD (2008a) Optimization study of the effects of bowl geometry, spray targeting, and swirl ratio for a heavy-duty diesel engine operated at low and high load. Int J Engine Res 9:325–346. https://doi.org/10.1243/14680874JER00808

Shu J, Fu J, Liu J, Ma Y, Wang S, Deng B, Zeng D (2019) Effects of injector spray angle on combustion and emissions characteristics of a natural gas (NG)-diesel dual fuel engine based on CFD coupled with reduced chemical kinetic model. Appl Energy 233–234:182–195. https://doi.org/10.1016/j.apenergy.2018.10.040

Shuai S, Abani N, Yoshikawa T, Reitz RD, Park S-W (2009) Evaluation of the effects of injection timing and rate-shape on diesel low temperature combustion using advanced CFD modeling. Fuel 88:1235–1244. https://doi.org/10.1016/j.fuel.2009.01.012

Shundoh S, Komori M, Tsujimura K, Kobayashi S (1992) NOx Reduction from diesel combustion using pilot injection with high pressure fuel injection. SAE Technical Paper, 920461. https://doi.org/10.4271/920461

Sindhu R, Rao APG, Murthy KM (2017) Effective reduction of NOx emissions from diesel engine using split injections. Alexandria Eng J 57(3):1379–1392. https://doi.org/10.1016/j.aej.2017.06.009

Som S, Aggarwal SK (2010) Effects of primary breakup modeling on spray and combustion characteristics of compression ignition engines. Combust Flame 157(6):1179–1193. https://doi.org/10.1016/j.combustflame.2010.02.018

Stanton D, Lippert A, Reitz R, Rutland C (1998) Influence of spray-wall interaction and fuel films on cold starting in direct injection diesel engines. SAE Technical Paper, 982584. https://doi.org/10.4271/982584

Stiesch G (2003) Modeling engine spray and combustion processes. Springer, Berlin

Tan Z, Reitz R (2004) Development of a universal turbulent combustion model for premixed and direct injection spark/compression ignition engines. SAE Technical Paper, 2004-01-0102. https://doi.org/10.4271/2004-01-0102

Tannehill JC, Anderson DA, Pletcher RH (1997) Computational fluid mechanics and heat transfer, 2nd edn. Taylor & Francis, Washington, DC

Tennison PJ, Reitz RD (2001) An experimental investigation of the effects of common-rail injection system parameters on emissions and performance in a high-speed direct-injection diesel engine. ASME J Eng Gas Turbines Power 123(1):167–174. https://doi.org/10.1115/1.1340638

Turns SR (2000) An introduction to combustion: concepts and applications, 2nd edn. McGraw-Hill

Vishwanathan G, Reitz RD (2010) Development of a practical soot modeling approach and its application to low-temperature diesel combustion. Combust Sci Technol 182(8):1050–1082. https://doi.org/10.1080/00102200903548124

Vishwanathan G, Reitz RD (2015) Application of a semi-detailed soot modeling approach for conventional and low temperature diesel combustion—Part I: model performance. Fuel 139:757–770. https://doi.org/10.1016/j.fuel.2014.08.026

Warsi ZUA (1993) Fluid dynamics: theoretical and computational approaches. CRC Press Inc.

White FM (1999) Fluid mechanics, 4th edn. McGraw-Hill, New York

Woschni G (1979) Prediction of thermal loading of supercharged diesel engines. SAE Technical Paper 790821

Xin J, Montgomery DT, Han Z, Reitz RD (1997) Multidimensional modeling of combustion for a six-mode emissions test cycle on a DI diesel engine. J Eng Gas Turbines Power 119(3):683–691. https://doi.org/10.1115/1.2817041

Xin J, Ricart L, Reitz RD (1998) Computer modeling of diesel spray atomization and combustion. Combust Sci Technol 137(1–6):171–194. https://doi.org/10.1080/00102209808952050

Yu R, Shahed S (1981) Effects of injection timing and exhaust gas recirculation on emissions from a DI diesel engine. SAE Technical Paper, 811234. https://doi.org/10.4271/811234

Zeng K, Huang Z, Liu B, Liu L, Jiang D, Ren Y, Wang J (2006) Combustion characteristics of a direct-injection natural gas engine under various fuel injection timings. Appl Therm Eng 26(8):806–813. https://doi.org/10.1016/j.applthermaleng.2005.10.011

Zhu Y, Zhao H, Melas DA, Ladommatos N (2004) Computational study of the effects of the re-entrant lip shape and toroidal radii of piston bowl on a HSDI diesel engine's performance and emissions. SAE Technical Paper, 2004-01-0118. https://doi.org/10.4271/2004-01-0118

Chapter 6
Physical and Data-Driven Models Hybridisation for Modelling the Dynamic State of a Four-Stroke Marine Diesel Engine

Andrea Coraddu, Miltiadis Kalikatzarakis, Gerasimos Theotokatos, Rinze Geertsma, and Luca Oneto

Abstract Accurate, reliable, and computationally inexpensive models of the dynamic state of combustion engines are a fundamental tool to investigate new engine designs, develop optimal control strategies, and monitor their performance. The use of those models would allow to improve the engine cost-efficiency trade-off, operational robustness, and environmental impact. To address this challenge, two state-of-the-art alternatives in literature exist. The first one is to develop high fidelity physical models (e.g., mean value engine, zero-dimensional, and one-dimensional models) exploiting the physical principles that regulate engine behaviour. The second one is to exploit historical data produced by the modern engine control and automation systems or by high-fidelity simulators to feed data-driven models (e.g., shallow and deep machine learning models) able to learn an accurate digital twin of the system without any prior knowledge. The main issues of the former approach are its complexity and the high (in some case prohibitive) computational require-

M. Kalikatzarakis · G. Theotokatos
Department of Naval Architecture, Ocean and Marine Engineering, University of Strathclyde, Glasgow, UK
e-mail: miltiadis.kalikatzarakis@strath.ac.uk

G. Theotokatos
e-mail: gerasimos.theotokatos@strath.ac.uk

A. Coraddu (✉) · R. Geertsma
Department of Maritime and Transport Technology, Delft University of Technology, Delft, The Netherlands
e-mail: a.coraddu@tudelft.nl

R. Geertsma
e-mail: r.d.geertsma@tudelft.nl;rd.geertsma@mindef.nl

L. Oneto
DIBRIS, University of Genova, Genoa, Italy
e-mail: luca.oneto@unige.it

R. Geertsma
Department of Military Systems & Technology, Netherlands Defence Academy, Breda, The Netherlands

A. K. Agarwal et al. (eds.), *Engine Modeling and Simulation*, Energy, Environment, and Sustainability, https://doi.org/10.1007/978-981-16-8618-4_6

ments. While the main issues of the latter approach are the unpredictability of their behaviour (guarantees can be proved only for their average behaviour) and the need for large amount of historical data. In this work, following a recent promising line of research, we describe a modelling framework that is able to hybridise physical and data driven models, delivering a solution able to take the best of the two approaches, resulting in accurate, reliable, and computationally inexpensive models. In particular, we will focus on modelling the dynamic state of a four-stroke diesel engine testing the performance (both in terms of accuracy, reliability, and computational requirements) of this solution against state-of-the-art physical modelling approaches on real-world operational data.

Keywords Marine diesel engine · Physical models · Data-driven models · Hybridisation

6.1 Introduction

In recent years the maritime industry has been challenged by several issues. The volatile bunker prices affect cargo transportation costs and the shipowners' competitiveness and operations viability (García-Martos et al. 2013). Strict regulations are in place to limit emissions (Committee 2011) with the aim of reducing CO_2 emissions from shipping by 40–50% (Commission 2013a, b). As a result of this combination, energy efficiency and environmental sustainability of maritime operations are currently the primary challenges to be faced by the maritime industry. Shipowners and operators are adopting several measures to lower fuel consumption and associated emissions (Chiong et al. 2021; Lion et al. 2020), and researchers in the field are developing innovative technologies and methods that can increase the environmental efficiency and cost-effectiveness of ship operations (Xing et al. 2020; Ni et al. 2020).

Improvements in energy efficiency can be obtained by a variety of design and retrofit measures (Lindstad et al. 2015; Gucwa and Schäfer 2013), such as hull design optimisation (Tillig et al. 2015; Lindstad and Eskeland 2015), adoption of alternative fuels (Chiong et al. 2021), and alternative energy sources (Bouman et al. 2017), as well as operational measures (Zis and Psaraftis 2018; Xing et al. 2020), including speed optimisation (Wen et al. 2017; Fagerholt and Psaraftis 2015; Psaraftis and Kontovas 2014), better capacity utilisation (Xing et al. 2020), and advanced route planning execution methods (Yu et al. 2021; Zis et al. 2020).

Nevertheless, it is widely known that for the majority of the vessels operating today, the main engines and to a lesser extent the auxiliary engines are the main factors of energy losses (Baldi et al. 2014, 2015). For this reason, engine manufacturers are focusing on further increasing Diesel Engine (DE) power density and enhancing operating performance (Lion et al. 2020).

Identifying and adopting new technologies in marine DEs is limited by the expensive design (Baldi et al. 2015), prototyping and experimentation processes (Kamal and Hui 2013). To mitigate these issues, computational models have been developed to simulate the physical processes occurring in a DE under steady-state (Larsen et al. 2015) and transient conditions (Theotokatos and Kyrtatos 2003) as a rapid and cost-effective prototyping tool (Merker et al. 2005). A common effort in the development of all computational models is towards improving their accuracy and reliability (Oberkampf and Trucano 2002). As such, the model verification, calibration, and validation processes occupy a central role, based on which the predictive capabilities of the model are assessed to justify its use for the application of interest (Oberkampf et al. 2004).

In the literature it is possible to find three main approaches to the DEs: Physical Models (PMs), Data-Driven Models (DDMs), and Hybrid Models (HMs). PM requires a deep knowledge of the physical phenomena (Grimmelius 2003; Grimmelius et al. 2007). The higher is the detail in the modelling of the equations which describe the physical phenomena, the higher is the expected accuracy of the results and the computational time required for the simulation. The second approach, instead, infers the desired model directly from historical data collected by on board machinery and does not require any a-priori knowledge of the underlying physical phenomena (Coraddu et al. 2021a). However, since these models are not supported by any physical interpretation, they need a significant amount of data to be built. The third approach is a combination of the previous ones and it is based on the integration of a PM and DDM into a single model. The DDM model compensates for the secondary effects not modelled by the PM and the PM helps the DDM in reducing the amount of historical data required to train it (Oneto et al. 2016).

A literature review on PMs, DDMs, and HMs for DE is reported in Sect. 6.2 showing that a complete and clear description of a modelling framework able to hybridise PMs and DDMs is not yet readily available. For this reason, this chapter is devoted to the presentation of this framework combining the computational efficiency of the 0D PM with the level of accuracy of the DDM.

With this goal in mind, first a 0D DE model will be proposed and validated for both steady-state and transient operations. This analysis will underline the high predictive capabilities of the PM in steady-state operations, and will also reveal the limited predictive capabilities under transient operation. Subsequently, different DDMs will be developed, tested, and compared. These models will leverage the information encapsulated in historical data to produce accurate predictions on a set of performance parameters of the DE. Finally, we will present the hybridisation framework where a set of HMs will be proposed, leveraging on both the DDMs and the PM previously developed. The authors will showcase the performance in terms of accuracy, reliability, and computational requirements of the HMs, clearly demonstrating the superiority of the proposed hybridisation framework on a comprehensive dataset containing real operational data from a marine DE for a time period of approximately 3 years.

The rest of the paper is organised as follows. Section 6.2 presents the state-of-the-art in PMs, DDMs, and HMs for marine DEs. Section 6.3 describes the hybridisation framework, starting from the theory of the PMs and DDMs and then going into the details of the HMs. Section 6.4 presents a specific DE case study and the related historical data available. Section 6.5 demonstrates the advantages and disadvantages of the different approaches presented in Sect. 6.3. Section 6.6 concludes this work.

6.2 Related Work

This section is devoted to the review of PMs, DDMs, and HMs for marine DEs. Note that the models can be categorised also according to other criteria (Grimmelius 2003). The choice that we made reflects the experience and knowledge on the subject.

6.2.1 Physical Models

PMs are the most traditional models for DE development, design, optimisation, and performance evaluation with extensive work carried out over several decades (Grimmelius et al. 2007). Many insightful reviews on this subject can be found (Grimmelius et al. 2007; Geertsma et al. 2017; Merker et al. 2005). All these reviews agree on the fact that the choice of the most suitable model strongly depends on the scope of work, the application requirements, and the available computational tools (Johnson et al. 2010). In general, to the best of the authors knowledge, PMs achieve errors well within the tolerance margins provided by engine manufacturers in steady state conditions, whereas reported results during transient operations tend to be less accurate.

Among PMs, Mean Value Engine Models (MVEMs) are a common choice when low computational effort is required (Maroteaux and Saad 2015; Guzzella and Onder 2009; He and Lin 2007; Lee et al. 2013; Miglianti et al. 2020, 2019). MVEMs are approximate first-principle models that adequately predict engine performance parameters, and are prevalent in applications in which the engine is considered as just one component of a wider system, or for control strategies development (Malkhede et al. 2005; Guan et al. 2014; Theotokatos 2010; Grimmelius et al. 2010; Theotokatos 2008; Nikzadfar and Shamekhi 2015; Geertsma et al. 2017; Theotokatos et al. 2018; Geertsma et al. 2018; Guzzella and Onder 2009).

MVEM can be classified in the following two categories: (a) not considering the engine manifolds dynamics; (b) including the engine manifold dynamics. Category (a) employs only two differential equations for calculating the engine and TC rotational speeds. Category (b) employs additional equations to calculate the pressure and temperature of the engine manifolds. MVEMs employ simpler approaches to estimate the engine parameters on an engine cycle basis. These models are employed in cases where calculations of low computational cost are required, with the most common example being the engine control system design.

0D models employ a more detailed formulation and are capable of predicting the in-cycle variations of the engine parameters. The latter are employed in the cases where the in-cylinder parameters (pressure, temperature, composition) variations are required leading to a more accurate estimation of the engine performance parameters. For instance, Llamas and Eriksson (2018) developed a control-oriented MVEM for a large two-stroke engine with Exhaust Gas Recirculation (EGR), with relative errors under 3.35% for steady state operations. A similar study was performed by the authors of Dahl et al. (Dahl et al. 2018), who also developed an MVEM for control strategy design, reporting errors smaller than 5% for all engine parameters under steady state conditions. In both these studies (Llamas and Eriksson 2018; Dahl et al. 2018), the authors did not report results under transient operation, nevertheless general trends of the real engine parameters were adequately captured. The authors of Sui et al. (2017) exploited an MVEM to investigate the performance of a medium-speed DE during preliminary design. The results of the in-cylinder process showed good correspondence with the test data across all process parameters, including in-cylinder temperatures, with satisfactory accuracy and adaptability to variable operating conditions. Authors of Sapra et al. (2017) studied back pressure effects on the performance of a marine DE, by means of an MVEM. Although quantitative performance metrics for the model are not reported, the graphical representation of the results indicates average relative percentage errors of around 4% for the turbine inlet temperature across all operating conditions.

More sophisticated approaches, with respect to MVEMs, are zero-dimensional (0D), one-dimensional (1D), and three-dimensional (3D) models that operate on a per-crank basis (Merker et al. 2005). These approaches are more computationally demanding compared to MVEMs, however they can predict the detailed gas processes inside the cylinders (Mohammadkhani et al. 2019; Stoumpos et al. 2018, 2020). For instance, authors of Guan et al. (2015) investigated a two-stroke marine DE with emphasis on part load operating conditions using a 0D model validated against experimental data obtained from engine shop tests in steady state operations. Very small errors were reported for loads between 25–100%, ranging between 0.6–2%. Authors of Larsen et al. (2015) also developed a 0D model, with a similar validation approach, reporting a root mean square deviation of approximately 1% across all engine performance parameters. Authors of Sapra et al. (2020) compared the capabilities of Seiliger-based and double Wiebe function-based 0D models to capture hydrogen-natural gas combustion in a marine engine for different fuel blends and engine loads. In this study, very small errors were reported for in-cylinder process parameters, ranging between 0.1–2.4%.

Several attempts to combine MVEM and 0D, 1D, or 3D models have been proposed, enhancing the predictive abilities of MVEMs with lower computational requirements than their 0D, 1D, or 3D counterparts (Livanos et al. 2007; Ding et al. 2010; Baldi et al. 2015; Maroteaux and Saad 2015; Tang et al. 2017).

In fact, in cases where low computational cost is required whilst the predictive capability for the engine in-cylinder parameters must be retained, approaches combining MVEMs and 0D models have been employed. In such cases, one engine cylinder (typically the closed cycle) is modelled by employing the 0D approach and interfaced with the MVEM. This provides the advantage of negating the extensive calibration process of the MVEM (which is typically based on data derived from either experiments or simulations with more detailed models) required to predict the engine performance parameters. For instance, authors of Baldi et al. (2015) combined MVEM and 0D models to investigate the propulsion behaviour of a handymax-size product carrier under constant and variable engine speed operations. The model was validated against experimental data from the engine's shop tests, for loads ranging between 50–110%. High accuracy was reported, with errors ranging between 1.9–2.7%. Furthermore, the authors concluded that their proposal provides a favourable time-accuracy trade-off and it can be used in cases where information, not provided by an MVEM, is required.

The integration of a 1D with a Computational Fluid Dynamics (CFD) based 3D model was proposed by the authors of Wang et al. (2020). The authors conducted a parametric investigation of a large four-stroke dual-fuel marine DE to identify the pre-injection effects on the engine combustion, knocking, and emission parameters. This modelling approach was validated under steady-state conditions in four operating points within a range of 25–100% of the nominal load. Near-zero deviation was reported for most parameters, whereas the maximum deviation for NOx emissions was about 2%. Finally, the authors of Hao et al. (2021) aimed to improve the in-cylinder fuel/air mixing process of heavy-duty DEs, which they modelled utilising CFD methods. Their simulations were validated in terms of the spray liquid/vapour penetration, heat release rate, and in-cylinder pressures, at various operational, and environmental conditions, reporting very low discrepancies.

When it comes to computational requirements, MVEMs are calibrated by using data from measurements or more detailed models. They typically exhibit absolute errors of less than 2%. However, outside the calibration envelope, their performance deteriorates. For 0D/1D models (where 0D refers to modelling the in-cylinder processes or volumes, whereas 1D refers to the modelling of pipes and manifolds), the expected typical errors are in the range of $\pm 5\%$. The more significant errors are exhibited for predicting the exhaust gas temperature and emissions, whereas errors for the other engine performance parameters are typically in the range $\pm 3\%$. Finally, evaluating the error is challenging for CFD models as the available measured parameters are pretty limited in most cases (typically the in-cylinder pressure diagram). However, the model set-up process consists of several steps (geometry, injection and mixing process, meshing studies in fixed combustion chambers (bombs), testing of various existing sub models) culminating in the set up of the final model version. In addition, several quantitative verification studies are typically carried out, thus providing confidence to the users that the model performance is appropriate.

We can conclude then that PMs can adequately capture most process parameters of a DE under a broad range of operating conditions. However, there is a clear trade-off between accuracy and computational requirements. In fact, the most detailed 3D models cannot run in real-time, whereas MVEMs lack accuracy, especially during transient operations.

6.2.2 Data-Driven Models

DDMs have been successfully applied in a variety of maritime applications, provided that the necessary quality and quantity of historical data is actually available (Coraddu et al. 2016, 2017, 2019, 2020, 2019, 2021a, b; Cipollini et al. 2018a, b; Stoumpos et al. 2020).

For instance, the Nikzadfar and Shamekhi (2014) developed an Artificial Neural Network (ANN) to study the relative contribution of several operating parameters to the performance of a DE. The authors utilised 4000 steady-state operating points to train the ANN, generated by means of numerical simulation, covering the entire envelope of the DE. Although the authors did not provide quantitative results regarding the model accuracy, graphical representation showed a relative difference of approximately 5%. Authors of Shin et al. (2020) developed an ANN to predict the NO_x emissions of a DE under transient operation. Highly accurate results were reported, with the developed ANN being able to predict NO_x with errors around 1.6%, comparable to the accuracy of physical NO_x measurement devices, with typical error margins of 1%. The ability of ANNs to predict performance parameters of a DE was also demonstrated by the authors of Ozener et al. (2013) predicting a variety of performance parameters and emissions with coefficient of determination (R^2) values over 0.95. A hydrogen dual-engine for automotive applications was the case study of the authors of Syed et al. (2017): ANNs proved to be extremely efficient, with near-zero errors being reported for the prediction of specific fuel consumption and a variety of emissions. A similar study was conducted by the authors of Liu et al. (2018) employing an ANN to predict the emissions and various performance parameters of a spark ignition engine fueled with butanol-gasoline blends. In this study, ANNs were able to predict the performances of the engine with very high R^2 values.

DDMs are frequently used in the field of condition monitoring of DEs, as they are extremely efficient at detecting and isolating faults and have proven to increase the reliability and decrease the probability of producing false alarms (Ahmed et al. 2015). For instance, authors of Namigtle-Jimenez (2020) developed 3 ANNs for fault detection and diagnosis of a DE, utilising only the pressure signal of the injection rail. They demonstrated that the proposed ANNs were able to detect and isolate the faulty injector of the electronic fuel system with 100% classification accuracy in offline training scenarios and near 100% accuracy in online scenarios. The authors of Wang et al. (2021) proposed a random convolutional ANN for health monitoring of DEs, relying on vibration measurements. Considering the very high accuracy of the developed ANN on this task, the authors concluded that, compared to traditional methods based on signal analysis techniques and shallow classifiers, their approach can automatically learn high-level representative features from the raw vibration signals and eliminate the necessity of the time-consuming manual feature extraction. A similar study was performed by the authors of Wang et al. (2019) presenting a Bayesian ANN-based approach for fault isolation in a DE fuel injection system under the presence of uncertainties. With the proposed approach, the authors demonstrated that symptoms under multiple faults could be decoupled into symptoms corresponding to each individual fault. This greatly reduced the prior knowledge needed for the diagnosis, decreased the complexity of the application, and significantly improved computational efficiency. The authors of Palmer and Bollas (2019) demonstrated the ability of DDMs to predict and isolate faults. More specifically, the authors showed that when proper fault detection and isolation test designs are selected, even the relatively simple combination of principal component analysis and k-nearest neighbours classifier could provide satisfactory results in fault detection. Likewise, the authors of Ahmed et al. (2015) trained an ANN employing a new estimation strategy known as the smooth variable structure filter to detect the engine's faults. This approach demonstrated stability and generalisation accuracy exhibiting improved performance compared with the first order back propagation algorithm and similar performance compared with the extended Kalman filter. Fault detection was also the scope of the work of the authors of Wang et al. (2020) where a noise-based method based on ANNs and wavelet packet analysis was presented. The authors concluded that ANNs are effective for feature extraction of engine fault noises in time and frequency domains, and are powerful for sound feature classification and recognition of the engine's faults. A further example is reported by the authors of Coraddu et al. (2021b) where multiple DDMs for weakly supervised marine dual fuel engines health monitoring were designed, presented, and tested. The proposed framework relied on a digital twin of the engine or on novelty detection algorithms which were compared against state-of-the-art fully supervised approaches. Utilising data from the validated simulation model of Stoumpos et al. (2020), the authors demonstrated that their approach can overcome the problematic requirement for a large amount of labelled samples, that are rarely available, with a decrease in performance of less than 1% compared to state-of-the-art fully-supervised approaches.

6.2.3 Hybrid Models

HMs are a quite recent modelling approach, especially in the maritime field, and just very few works showed the advantages of a hybrid approach, with respect to pure PMs and DDMs (Coraddu et al. 2021a, 2018; Miglianti et al. 2019, 2020). For instance, in Coraddu et al. (2017) the authors show that it is possible to effectively predict fuel consumption with HMs. Moreover, in Coraddu et al. (2018, 2021a), the authors also attempted to model the engine exhaust gas temperature with HMs under steady state and transient conditions.

Authors of Mishra and Subbarao (2021) compared the performance of a PM, a DDM, and an HM to predict dynamic combustion control parameters of a Reactivity Controlled Compression Ignition engine across five engine loads. The parameters included the start of combustion, the 50% mass fraction burnt crank angle, and combustion peak pressure. The authors compared the model predictions with measured data from experiments, concluding that the prediction capability of the HM was far superior to the DDM and PM across all parameters.

Authors of Bidarvatan et al. (2014) developed an HM to predict several performance parameters of Homegeneous Charge Compression Ignition (HCCI) engines. Namely, the the 50% mass fraction burnt crank angle, the indicated mean effective pressure, exhaust temperature, and concentration of CO, total unburned hydrocarbons and NO_x. The proposed HM combined a PM and 3 ANNs, designed to minimise computational time requirements, with minimal sacrifice in accuracy. The authors compared the predictions of the proposed HM with experimental data at 309 steady state and transient conditions for two HCCI engines, concluding that the HM offered approximately 80% better accuracy compared to the PM, or 60% compared to the DDM.

As expected, the amount of literature available on the HMs is limited, being this a relatively new and still partially explored research field. Moreover, focusing on the marine DE applications, to the authors best knowledge, a complete and clear description of a modelling framework for marine DEs able to hybridise PMs and DDMs is not yet readily available.

6.3 Modelling

This section is devoted first to the presentation of a state-of-the-art 0D PM (see Sect. 6.3.1) and DDM (see Sect. 6.3.2). Then the hybridisation framework is presented together with the proposed HM (see Sect. 6.3.3).

6.3.1 Zero Dimensional Diesel Engine Model

The DE PM has been developed utilising a modular approach, in which the engine scavenging air and exhaust gas receivers are modelled as control volumes, whereas the compressor and turbine are modelled as flow elements. The exhaust receiver contains states for pressure p, temperature T, and exhaust gas composition $\boldsymbol{g}$. The gas composition of the air path has been assumed constant. The engine boundaries are modelled using fixed fluid elements of constant pressure and temperature, and shaft elements are utilised to compute the rotational speed of the turbocharger ω_{tc} and crankshaft ω_e. Finally, for the in-cylinder process we exploited a two-zone 0D approach, with state quantities being the in-cylinder pressure, temperature, gas composition, and air-fuel equivalence ratio λ for each zone. The layout and state variables of the developed model are presented in Fig. 6.1. The governing equations of all relevant components is discussed in the the following.

Gas composition and properties—The working fluid of the engine is considered to be a mixture of the following 11 species: CO_2, H_2O, N_2, O_2, CO, H_2, O, H, OH, NO, and N. As such, the concentration vector in an arbitrary control volume is defined as

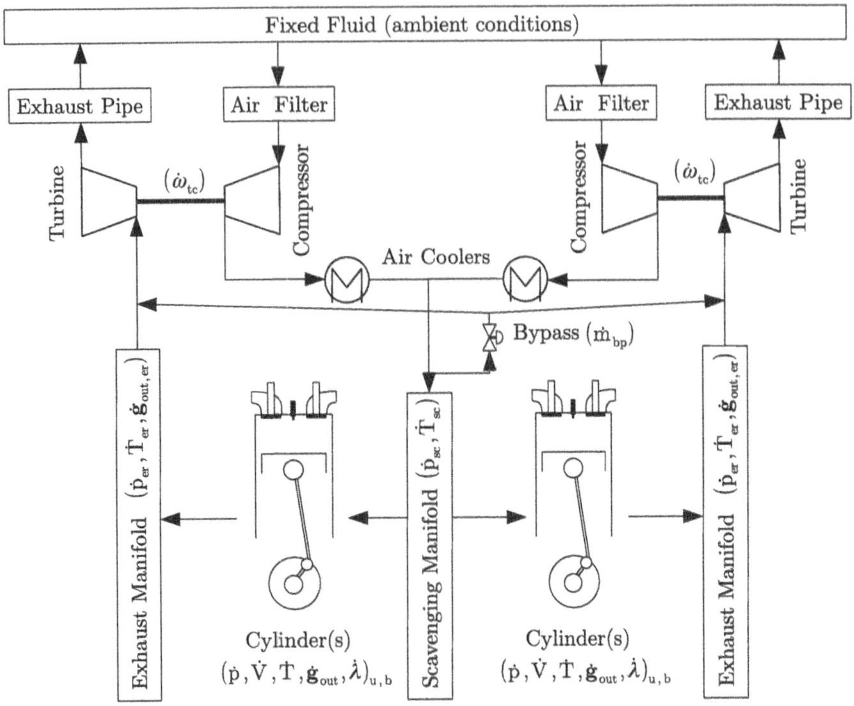

Fig. 6.1 Layout of the PM with all relevant components and state variables

$$g = \frac{\left[m_{CO_2}, m_{H_2O}, m_{N_2}, m_{O_2}, m_{CO}, m_{H_2}, m_O, m_H, m_{OH}, m_{NO}\right]}{m_{tot}}, \quad (6.1)$$

where m_i is the mass flow of species i and $m_{tot} = \sum_{i=1}^{11} m_i$. Arbitrarily, N is not included in the concentration vector since it can be computed as the remaining part, to ensure that the mass conservation law is not violated and to act as a sanity check of the numerical integration.

The thermodynamic properties of the gas are calculated using the NASA polynomials (McBride et al. 2002), under the assumption of a thermally perfect gas. The concentration along the air path is assumed constant and equal to the standard air concentration.

Control Volumes—The control volumes are modelled using the open thermodynamic system concept (Watson and Janota 1982; Heywood 1988) which use as inputs p, T, and the composition of the working medium contained in the adjacent elements. Considering the control volumes as cylindrical solenoids (neglecting dissociation effects and the kinetic energy of the flows entering/exiting the receivers) and assuming ideal gas, the change of rate in the mass stored in the volume m can be expressed as

$$\frac{\mathrm{d}m}{\mathrm{d}t} = \sum_i \dot{m}_{\text{in},i} - \sum_j \dot{m}_{\text{out},j}, \tag{6.2}$$

where $\dot{m}_i$ is the mass flow rate of stream i and the subscripts in, out refer to flows entering and exiting the control volume respectively. Note that, the mass rate of change does not need to be integrated since it only represents the difference between entering and exiting mass flow rates.

The temperature rate of change is derived from the energy conservation as

$$\frac{\mathrm{d}T}{\mathrm{d}t} = \frac{\dot{Q} + \sum_i mh_{\text{in},i} - \sum_j mh_{\text{out},j} - u\frac{\mathrm{d}m}{\mathrm{d}t}}{mc_v}, \tag{6.3}$$

where mh refers to the enthalpy flow rate (entering $mh_{\text{in},i}$ and exiting $mh_{\text{out},i}$ for the stream i), $\dot{Q} = kA(T - T_{\text{amb}})$ represents the heat transfer from the control volume to the surrounding environment, with k being the heat transfer coefficient, $A = \pi d^2/4$ being the heat transfer area, and u is the internal energy.

The heat transfer coefficient can be computed as (Merker et al. 2005)

$$k = 0.024\frac{\kappa}{d}\left[1 - \left(\frac{d}{l}\right)^{\frac{2}{3}}\right]\mathrm{Re}^{0.786}\mathrm{Pr}^{0.45}, \tag{6.4}$$

where l and d refer to the length and inner diameter of the control volume and κ being the heat conductivity, evaluated as

$$\kappa = 3.65182\ 10^{-4}T^{0.748}. \tag{6.5}$$

The Prandtl number Pr has been kept constant and equal to 0.731 and the Reynolds number Re is evaluated on the average mass flow rate of the input and output streams $\dot{m}_{\text{avg}}$ as

$$Re = \frac{\dot{m}_{\text{avg}}d}{A\nu}, \tag{6.6}$$

with ν being the kinematic viscosity computed as

$$\nu = 5.17791\ 10^{-7}T^{0.62}. \tag{6.7}$$

Due to to the lack of information regarding the geometry of the control volumes, their inner diameters d and lengths l are considered as calibration parameters.

Moreover, considering the control volume as a well-stirred mixer, the dynamics of the concentration states can be derived as (Llamas and Eriksson 2019)

$$\frac{\partial g_{\text{out}}}{\partial t} = \frac{RT}{pV}\sum_j \left(g_{\text{in},j} - g_{\text{out}}\right)\dot{m}_{\text{in},j}, \tag{6.8}$$

where $R = R(\mathbf{g}, T)$ is the gas mixture constant, V is the control volume, $\mathbf{g}_{\mathrm{in},j}$ is the composition of gas of the input stream j, and $\mathbf{g}_{\mathrm{out}}$ is the composition of gas of the output stream. Note that in the case of the scavenging air receiver, under the simplification of constant fresh air composition, these derivatives have been set to zero.

The pressure of the working medium contained in each control volume is calculated using the ideal gas law equation, from which the pressure's rate of change can be computed as

$$\frac{\mathrm{d}p}{\mathrm{d}t} = \frac{R}{V}\left(T\frac{\mathrm{d}m}{\mathrm{d}t} + m\frac{\mathrm{d}T}{\mathrm{d}t}\right). \tag{6.9}$$

Valves—The mass flow rate through a valve is computed assuming subsonic or sonic flow through a flow restriction (Heywood 1988), with the valve opening signal (lift) u_v as input

$$\dot{m}_v = A_{\mathrm{eff}}(u_v)\frac{p}{\sqrt{RT}}\Psi, \quad \Psi = \begin{cases} \Pi^{\frac{1}{\gamma}}\sqrt{\frac{2\gamma}{\gamma-1}\left(1-\Pi^{\frac{\gamma-1}{\gamma}}\right)}, & \text{if } \Pi > \left(\frac{2}{\gamma+1}\right)^{\frac{\gamma}{\gamma-1}} \\ \sqrt{\gamma}\left(\frac{2}{\gamma+1}\right)^{\frac{\gamma+1}{2(\gamma-1)}}, & \text{otherwise} \end{cases}, \tag{6.10}$$

where u_v is chosen as the percentage lift of the valve, Π refers to the ratio of the static pressure downstream of the restriction to the upstream stagnation pressure, $\gamma = \gamma(\mathbf{g}, T)$ is the ratio of specific heats of the medium, and $A_{\mathrm{eff}} = \mu(u_v)A_{\mathrm{ref}}$ refers to the effective area of the valve computed as the product of a reference area A_{ref} with the appropriate flow coefficient $\mu(u_v)$ (Heywood 1988).

The bypass valve is assumed to be activated if the pressure on the compressor exceeds 90% of the surge limit for the instantaneous flow rate. Furthermore, we have assumed a linear opening characteristic, with the reference area considered as a calibration parameter. For the intake and exhaust valves, we utilised a non-linear characteristic, parameterised on the basis of the maximum cross-flow area of the inlet A_{iv} and exhaust valves A_{ev} and the crank-angle duration that the valve stays at its maximum lift denoted as $\mathrm{vmd}_{\mathrm{iv}}$ and $\mathrm{vmd}_{\mathrm{ev}}$ (Merker et al. 2005).

Air cooler, Air Filter, and Exhaust Pipe—Pressure losses in all these components, as well as air cooler effectiveness, have been modelled as quadratic functions of their corresponding input mass flow rate (Theotokatos 2010; Theotokatos et al. 2018)

$$\begin{aligned} \Delta p_{\mathrm{ac}} &= \sum_{i=0}^{2} \dot{m}_{\mathrm{ac}}^{i} c_{\mathrm{ac},i}, \quad \Delta p_{\mathrm{af}} = \sum_{i=0}^{2} \dot{m}_{\mathrm{af}}^{i} c_{\mathrm{af},i}, \\ \Delta p_{\mathrm{ep}} &= \sum_{i=0}^{2} \dot{m}_{\mathrm{ep}}^{i} c_{\mathrm{ep},i}, \quad \epsilon_{\mathrm{ac}} = \sum_{i=0}^{2} \dot{m}_{\mathrm{ac}}^{i} c_{\mathrm{ac}_e,i}, \end{aligned} \tag{6.11}$$

where the subscripts ac, af, and ep refer to the air cooler, air filter, and exhaust pipe, respectively. $c_{\mathrm{ac},i}$, $c_{\mathrm{af},i}$, c_{ep_i}, and $c_{\mathrm{ac}_e,i}$ are considered calibration parameters.

As a consequence, the temperature at the air cooler outlet is given by

$$T_{\text{ac}}^{\text{out}} = \epsilon_{\text{ac}} T_w + (1 - \epsilon_{\text{ac}}) T_{\text{ac}}^{\text{in}}, \tag{6.12}$$

where ϵ_{ac} refers to the heat exchanger effectiveness, T_w corresponds to the temperature of the cooling water, and $T_{\text{ac}}^{\text{in}}$ is the temperature of the working medium at the inlet.

Turbocharger—The engine is equipped with two turbocharging units operating in parallel to supply the engine with sufficient air mass flow. We modelled the compressor using its steady state performance map, which provides the relations between the compressor performance variables: reference flow rate, pressure ratio, reference speed, and isentropic efficiency. The rotational speed and pressure ratio are considered as inputs to the model, which allows the computation of the reference flow rate $\dot{m}$ and isentropic efficiency η_c as reported in (Theotokatos and Tzelepis 2015; Baldi et al. 2015)

$$\dot{m} = f_{m_c}\left(\Pi_c, T_c^{\text{in}}, \omega_{\text{tc}}\right), \quad \eta_c = f_{\eta_c}\left(\Pi_c, T_c^{\text{in}}, \omega_{\text{tc}}\right), \tag{6.13}$$

where f_{m_c} and f_{η_c} are functions that have been interpolated based on the the compressor maps, representing the compressor mass flow rate and efficiency, respectively. Π_c is the pressure ratio of the compressor, and T_c^{in} is the compressor inlet temperature. After accounting for pressure losses in the air cooler and air filter, Π_c can be computed as

$$\Pi_c = \frac{p_{\text{sc}} + \Delta p_{ac}}{p_{\text{amb}} - \Delta p_{\text{af}}}, \tag{6.14}$$

where p_{sc} is the pressure of the scavenging receiver and p_{amb} is the ambient pressure. The temperature at the outlet of the compressor is given by (Watson and Janota 1982)

$$T_c^{\text{out}} = T_c^{\text{in}}\left(1 + \frac{\Pi_c^{\frac{\gamma-1}{\gamma}} - 1}{\eta_c}\right). \tag{6.15}$$

For the turbine, we exploited its swallowing capacity and efficiency maps, which allow the calculation of the turbine flow rate $\dot{m}_t$ and isentropic efficiency η_t as

$$\dot{m}_t = f_{m_t}\left(\Pi_t, T_t^{\text{in}}, \omega_{\text{tc}}\right), \quad \eta_t = f_{\eta_t}\left(\Pi_t, T_t^{\text{in}}, \omega_{\text{tc}}\right). \tag{6.16}$$

where f_{m_t} and f_{η_t} are functions that have been interpolated based on the the turbine maps, representing the turbine mass flow rate and efficiency, respectively. Π_t is the pressure ratio of the turbine, and T_t^{in} is the turbine inlet temperature. The turbine pressure ratio is computed by taking the exhaust pipe pressure losses Δp_{ep} into account as

$$\Pi_t = \frac{p_{\text{er}}}{p_{\text{amb}} + \Delta p_{\text{ep}}}, \tag{6.17}$$

where p_{er} is the pressure of the exhaust receiver. The temperature at the turbine outlet T_t^{out} can be computed from the turbine isentropic efficiency definition as

$$T_t^{\text{out}} = T_t^{\text{in}} \eta_t \left(1 - \Pi_t^{\frac{\gamma-1}{\gamma}}\right), \tag{6.18}$$

The rotational speed of the turbochargers ω_{tc} is a model state, defined by the power balance between compressor and turbocharger as

$$\frac{d\omega_{\text{tc}}}{\text{d}t} = \frac{P_t \eta_{\text{tc}}^{\text{mech}} - P_c}{J_{\text{tc}}\omega_{\text{tc}}}, \tag{6.19}$$

where $P_t = \dot{m}_t \Delta h_t$, $P_c = \dot{m}_c \Delta h_c$ refer to the turbine and compressor power, respectively, with Δh_t, Δh_c being the enthalpy difference between the inlet and outlet of the turbine and compressor. J_{tc} refers to the turbocharger shaft inertia and $\eta_{\text{tc}}^{\text{mech}}$ corresponds to the mechanical efficiency of the turbocharger unit, accounting for friction losses. J_{tc} and $\eta_{\text{tc}}^{\text{mech}}$ are considered calibration parameters.

Cylinder—For the in-cylinder process, apart from the assumptions on the working medium described before, we have further neglected valve leakage and blow-by. Furthermore, the temperatures at the cylinder wall T_{cw}, head T_{hw}, piston wall T_{pw}, liner wall T_{lw}, exhaust valve wall T_{evw}, as well as the injected fuel temperature T_f are considered all uniform and constant. In fact, the temperature variations of the inner cylinder surface during the thermodynamic cycle are trivial compared to the temperature variations of the combustion gases (Descieux and Feidt 2007; Rakopoulos et al. 2004). Moreover, we have assumed a uniform cylinder pressure and that the combustion chamber volume consists of two zones. The burned zone contains incompletely oxidised fuel (denoted with the subscript b), whereas the unburned zone contains air and fuel (denoted with the subscript u). Each zone is spatially homogeneous, separated by a massless and infinitesimally thin flame, and no heat transfer takes place between the two zones. A schematic of the two combustion zone model is presented in Fig. 6.2. The main equations governing the two combustion zone model include the conservation of mass

$$\frac{\text{d}m}{\text{d}\theta} = \frac{\text{d}m_u}{\text{d}\theta} + \frac{\text{d}m_b}{\text{d}\theta} = \frac{\text{d}m_f}{\text{d}\theta} + \frac{\text{d}m_a}{\text{d}\theta} \tag{6.20}$$

where the subscripts u and b refer the the unburned and burned zone, respectively, and the subscripts a, f refer to air and fuel, respectively. The equations of state for the working medium in each zone can be described as

$$pV_u = m_u R_u T_u, \quad pV_b = m_b R_b T_b, \tag{6.21}$$

and the evolution of volume derivative with respect to crank angle in the two zones as

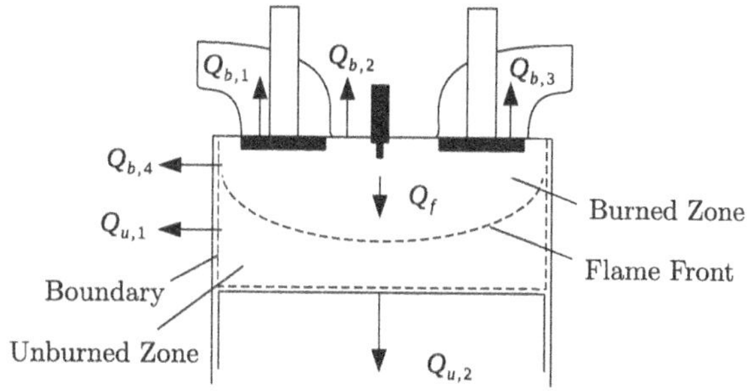

Fig. 6.2 Energy flow of the two combustion zone model

$$\frac{dV}{d\theta} = \frac{dV_u}{d\theta} + \frac{dV_b}{d\theta}, \tag{6.22}$$

in which the instantaneous cylinder volume can be expressed as a function of crank-angle θ according to the kinematics of the crankshaft as (Merker et al. 2005; Heywood 1988)

$$V(\theta)=V_c+B^2\frac{\pi}{4}(l_r+a-s(\theta)), \quad s(\theta)=a+\cos\theta+\sqrt{l_r^2 - a^2\sin\theta^2}, \tag{6.23}$$

where $V_c = V_s/(CR - 1)$ is the combustion chamber volume, V_s is the stroke volume, CR is the geometric compression ration, B is the bore diameter, l_r is the connecting rod length, and a is the crank radius.

Finally, the energy conservation equations for each combustion zone can be expressed as

$$\frac{d(m_u u_u)}{d\theta} = -p\frac{dV_u}{d\theta} - \sum_i \frac{dQ_{u_i}}{d\theta} - h_u\frac{dm_b}{d\theta} \tag{6.24}$$

$$\frac{d(m_b u_b)}{d\theta} = -p\frac{dV_b}{d\theta} - \sum_i \frac{dQ_{b_i}}{d\theta} + h_u\frac{dm_b}{d\theta} + \frac{dQ_f}{d\theta} \tag{6.25}$$

where $\sum_i \frac{dQ_i}{d\theta}$ for each one of the combustion zones refers to the summation of heat transfer rates through the engine's different parts surfaces in contact with the cylinder gases and $\frac{dQ_f}{d\theta}$ refers to the heat release rate.

Burn fraction and heat release rate in internal combustion engines are mostly governed by functions based on the law of Normal distribution of continuous random variables. In this context, one of the most popular functions is the Wiebe function (Ghojel 2010; Galindo et al. 2011). While the Wiebe function by no means describes the complex fuel air mixing in the diesel combustion process, it can provide valuable thermodynamic input for the model in terms of a realistic shape of the heat release. In this work, we have exploited a double-Wiebe profile for the premixed and diffusive combustion processes (Miyamoto et al. 1985; Kökkülünk et al. 2016) as reported in (Merker et al. 2005).

$$\frac{\mathrm{d}Q_{f,1}}{\mathrm{d}\theta}=Q_{f,1}\alpha(m_1+1)\left(\frac{\theta-\theta_{\mathrm{SOC}_1}}{\Delta\theta_{\mathrm{CD},1}}\right)^{m_1}\mathrm{e}^{-\alpha_l\left(\frac{\theta-\theta_{\mathrm{SOC}_1}}{\Delta\theta_{\mathrm{CD}_1}}\right)^{m_1+1}}, \tag{6.26}$$

$$\frac{\mathrm{d}Q_{f,2}}{\mathrm{d}\theta}=Q_{f,2}\alpha(m_2+1)\left(\frac{\theta-\theta_{\mathrm{SOC}_2}}{\Delta\theta_{\mathrm{CD},2}}\right)^{m_2}\mathrm{e}^{-\alpha_2\left(\frac{\theta-\theta_{\mathrm{SOC}_2}}{\Delta\theta_{\mathrm{CD}_2}}\right)^{m_2+1}}, \tag{6.27}$$

$$Q_{f,1}=qQ_{f,tot},\quad Q_{f,2}=(1-q)Q_{f,tot},\quad \frac{\mathrm{d}Q_{f,\mathrm{tot}}}{\mathrm{d}\theta}=\frac{\mathrm{d}Q_{f,1}}{\mathrm{d}\theta}+\frac{\mathrm{d}Q_{f,2}}{\mathrm{d}\theta}, \tag{6.28}$$

where α is a factor related to the percentage of the total energy added with the fuel at the end of combustion $\eta_{\mathrm{conv,total}}$ as

$$\alpha=-\ln\left(1-\eta_{\mathrm{conv,total}}\right), \tag{6.29}$$

its value is related to the combustion efficiency and it was assumed equal to 6.9 as suggested in Theotokatos et al. (2018) and Gogoi and Baruah (2010). Moreover, $Q_{f,tot}=m_f LHV_f$ represents the total thermal energy of the fuel released during combustion, q is the fraction of heat release of the first Wiebe profile, $\theta_{\mathrm{SOC},i}$ corresponds to the start of combustion, m_1 and m_2 are the Vibe form factors, and $\Delta\theta_{\mathrm{CD},i}$ denotes the combustion duration for each Wiebe profile. For the latter, the estimation of the ignition delay $\Delta\theta_{id}$ is required, for example according to Sitkei (1963) as

$$\Delta\theta_{id}=6n_e10^{-3}\left(a_{\mathrm{IGD}}+(p^{-0.7}b_{\mathrm{IGD}}+p^{-1.8}c_{\mathrm{IGD}})\,\mathrm{e}^{\frac{7800}{6.9167RT}}\,1.0197\right), \tag{6.30}$$

where n_e is the engine rotational speed (expressed in Hz). The constants a_{IGD}, b_{IGD}, c_{IGD} are treated as calibration parameters.

In Eqs. (6.26) and (6.27), the constants m_1, m_2, $\Delta\theta_{\mathrm{CD}_1}$, and $\Delta\theta_{\mathrm{CD}_2}$ for each Wiebe profile are calibrated at the engine reference point (subscript ref) and updated for other operating points according to Woschni and Anisits (1973)

$$m=(m_{\mathrm{ref}}+\Delta m)\left(\frac{\theta_{\mathrm{IGD,ref}}}{\theta_{\mathrm{IGD}}}\right)^{a_{\mathrm{VM}}}\left(\frac{n_{e,\mathrm{ref}}}{n_e}\right)^{b_{\mathrm{VM}}}\left(\frac{p_{\mathrm{IVC}}}{p_{\mathrm{IVC,ref}}}\cdot\frac{V_{\mathrm{IVC}}}{V_{\mathrm{IVC,ref}}}\cdot\frac{T_{\mathrm{IVC,ref}}}{T_{\mathrm{IVC}}}\right)^{c_{\mathrm{VM}}}-\Delta_m, \tag{6.31}$$

$$\Delta\theta_{\mathrm{CD}}=\Delta\theta_{\mathrm{CD,ref}}\left(\frac{\lambda_{\mathrm{ref}}}{\lambda}\right)^{a_{\mathrm{CD}}}\left(\frac{n_e}{n_{e,\mathrm{ref}}}\right)^{b_{\mathrm{CD}}}, \tag{6.32}$$

The constants a_{CD}, b_{CD}, a_{VM}, b_{VM}, c_{VM} are regarded as calibration parameters, since their values are known to vary between engines of different types and sizes as reported in Merker et al. (2005). The subscripts IVC,ref and IVC refer to the inlet valve closing point at engine reference speeds $n_{e,\mathrm{ref}}$ and operating speed n_e, respectively.

The heat transfer between the mass trapped in the cylinder and the surrounding areas is calculated according to the standard Newtonian relation for convective heat transfer, as

$$\sum_i\frac{\mathrm{d}Q_{u_i}}{\mathrm{d}\theta}=k_{\mathrm{csa}}\sum_i A_i(T(\theta)-T_i), \tag{6.33}$$

Table 6.1 Coefficients for Woschni's heat transfer model of Eqs. (6.34), (6.35)

Phase	$c_1[-]$	c_2[m/s K]
Intake-exhaust	$6.18 + 0.417 c_c/c_m$	0
Compression	$2.28 + 0.308 c_c/c_m$	0
Combustion-expansion		$3.24 \cdot 10^{-3}$

with $i = \{\text{cw, hw, pw, lw, evw}\}$ referring to cylinder, head, piston, liner, and exhaust valve wall. The heat transfer coefficient k_{csa} is evaluated according to Woschni (1968), Merker et al. (2005)

$$k_{\text{csa}} = 127.93 p^{0.8} v^{0.8} B^{-0.2} T^{-0.53}, \tag{6.34}$$

where v is a representative velocity evaluated taking into account the mean piston speed $c_m = 2L_s n_e$ (L_s is the stroke length), and the compression induced turbulence as Merker et al. (2005)

$$v = c_1 c_m + c_2 \frac{V_c T_{\text{IC}}}{p_{\text{IC}} V_{\text{IC}}} (p - p_0), \tag{6.35}$$

where p_0 is the cylinder pressure during motored operation, computed over a polytropic relation from the cylinder volume according to Merker et al. (2005). The constants c_1, c_2 are functions of the intake swirl c_c/c_m, according to Table 6.1. c_c/c_m is considered as a calibration parameter. The subscript IC refers to the cylinder conditions at the start of compression, when the intake valve closes.

We exploited the method of Chen and Flynn (1965) for the evaluation of the friction losses, according to which the friction mean effective pressure FMEP accounts for the effect of engine speed through a quadratic law. The effect of engine load is represented through the maximum in-cylinder pressure p_{max}, while the energy drawn by accessories and all the other invariable factors is accounted for by a constant term, as

$$\text{FMEP} = c_{f,1} + c_{f,2} p_{\text{max}} + c_{f,3} n_e + c_{f,4} n_e^2 \tag{6.36}$$

where coefficients $c_{f,1} - c_{f,4}$ are considered as calibration parameters.

The instantaneous cylinder torque is then computed by using the gross cylinder torque, through the cylinder indicated work, and cylinder torque due to friction. Since all cylinders are considered to operate under the same conditions, the brake power is computed using the current engine speed, multiplied by the cylinder torque and the number of cylinders in the engine.

The combustion products are evaluated exploiting the method of Rakopoulos et al. (1994), due to its minimal computational time requirements and reasonable agreement with experiments. For the burning zone, given its volume, temperature, mass of fuel burnt and mass of air entrained, the concentration of each species can be evaluated by solving a 11×11 non-linear system obtained from 7 non-linear equilibrium equations and 4 linear atom balance equations. This system is converted to a 4×4

non-linear system which can be solved efficiently by any root-finding algorithm [in this study, the Newton-Raphson method (Stoer and Bulirsch 2013)]. Finally, thermal NO has been evaluated according to the extended Zeldovich mechanism, for which the reaction rates were selected according to Hanson and Salimian (1984).

Equations (6.20)–(6.28) form a system of first order differential equations of the form $F(\theta, y, \dot{y}) = 0$ that is solved for each crank angle step by using the classic Runge-Kutta method (Stoer and Bulirsch 2013).

Sensor Dynamics—The thermal inertia of the temperature sensors which are mounted on the outer surface of the exhaust pipes, is modelled according to Llamas and Eriksson (2019) as

$$\frac{dT_s}{dt} = \frac{T - T_s}{\tau_s}, \tag{6.37}$$

where T_s refers to the temperature including the sensor dynamics, and T is the temperature of the working medium in the engine. The time constant τ_s is considered as a calibration parameter. Dynamic response for all other sensors has been neglected, as it is known to be in the order of milliseconds (Wahlström and Eriksson 2011).

Calculation Procedure—Inputs towards the cylinder model include the pressure, temperature, the medium composition from the scavenging manifold, the pressure of the exhaust manifold, engine rotational speed, and fuel injected per cycle from the governor. Subsequently, the cylinders air and exhaust gas mass flow rates, pressures, temperatures, the composition of the exhaust gas, and the equivalence ratio of the exhaust gas exiting the cylinders are calculated. Additional outputs include the energy flow of the exhaust gas exiting cylinders, the indicated power, the friction power, brake power torque, brake specific fuel consumption, and engine brake efficiency.

For all the other components (i.e., control volumes or flow elements) the following structure is employed. Inputs required for the flow elements are utilised from the adjacent flow receiver or fixed fluid structures for the engine boundaries which include the necessary parameters to fully characterise the working medium state (temperature, pressure, composition). Subsequently, mass and energy flows through the flow elements are computed and provided to the adjacent control volumes. In addition, the absorbed compressor torque and produced turbine torque are calculated and used as inputs to the turbocharger shaft element, which derives the turbocharger speed, which, in turn, is provided to the turbine and compressor blocks.

This framework forms an additional system of first order differential equations in the form $F(t, y, \dot{y}) = 0$, that is solved for each time step by using the classic Runge-Kutta method (Stoer and Bulirsch 2013). In total, the model contains 50 states over two major integration steps. Parameters include the geometric data of the engine, the intake and exhaust valves profiles, the compressor and turbine performance maps, the bypass valve geometric and control details, constants present in any sub-model, and the ambient conditions for the engine boundaries. Finally, initial values are also required for the engine and turbocharger rotational speeds, and the temperature, pressure and composition of the working medium contained in the scavenging and exhaust manifolds.

Model Parameterisation—The proposed model contains internal feedback systems, hence the modelling errors of any subsystems will be propagated and amplified towards the rest of the model. For instance, errors in the exhaust temperature will affect the turbine power production, which will in turn alter the scavenging pressure, and this finally will modify the exhaust temperature prediction anew. Due to this fact, balancing out the complete model by readjusting the model parameters is essential to obtain an overall accurate model (Llamas and Eriksson 2019).

The calibration process is treated as an optimisation problem, with the objective being to minimise the error between the model estimated outputs and the available measurements.

Formally, we seek the solution of the following continuous, non-convex problem

$$\begin{aligned} \arg\min_{\phi} \quad & \hat{L}(\phi, \mathcal{D}_n) = \sum_{i=1}^{M} l(h(\boldsymbol{x}_i, \phi), \boldsymbol{y}_i), \\ \text{s.t.} \quad & \phi_{\min} \le \phi \le \phi_{\max} \end{aligned} \tag{6.38}$$

where h refers to the outputs of the PM, ϕ is the set of parameters that needs to be estimated from a given bounded space Φ, $\boldsymbol{x}_i$ refers to the measurements corresponding with the model inputs, and $\boldsymbol{y}_i$ refers to the measurements corresponding with the model outputs. $\hat{L}(\phi, \mathcal{D}_n)$ is the empirical error of the model h on the dataset $\mathcal{D}_n = \{(\boldsymbol{x}_1, \boldsymbol{y}_1), \ldots, (\boldsymbol{x}_n, \boldsymbol{y}_n)\}$, measured according to a loss function $\ell(h(\boldsymbol{x}, \boldsymbol{\theta}), \boldsymbol{y})$.

We have adopted the absolute relative error, given by

$$\ell\left(h\left(\boldsymbol{x}_i, \boldsymbol{\theta}\right), \boldsymbol{y}_i\right) = \sum_{j=1}^{S} \left| \frac{y_i^j - h^j(\boldsymbol{x}_i, \boldsymbol{\theta})}{y_i^j} \right| \tag{6.39}$$

Note that, because $\boldsymbol{y}$ is a vector, the loss function of Eq. (6.39) refers to the sum of relative errors of all model outputs $j = (1, \ldots, S)$ and their corresponding measured values.

Given the nature of the problem, a Derivative-Free Optimisation (DFO) method must be exploited, as obtaining or estimating the derivatives of the physical models with respect to the parameters is a computationally and time-intensive procedure. The literature on DFO methods is quite large, with a variety of algorithms that can solve different classes of problems (Conn et al. 2009; Galinier et al. 2013; Floudas and Pardalos 2008).

In this work, we leveraged an algorithm from the class of directional direct search methods: the Mesh-Adaptive Direct Search (MADS) algorithm, which is a local optimisation technique with established convergence theory under some mild assumptions (Audet and Dennis 2006; Audet et al. 2008). MADS is an extension of the Generalized Pattern Search algorithm (Torczon 1997; Lewis et al. 2000), specifically developed to handle non-smooth black-box problems. It is an iterative method that uses a discretisation of the solution space, called the mesh, to select and evaluate new trial points, given an initial iterate. Each iteration consists of two steps:

the search and the poll, followed by a parameter update step. If the search step succeeds (i.e., the selected trial point improves upon the current iterate), then this trial point becomes the new iterate and the poll step is skipped. If the search step fails, the poll step becomes mandatory. The poll is used to choose mesh points near the current iterate and to evaluate their objective and constraint values. If the poll fails to find a better solution, the update step will reduce the mesh size and the poll size, to concentrate near the current iterate. The mesh size is the parameter that scales the space discretization and the poll size is the maximum distance allowed between a trial point and the current iterate. On the other hand, once a better solution is found, the pol step terminates and the update step increases the mesh size. This process is repeated until a stopping condition is satisfied.

Given that our problem is highly non-linear and non-convex, there is no guarantee that the solution obtained corresponds to the global optima (Floudas and Pardalos 2008). In order to ensure the quality of the final solution the algorithm has been started from a number of different initial points, and from all solutions obtained, the best one has been chosen (Martí 2003).

The dynamic behaviour of the model is largely defined by the turbocharger inertias, the control volume sizes, and the time constants of the temperature sensors. The optimisation algorithm was initialised by a single starting point with suitable values of the turbocharger inertia corresponding to other engines of this size, whereas the values of the control volumes have been set to reasonable values based on the real pipe volume sizes from the engine design drawings.

6.3.2 *Data Driven Models*

In the proposed context of developing a fast yet accurate dynamic model of a four-stroke marine DE, a general modelisation framework can be defined, characterised by an input space $\mathcal{X} \subseteq \mathbb{R}^d$, an output space $\mathcal{Y} \subseteq \mathbb{R}^b$, and an unknown relation $\mu : \mathcal{X} \to \mathcal{Y}$ to be learned (Shalev-Shwartz and Ben-David 2014; Hamilton 2020). For what concerns this work, $\mathcal{X}$ is composed by the measurements available from the engine monitoring system (see Sect. 6.4), while the output space $\mathcal{Y}$ refers to the target features accounting for the engine fuel consumption, turbocharger rotational speed, turbine outlet temperature, and exhaust manifold temperature (see Sect. 6.4).

In this context, the authors define the model $h : \mathcal{X} \to \mathcal{Y}$ as an artificial simplification of μ. Analogously to what has been done in Sect. 6.3.1 we will assume to know all the information until time t_0 to make a prediction of the quantity of interest. In particular, we will consider all the information in $[t_0 - \Delta, t_0]$ (see Fig. 6.3). Δ represents how much history of the different available data we want to exploit to make predictions. Δ is an hyperparameter for which an optimal value exists: too much history (too large Δ) will make us face with the curse of dimensionality while too little history (too small Δ) will limit our ability to make accurate predictions (Shalev-Shwartz and Ben-David 2014; Oneto 2020; Hamilton 2020).

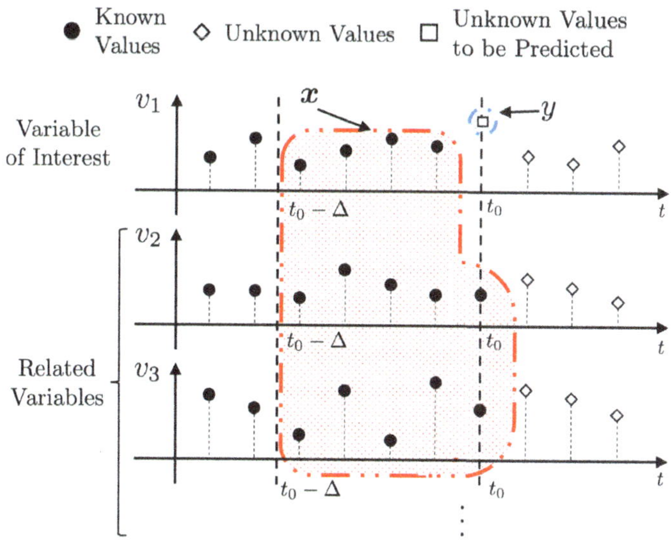

Fig. 6.3 Input and output variables of the DDMs

The model h, as described in Sect. 6.1 can be obtained with different kinds of techniques, for example, requiring some physical knowledge of the problem, as in PMs (see Sect. 6.3.1), or the acquisition of large amounts of data, as in DDMs or using both information (see Sect. 6.3.3). In this section we will use a state-of-the-art DDM. Between the DDMs it is possible to identify two families of approaches (Shalev-Shwartz and Ben-David 2014; Goodfellow et al. 2016). The first one, comprising traditional Machine Learning methods, needs an initial phase where the features must be defined a-priori from the data via feature engineering or implicit or explicit feature mapping (Shalev-Shwartz and Ben-David 2014; Zheng and Casari 2018; Shawe-Taylor and Cristianini 2004). The second family, which includes deep learning methods, automatically learns both the features and the models from the data (Goodfellow et al. 2016). For small cardinality datasets and outside particular applications (e.g., computer vision and natural language processing) Deep Learning does not perform well since they require huge amount of data to be reliable and to outperform traditional Machine Learning models (Fernández-Delgado et al. 2014; Wainberg et al. 2016).

In the Machine Learning maps the problem our problem can be easily mapped in a typical regression problem (Vapnik 1998; Shawe-Taylor and Cristianini 2004). In fact, ML techniques aim at estimating the unknown relationship μ between input and output through a learning algorithm $\mathscr{A}_{\mathcal{H}}$ which exploits some historical data to learn h and where $\mathcal{H}$ is a set of hyperparameters which characterises the generalisation performance of $\mathscr{A}$ (Oneto 2020). The historical data consists on a series of n examples of the input/output relation μ and are defined as $\mathcal{D}_n = \{(\boldsymbol{x}_1, y_1), \dots, (\boldsymbol{x}_n, y_n)\}$ where $\boldsymbol{x} \in \mathcal{X}$ and $y \in \mathcal{Y}$.

In this paper we will leverage on a Machine Learning model coming from the Kernel Methods family called Kernel Regularised Least Squares (KRLS) (Vovk 2013). The idea behind KRLS can be summarised as follows. During the training phase, the quality of the learned function $h(\boldsymbol{x})$ is measured according to a loss function $\ell(h(\boldsymbol{x}), y)$ (Rosasco et al. 2004) with the empirical error

$$\hat{L}_n(h) = \frac{1}{n} \sum_{i=1}^{n} \ell(h(\boldsymbol{x}_i), y_i). \tag{6.40}$$

A simple criterion for selecting the final model during the training phase could then consist in simply choosing the approximating function that minimises the empirical error $\hat{L}_n(h)$. This approach is known as Empirical Risk Minimization (ERM) (Vapnik 1998). However, ERM is usually avoided in Machine Learning as it leads to severe overfitting of the model on the training dataset. As a matter of fact, in this case the training process could choose a model, complicated enough to perfectly describe all the training samples (including the noise, which afflicts them). In other words, ERM implies memorisation of data rather than learning from them. A more effective approach is to minimise a cost function where the trade-off between accuracy on the training data and a measure of the complexity of the selected model is achieved (Tikhonov and Arsenin 1979), implementing the Occam's razor principle

$$h^*: \min_h \quad \hat{L}_n(h) + \Lambda\, C(h). \tag{6.41}$$

In other words, the best approximating function h^* is chosen as the one that is complicated enough to learn from data without overfitting them. In particular, $C(\cdot)$ is a complexity measure: depending on the exploited Machine Learning approach, different measures are realised. Instead, $\Lambda \in [0, \infty)$ is a hyperparameter, that must be set a-priori and is not obtained as an output of the optimisation procedure: it regulates the trade-off between the overfitting tendency, related to the minimisation of the empirical error, and the underfitting tendency, related to the minimisation of $C(\cdot)$. The optimal value for Λ is problem-dependent, and tuning this hyperparameter is a non-trivial task, as will be discussed later in this section. In KRLS, models are defined as

$$h(\boldsymbol{x}) = \boldsymbol{w}^T \boldsymbol{\varphi}(\boldsymbol{x}), \tag{6.42}$$

where $\boldsymbol{\varphi}$ is an a-priori defined Feature Mapping (FM) (Shalev-Shwartz and Ben-David 2014) allowing to keep the structure of $h(\boldsymbol{x})$ linear. The complexity of the models, in KRLS, is measured as

$$C(h) = \|\boldsymbol{w}\|^2, \tag{6.43}$$

i.e., the Euclidean norm of the set of weights describing the regressor, which is a standard complexity measure in ML (Shalev-Shwartz and Ben-David 2014; Vovk

2013). Regarding the loss function, the square loss is typically adopted because of its convexity, smoothness, and statistical properties (Rosasco et al. 2004)

$$\hat{L}_n(h) = \frac{1}{n}\sum_{i=1}^{n} \ell(h(\boldsymbol{x}_i), y_i) = \frac{1}{n}\sum_{i=1}^{n} [h(\boldsymbol{x}_i) - y_i]^2 . \tag{6.44}$$

Consequently, Problem (6.41) can be reformulated as

$$\boldsymbol{w}^*: \min_{\boldsymbol{w}} \quad \sum_{i=1}^{n} \left[\boldsymbol{w}^T \boldsymbol{\varphi}(\boldsymbol{x}) - y_i\right]^2 + \Lambda \|\boldsymbol{w}\|^2. \tag{6.45}$$

By exploiting the Representer Theorem (Schölkopf et al. 2001), the solution h^* of the Problem (6.45) can be expressed as a linear combination of the samples projected in the space defined by $\boldsymbol{\varphi}$

$$h^*(\boldsymbol{x}) = \sum_{i=1}^{n} \iota_i \boldsymbol{\varphi}(\boldsymbol{x}_i)^T \boldsymbol{\varphi}(\boldsymbol{x}). \tag{6.46}$$

It is worth underlining that, according to the kernel trick, it is possible to reformulate $h^*(\boldsymbol{x})$ without an explicit knowledge of $\boldsymbol{\varphi}$, and consequently avoiding the curse of dimensionality of computing $\boldsymbol{\varphi}$, by using a proper kernel function $K(\boldsymbol{x}_i, \boldsymbol{x}) = \boldsymbol{\varphi}(\boldsymbol{x}_i)^T \boldsymbol{\varphi}(\boldsymbol{x})$

$$h^*(\boldsymbol{x}) = \sum_{i=1}^{n} \iota_i K(\boldsymbol{x}_i, \boldsymbol{x}). \tag{6.47}$$

Several kernel functions can be retrieved in literature (Scholkopf 2001; Cristianini and Shawe-Taylor 2000), each one with a particular property that can be exploited based on the problem under exam. Usually the Gaussian kernel is chosen

$$K(\boldsymbol{x}_i, \boldsymbol{x}) = \mathrm{e}^{-\gamma \|\boldsymbol{x}_i - \boldsymbol{x}\|^2}, \tag{6.48}$$

because of the theoretical reasons described in Keerthi and Lin (2003), Oneto et al. (2015) and because of its effectiveness (Fernández-Delgado et al. 2014; Wainberg et al. 2016). γ is another hyperparameter, which regulates the nonlinearity of the solution that must be tuned as explained later. Basically the Gaussian kernel is able to implicitly create an infinite dimensional $\boldsymbol{\varphi}$ and thanks to this, the KRLS are able to learn any possible function (Keerthi and Lin 2003). The KRLS problem of Eq. (6.45) can be reformulated by exploiting kernels as

$$\boldsymbol{\iota}^*: \min_{\boldsymbol{\iota}} \quad \|Q\boldsymbol{\iota} - \boldsymbol{y}\|^2 + \Lambda \boldsymbol{\iota}^T Q \boldsymbol{\iota}, \tag{6.49}$$

where $y = [y_1, \ldots, y_n]^T$, $\iota = [\iota_1, \ldots, \iota_n]^T$, the matrix Q such that $Q_{i,j} = K(\boldsymbol{x}_j, \boldsymbol{x}_i)$, and the identity matrix $I \in \mathbb{R}^{n \times n}$. By setting the gradient equal to zero w.r.t. $\boldsymbol{\iota}$ it is possible to state that

$$(Q + \Lambda I)\,\iota^* = y, \tag{6.50}$$

which is a linear system for which effective solvers have been developed over the years, allowing it to cope with even very large sets of training data (Young 2003).

The problems that still have to be faced is how to tune the hyperparameters of the approach (Λ, γ, and Δ^- for the second DT) and to estimate the performance of the final model. Model Selection (MS) and Error Estimation (EE) deal exactly with these problems (Oneto 2020). Resampling techniques are commonly used by researchers and practitioners since they work well in most situations and this is why we will exploit them in this work (Oneto 2020). Other alternatives exist, based on the Statistical Learning Theory, but they tend to underperform resampling techniques in practice (Oneto 2020). Resampling techniques are based on a simple idea: the original dataset $\mathcal{D}_n$ is resampled once or many (n_r) times, with or without replacement, to build three independent datasets called learning, validation and test sets, respectively $\mathcal{L}_l^r$, $\mathcal{V}_v^r$, and $\mathcal{T}_t^r$, with $r \in \{1, \ldots, n_r\}$ such that

$$\mathcal{L}_l^r \cap \mathcal{V}_v^r = \oslash, \quad \mathcal{L}_l^r \cap \mathcal{T}_t^r = \oslash, \quad \mathcal{V}_v^r \cap \mathcal{T}_t^r = \oslash, \quad \mathcal{L}_l^r \cup \mathcal{V}_v^r \cup \mathcal{T}_t^r = \mathcal{D}_n. \tag{6.51}$$

Subsequently, to select the best hyperparameters' combination $\mathcal{H} = \{\Lambda, \gamma, (\Delta^-)\}$ in a set of possible ones $\mathfrak{H} = \{\mathcal{H}_1, \mathcal{H}_2, \ldots\}$ for the algorithm $\mathscr{A}_\mathcal{H}$ or, in other words, to perform the MS phase, the following procedure has to be applied:

$$\mathcal{H}^*\colon \arg\min_{\mathcal{H} \in \mathfrak{H}} \sum_{r=1}^{n_r} M(\mathscr{A}_\mathcal{H}(\mathcal{L}_l^r), \mathcal{V}_v^r), \tag{6.52}$$

where $h = \mathscr{A}_\mathcal{H}(\mathcal{L}_l^r)$ is a model built with the algorithm $\mathscr{A}$ with its set of hyperparameters $\mathcal{H}$ and with the data $\mathcal{L}_l^r$, and where $M(h, \mathcal{V}_v^r)$ is a desired metric. Since the data in $\mathcal{L}_l^r$ are independent from the data in $\mathcal{V}_v^r$, $\mathcal{H}^*$ should be the set of hyperparameters which allows achieving a small error on a data set that is independent from the training set. Then, to evaluate the performance of the optimal model which is $h^*_\mathscr{A} = \mathscr{A}_{\mathcal{H}^*}(\mathcal{D}_n)$ or, in other words, to perform the EE phase, the following procedure has to be applied:

$$M(h^*_\mathscr{A}) = \frac{1}{n_r} \sum_{r=1}^{n_r} M(\mathscr{A}_{\mathcal{H}^*}(\mathcal{L}_l^r \cup \mathcal{V}_v^r), \mathcal{T}_t^r). \tag{6.53}$$

Since the data in $\mathcal{L}_l^r \cup \mathcal{V}_v^r$ are independent from the ones in $\mathcal{T}_t^r$, $M(f^*_\mathscr{A})$ is an unbiased estimator of the true performance, measured with the metric M, of the final model (Oneto 2020). In this work we will rely on Complete k-fold cross validation which means setting $n_r \leq \binom{n}{k}\binom{n-\frac{n}{k}}{k}$, $l = (k-2)\frac{n}{k}$, $v = \frac{n}{k}$, and $t = \frac{n}{k}$ and the resampling must be done without replacement (Oneto 2020). Notc that, in our application, we have a further constraint in terms of dependence in time between the samples. For this reason, when resampling the data form $\mathcal{D}_n$ we actually keep data of different periods in $\mathcal{L}_l^r$, $\mathcal{V}_v^r$, and $\mathcal{T}_t^r$ (Hamilton 2020).

For what concerns the metric M that we will use in our paper we will rely on the Mean Absolute Error (MAE), the Mean Absolute Percentage of Error (MAPE) and the Pearson Product-Moment Correlation Coefficient PPMCC (Willmott and Matsuura 2005). Since in regression it is quite hard to synthesise the quality of a predictor in a single metric we will also rely on visualisation techniques like the scatter plot and histograms (Shao et al. 2017).

6.3.3 Hybrid Models

In this section we would like to depict a framework able to take into account both the physical knowledge about the problem encapsulated in the PMs of Sect. 6.3.1 and the information hidden in the available data as the DDMs of Sect. 6.3.2, into account. For this purpose authors will start from a simple observation: a HM, based on the previous observation, should be able to learn from the data without being too different, or too far away, from the PM.

From the Data Science and Machine Learning point of view, this requirement can be straightforwardly mapped in a typical ML Multi Task Learning (MTL) problem (Baxter 2000; Caruana 1997; Evgeniou and Pontil 2004; Bakker and Heskes 2003; Argyriou et al. 2008). MTL aims at simultaneously learning two concepts, in this case the PM and the available data, through a learning algorithm $\mathscr{A}_{\mathcal{H}}$ which exploits the data in $\mathcal{D}_n$ to learn a function h which is both close to the observation, the data $\mathcal{D}_n$ and the PM, namely its forecasts.

Consequently, in this case a slightly different scenario is presented where the dataset is composed by a triple of points $\mathcal{D}_n = \{(\boldsymbol{x}_1, y_1, p_1), \dots, (\boldsymbol{x}_n, y_n, p_n)\}$ where p_i is the output of the PM in the point $\boldsymbol{x}_n$ with $i \in \{1, \dots, n\}$. The target is to learn a function able to approximate both μ, namely the relation between the input $\boldsymbol{x} \in \mathcal{X}$ and the output $y \in \mathcal{Y}$, and the PM, namely the relation between the input and the output of the PM. Two tasks have to be learned. For this purpose there are two main approaches: the first approach is called Shared Task Learning (STL) and the second Independent Task Learning (ITL). While the latter independently learns a different model for each task, the former aims to learn a model that is common between all tasks. A well-known weakness of these methods is that they tend to generalise poorly on one of the two tasks (Baxter 2000). In this work, authors show that an appealing approach to overcome such limitations is provided by MTL (Baxter 2000; Caruana 1997; Evgeniou and Pontil 2004; Bakker and Heskes 2003; Argyriou et al. 2008). This methodology leverages on the information between the tasks to learn more accurate models.

In order to apply the MTL approach to this case, it is possible to modify the KRLS problem of Eq. (6.45) to simultaneously learn a shared model and a task specific model which should be close to the shared model. In this way, authors obtain a model which is able to simultaneously learn the two tasks. The model that authors are interested in is the shared model, while the task specific models are just used as a tool. A shared model is defined as

$$h(\boldsymbol{x}) = \boldsymbol{w}^T \varphi(\boldsymbol{x}), \tag{6.54}$$

and two task specific models as

$$h_i(\boldsymbol{x}) = \boldsymbol{w}_i^T \varphi(\boldsymbol{x}), \quad i \in \{y, p\}. \tag{6.55}$$

Then, it is possible to state the MTL version of Eq. (6.45), as follows

$$\begin{aligned} \boldsymbol{w}^*, \boldsymbol{w}_y^*, \boldsymbol{w}_p^* : \min_{\boldsymbol{w}, \boldsymbol{w}_y, \boldsymbol{w}_p} \quad & \sum_{i=1}^{n} \left[\boldsymbol{w}^T \varphi(\boldsymbol{x}) - y_i\right]^2 + \left[\boldsymbol{w}^T \varphi(\boldsymbol{x}) - p_i\right]^2 \\ & + \sum_{i=1}^{n} \left[\boldsymbol{w}_y^T \varphi(\boldsymbol{x}) - y_i\right]^2 + \left[\boldsymbol{w}_p^T \varphi(\boldsymbol{x}) - p_i\right]^2 \\ & + \Lambda \|\boldsymbol{w}\|^2 + \varkappa(\|\boldsymbol{w} - \boldsymbol{w}_y\|^2 + \|\boldsymbol{w} - \boldsymbol{w}_p\|^2), \end{aligned} \tag{6.56}$$

where Λ is the usual regularization of KRLS and $\varkappa \in [0, \infty)$, instead, is another hyperparameter that forces the shared model to be close to the task specific models. Basically the MTL problem of Eq. (6.56) is a concatenation of three learning problems solved with KRLS plus a term which tries to keep a relation between all the three different problems.

By exploiting the kernel trick as in KRLS, it is possible to reformulate Problem (6.56), as follows

$$\begin{aligned} \boldsymbol{\iota}^* : \min_{\boldsymbol{\iota}} \quad & \left\| \begin{bmatrix} Q & Q & 0 & 0 \\ Q & Q & 0 & 0 \\ 0 & 0 & Q & 0 \\ 0 & 0 & 0 & Q \end{bmatrix} \boldsymbol{\iota} - \begin{bmatrix} \boldsymbol{y} \\ \boldsymbol{p} \\ \boldsymbol{y} \\ \boldsymbol{p} \end{bmatrix} \right\|^2 \\ & + \boldsymbol{\iota}^T \begin{bmatrix} (\Lambda + 2\varkappa)Q & (\Lambda + 2\varkappa)Q & -\varkappa Q & -\varkappa Q \\ (\Lambda + 2\varkappa)Q & (\Lambda + 2\varkappa)Q & -\varkappa Q & -\varkappa Q \\ -\varkappa Q & -\varkappa Q & \varkappa Q & 0 \\ -\varkappa Q & -\varkappa Q & 0 & \varkappa Q \end{bmatrix} \boldsymbol{\iota}, \end{aligned} \tag{6.57}$$

where $\boldsymbol{p} = [p_1, \ldots, p_n]^T$. The solution of this problem is again equivalent to solving a linear system

$$\begin{bmatrix} Q+(\Lambda+2\varkappa)I & Q+(\Lambda+2\varkappa)I & -\varkappa I & -\varkappa I \\ Q+(\Lambda+2\varkappa)I & Q+(\Lambda+2\varkappa)I & -\varkappa I & -\varkappa I \\ -\varkappa I & -\varkappa I & Q+\varkappa I & 0 \\ -\varkappa I & -\varkappa I & 0 & Q+\varkappa I \end{bmatrix} \iota^* = \begin{bmatrix} y \\ p \\ y \\ p \end{bmatrix}. \quad (6.58)$$

The function that the authors are interested in, the shared one, can be expressed as follows

$$h(\boldsymbol{x}) = \boldsymbol{w}^T \boldsymbol{\varphi}(\boldsymbol{x}) = \sum_{i=1}^{n} (\iota_i + \iota_{i+n}) K(\boldsymbol{x}_i, \boldsymbol{x}). \quad (6.59)$$

What changes here, with respect to the MS phase of the DDM described in Sect. 6.3.2, is the MS phase where just Λ, γ, and also $\varkappa$ need to be tuned.

6.4 Case Study and Dataset Description

In this work, as a case study, we have exploited data acquired from a naval vessel equipped with a MAN B&W V28-33D medium speed four-stroke DE. The engine has 12 cylinders with 0.28m bore and 0.33m stroke, with a Maximum Continuous Rating (MCR) of 5.4MW at 1000rpm and two turbochargers (TCs) operating in parallel to deliver the necessary air. The main characteristics of the engine are summarised in Table 6.2.

The engine is characterised by a power-to-weight and power-to-installation space ratios favourable to fast mono-hull and multi-hull vessels, offshore patrol vessels with either single or twin engine-gear-propeller systems or corvettes, frigates and destroyers with combined propulsion plant configurations, such as COmbined Diesel

Table 6.2 Main characteristics of the MAN 12 V28-33D engine

Feature	Value	Unit
Cylinders	V12, 16, 20	[–]
Bore diameter	280	[mm]
Stroke length	330	[mm]
Number of cylinders	12	[–]
Revolutions per cycle	2	[–]
Engine speed at MCR	1000	[rpm]
Brake power at 60% MCR	3240	[kW]
Brake power at 80% MCR	4320	[kW]
Brake power at MCR	5400	[kW]
Mean effective pressure	26.9	[bar]
Mean piston speed	11	[m/s]
Specific fuel consumption (100% load)	191	[g/kWh]

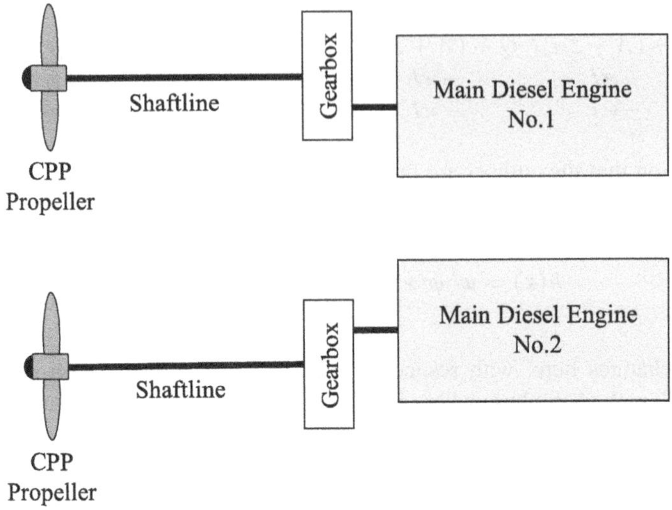

Fig. 6.4 Propulsion system layout for the holland class oceangoing patrol vessels

And Diesel (CODAD), COmbined Diesel Or Gas (CODOG), and COmbined Diesel And Gas (CODAG). The DE under investigation is installed on board one of the the Holland Class Oceangoing Patrol Vessels. The propulsion system of the vessel consists of two shafts with Controllable Pitch Propellers (CPP), a gearbox, and one DE per shaft, as shown in Fig. 6.4. This configuration is typical for multi-function ships that require silent, manoeuvrable, highly reliable and low emission propulsion.

The Patrol vessel is equipped with a data logging system which is used by the Royal Netherlands Navy both for on-board monitoring and control and for land-based performance analysis. The dataset utilised consists of two different data sources: standard measurements (steady-state) performed during Shop Trials (ST) that have been exploited to calibrate the PM model (see Sect. 6.3.1), and operational data originating from the vessel's data logging system, used by the ship operator for performance monitoring purposes, which has been exploited to evaluate the performance of the PM model in dynamic conditions (see Sect. 6.3.1), and to train, validate, and test the DDMs and HMs (see Sects. 6.3.2 and 6.3.3). Operational measurements are sampled at 3Hz for approximately 3 calendar years for a total of 7900 h of operations. A summary of the available measurements is reported in Table 6.3.

It should be noted that fuel consumption is measured using a mass flow meter which is known to be more accurate of the more common volume flow meters as it eliminates uncertainty on fuel density. However, measurements of fuel specific energy content are not available. Unfortunately, the energy content of a compound fuel can vary in quality among markets, a variation which is known to be in the order of ± 2 MJ/kg, or approximately $\pm 5\%$ (Coraddu et al. 2017). Moreover, measurements regarding the ambient conditions of the engine's surrounding environment are also not available, as well as the uncertainty of the measurements performed during the ST.

Table 6.3 Measurements available from the engine monitoring system

Variable name	Symbol	Unit
Time stamp	t	[hh:mm:ss]
Governor position	G_p	[–]
Engine rotational speed	n_e	[rpm]
Engine torque	M_e	[kNm]
Charge air temperature at scavenging receiver	T_{sc}	[°C]
Charge air temperature at compressor outlet	$T_{\mathrm{c,out}}$	[°C]
Charge air temperature at compressor inlet	$T_{\mathrm{c,in}}$	[°C]
Exhaust gas temperature at turbine outlet	$T_{\mathrm{t,out}}$	[°C]
Main bearing no. 1 temperature	$T_{\mathrm{b,1}}$	[°C]
Main bearing no. 2 temperature	$T_{\mathrm{b,2}}$	[°C]
Main bearing no. 3 temperature	$T_{\mathrm{b,3}}$	[°C]
Main bearing no. 4 temperature	$T_{\mathrm{b,4}}$	[°C]
Main bearing no. 5 temperature	$T_{\mathrm{b,5}}$	[°C]
Main bearing no. 6 temperature	$T_{\mathrm{b,6}}$	[°C]
Main bearing no. 7 temperature	$T_{\mathrm{b,7}}$	[°C]
Lube oil compartment no. 1 temperature	$T_{\mathrm{l,1}}$	[°C]
Lube oil compartment no. 2 temperature	$T_{\mathrm{l,2}}$	[°C]
Lube oil compartment no. 3 temperature	$T_{\mathrm{l,3}}$	[°C]
Lube oil compartment no. 4 lube oil temperature	$T_{\mathrm{l,4}}$	[°C]
Lube oil compartment no. 5 lube oil temperature	$T_{\mathrm{l,5}}$	[°C]
Lube oil engine inlet temperature	$T_{\mathrm{le,in}}$	[°C]
Lube oil engine outlet temperature	$T_{\mathrm{le,out}}$	[°C]
High-temperature sea cooling water—inlet	$T_{\mathrm{ht,in}}$	[°C]
High-temperature sea cooling water—outlet	$T_{\mathrm{ht,out}}$	[°C]
Low-temperature sea cooling water—inlet	$T_{\mathrm{lt,in}}$	[°C]
Low-temperature sea cooling water—outlet	$T_{\mathrm{lt,out}}$	[°C]
Fuel oil supply temperature	T_f	[°C]
Charge air temperature at compressor outlet—bank A	$T_{\mathrm{c,out}}^A$	[°C]
Charge air temperature at compressor outlet—bank B	$T_{\mathrm{c,out}}^B$	[°C]
Charge air temperature at compressor inlet—bank A	$T_{\mathrm{c,in}}^A$	[°C]
Charge air temperature at compressor inlet—bank B	$T_{\mathrm{c,in}}^B$	[°C]
Charge air engine inlet pressure	$p_{\mathrm{ca,in}}$	[Pa]
Charge air engine inlet temperature	$T_{\mathrm{ca,in}}$	[°C]
Fuel consumption	$\dot{m}_f$	[kg/h]
TC rotational speed	N_{tc}	[rpm]
Turbine outlet temperature	$T_{\mathrm{t,out}}$	[°C]
Exhaust receiver temperature	T_{er}	[°C]

6.5 Experimental Results

In this section, we exploited the data described in Sect. 6.4 to test the models developed in Sect. 6.3. To begin with, calibration results of the PM described in Sect. 6.5.1 are reported. Then a comparison of the performance of PMs, DDMs, and HMs in operational dynamic conditions is reported.

6.5.1 Zero Dimensional Diesel Engine Model Calibration Results

The PM model validation has been carried out taking into account both the standard (steady-state) measurements performed during ST and Stationary Operations (SO). To identify SO, the dataset was first split into a set of time intervals of continuous operation. Within each interval, operation under stationary conditions is defined as any continuous set of measurements for which the rotational speed and load of the engine vary by less than 1%, for a period of at least 3 hours. For each of these stationary conditions, the last 10 minutes of measurements were extracted and the median value of each signal was computed. This allowed us to summarise each stationary operation as one value per signal, for a total of 256 stationary operation points.

To perform the calibration and validation of the PM we exploited a subset of the data reported in Table 6.3. In particular, Table 6.4 reports the subset of the data source exploited as input, as validation on ST and SO, and comparison with the DDM and HM in transient analysis.

The performance metrics discussed in Sect. 6.3.2 are reported in Table 6.5 on the ST dataset.

The reported performances indicate that the model can capture all measurements well within 1% for engine loads ranging between 20 and 100%. The maximum combustion pressure shows the highest errors, with a MAPE equal to 0.98%. The lowest discrepancy between the PM and the measurements is reported for the temperature at the turbine outlet, with a MAPE of only 0.01%, well within the uncertainty of most conventional thermocouples used for this application. The prediction accuracy for all other variables is equally good regardless of the subsystem considered.

Regarding the validation of the model in SO, we have to consider that the performance of the PM have been assesses on a dataset representing the behaviour of the engine during sailing. Unfortunately, a holistic comparison on the performance of the PM on the ST and SO dataset is not possible as only a subset of signals is available for this second validation as reported in Table 6.6.

Nevertheless, the metrics reported in Table 6.6 reveal that the PM is still able to capture the performances of the DE in real operations. In fact, the MAPE on the compressor outlet temperature has increased from 0.638% on the ST to 2.752% on SO. Similarly, the scavenging manifold and turbine outlet temperatures have increased

Table 6.4 Subset of data source from Table 6.3 exploited as PM inputs, validation on ST and SO, and comparison with the DDM and HM in transient analysis

	Variable name	Symbol	Unit
Input	Engine rotational speed	n_e	[rpm]
	Governor position	G_p	[rpm]
Output for ST and SO validation	Compressor outlet pressure	$p_{c,\text{out}}$	[Pa]
	Compressor outlet temperature	$T_{c,\text{out}}$	[kg/h]
	Turbine outlet pressure	$p_{\text{t,out}}$	[Pa]
	Turbine outlet temperature	$T_{\text{t,out}}$	[°C]
	Scavenging receiver temperature	T_{sc}	[°C]
	TC Rotational speed	n_{tc}	[Hz]
	Maximum cylinder pressure	p_{max}	[Pa]
	Specific fuel consumption	sfc	[g/kWh]
Output for transient analysis	Fuel consumption	$\dot{m}_f$	[kg/h]
	TC rotational speed	N_{tc}	[Hz]
	Turbine outlet temperature	$T_{\text{t,out}}$	[°C]
	Exhaust receiver temperature	T_{er}	[°C]

Table 6.5 PM Performance metrics on the ST

Variable name	Symbol	Unit	MAE	MAPE [%]	PPMCC
Compressor outlet pressure	$p_{c,\text{out}}$	[Pa]	$1.2 \cdot 10^3$	0.363	1.000
Compressor outlet temperature	$T_{c,\text{out}}$	[K]	2.750	0.638	1.000
Turbine outlet pressure	$p_{t,\text{out}}$	[Pa]	62.354	0.060	1.000
Turbine outlet temperature	$T_{t,\text{out}}$	[K]	0.066	0.010	1.000
Scavenging manifold temperature	$T_{\text{s,in}}$	[K]	0.393	0.122	0.979
Turbocharger rotational speed	n_{tc}	[Hz]	4.302	0.041	0.989
Cylinder maximum pressure	p_{max}	[Pa]	$1.1 \cdot 10^5$	0.981	0.999
Specific fuel consumption	sfc	[g/kWh]	1.207	0.056	0.988

Table 6.6 Performance metrics on PM-SO dataset

Variable name	Symbol	Unit	MAE	MAPE [%]	PPMCC
Compressor outlet temperature	$T_{c,out}$	[K]	11.545	2.752	0.643
Turbine outlet temperature	$T_{t,out}$	[K]	3.798	1.199	0.342
Scavenging manifold temperature	$T_{sc,in}$	[K]	17.545	2.570	−0.145
Turbocharger rotational speed	n_{tc}	[Hz]	12.013	3.506	0.881
Specific fuel consumption	sfc	[g/kWh]	4.400	2.178	0.514

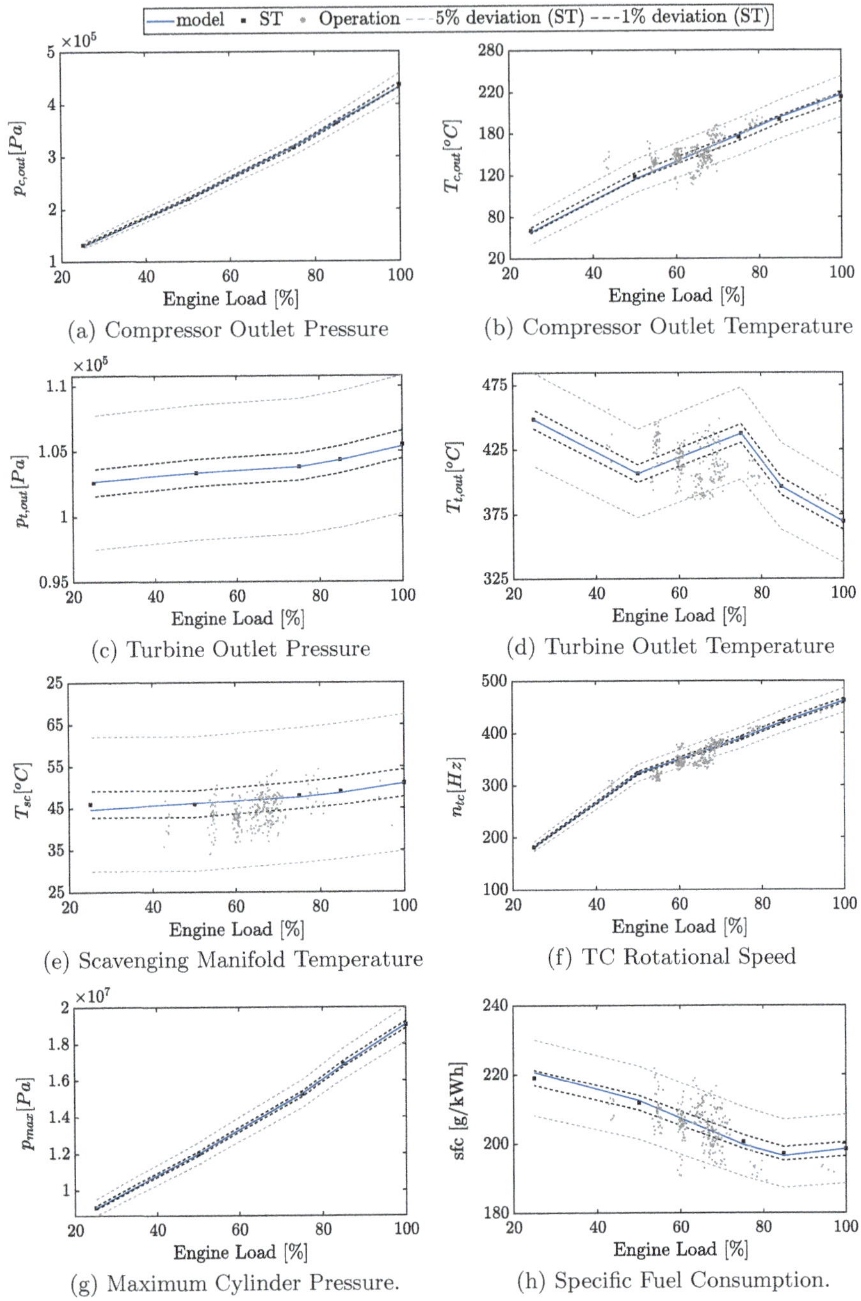

Fig. 6.5 PM verification in steady-state conditions (ST and SO)

Table 6.7 Subset of data source from Table 6.3 exploited as inputs and outputs of the DDM and HM

	Symbol					
Input variables for DDMs and HMs	n_e	$T_{l,4}$	T_{sc}	$T_{l,5}$	$T_{c,out}$	$T_{le,in}$
	$T_{c,in}$	$T_{le,out}$	$T_{t,out}$	$T_{ht,in}$	$T_{b,1}$	$T_{ht,out}$
	$T_{b,2}$	$T_{lt,in}$	$T_{b,3}$	$T_{lt,out}$	$T_{b,4}$	T_f
	$T_{b,5}$	$T^A_{c,out}$	$T_{b,6}$	$T^B_{c,out}$	$T_{b,7}$	$T^A_{c,in}$
	$T_{l,1}$	$T^B_{c,in}$	$T_{l,2}$	$p_{ca,in}$	$T_{l,3}$	$T_{ca,in}$
Output variables for DDMs and HMs	$\dot{m}_f$					
	N_{tc}					
	$T_{t,out}$					
	T_{er}					

from 0.122 and 0.01% to 1.2 and 2.6%, respectively. Moreover, a similar decrease in prediction capability can be observed for the specific fuel consumption and turbocharger rotational speed, with the MAPEs increasing from 0.041 and 0.056% to 3.5 and 2.2%, respectively. Finally, a visual impression of the results reported in Tables 6.5 and 6.6 is reported in Fig. 6.5 for various engine loading conditions.

6.5.2 *Physical Model, Data Driven Model, and Hybrid Model*

This section is devoted to the comparison between the PM, DDM, and HM. As a first step we have to detail the inputs and the outputs of the DDM and HM. For this purpose Table 6.7 reports the subset of Table 6.3 exploited ad inputs and outputs of the DDM and HM.

Then we have to report the hyperparametes ranges for the DDM and HM. For the DDM the set of hyperparameters tuned during the MS phase are $\mathcal{H} = \{\gamma, \Lambda\}$ chosen in $\mathfrak{H} = \{10^{-4.0}, 10^{-3.8}, \dots, 10^{+4.0}\} \times \{10^{-4.0}, 10^{-3.8}, \dots, 10^{+4.0}\}$. For the DDM the set of hyperparameters tuned during the MS phase are $\mathcal{H} = \{\gamma, \Lambda, \varkappa\}$ chosen in $\mathfrak{H} = \{10^{-4.0}, 10^{-3.8}, \dots, 10^{+4.0}\} \times \{10^{-4.0}, 10^{-3.8}, \dots, 10^{+4.0}\} \times \{10^{-4.0}, 10^{-3.8}, \dots, 10^{+4.0}\}$.

All the tests have been repeated 30 times, and the average results are reported together with their t-student 95% confidence interval, to ensure the statistical validity of the results.

Table 6.8 reports the performance (measured with the MAE, MAPE, and PPMCC) of the different models (PM, DDM, and HM) for different values of $\Delta \in \{0, 10, 20, 30\}$ for the different targets to to predict ($\dot{m}_f$, N_{tc}, $T_{t,out}$, and T_{er}). Note that $\Delta = 0$ means that the authors do not exploit time series information from the past, for $\Delta > 0$ there is no PM result. Moreover, to improve the readability and better understand the quality of the results, Figs. 6.6, 6.7, 6.8 and 6.9 reports for $\Delta = 0$ the scatter

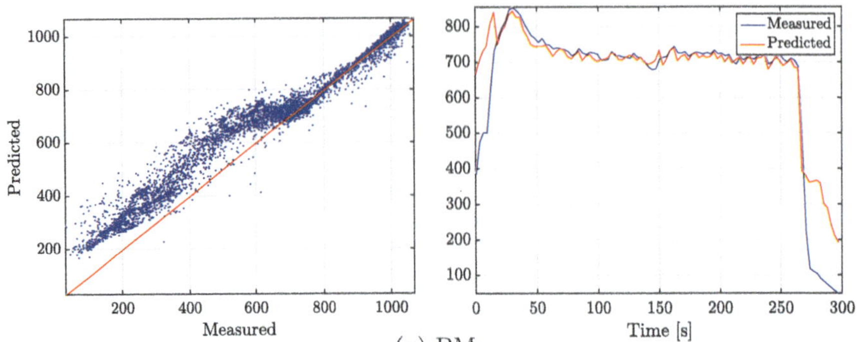

(a) PM

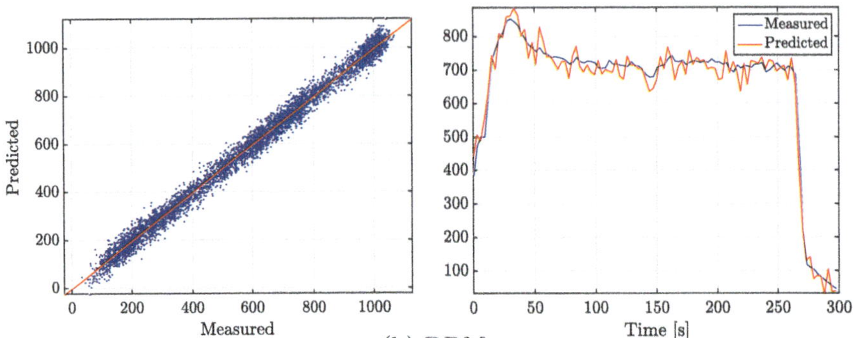

(b) DDM

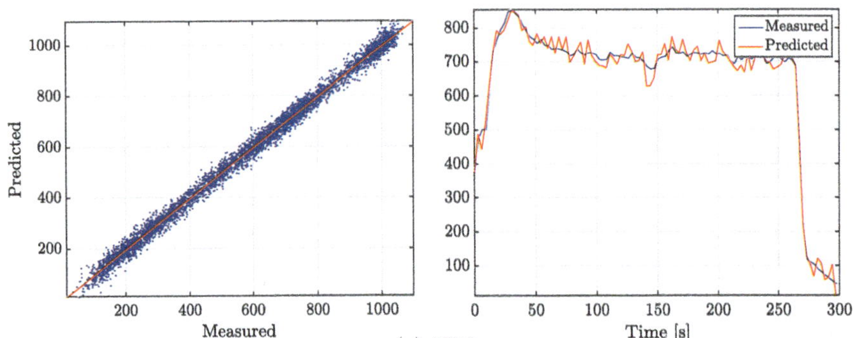

(c) HM

Fig. 6.6 Scatter plot (measured vs. predicted) and trend in time for the $\dot{m}_f$ (kg/h) output feature using the different models (PMs, DDMs, and HMs) with $\Delta = 0$

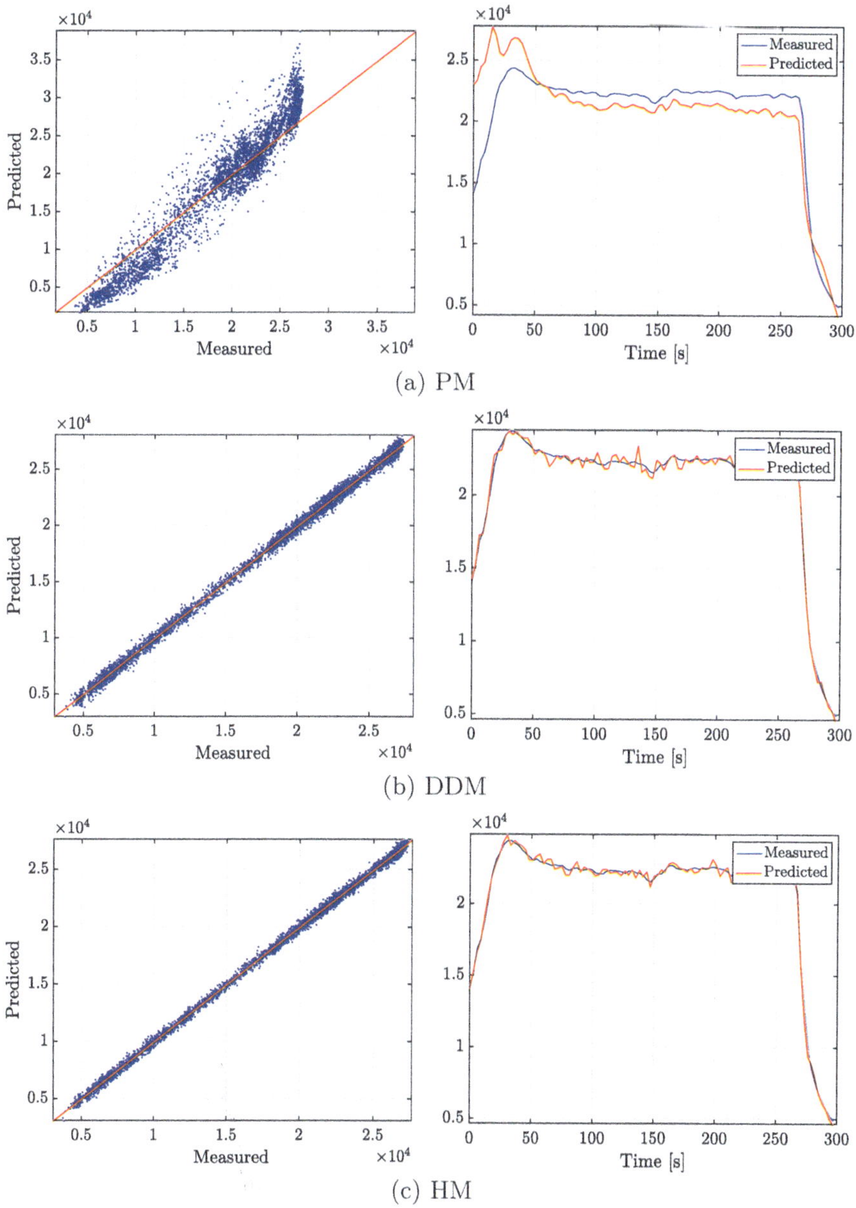

Fig. 6.7 Scatter plot (measured vs. predicted) and trend in time for the N_{tc} (rpm) output feature using the different models (PMs, DDMs, and HMs) with $\Delta = 0$

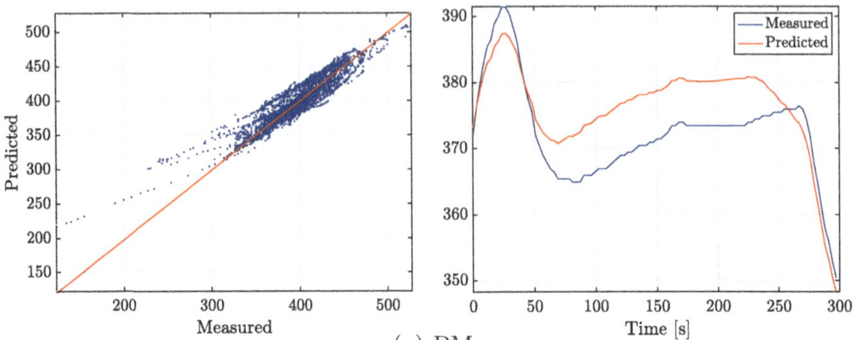

(a) PM

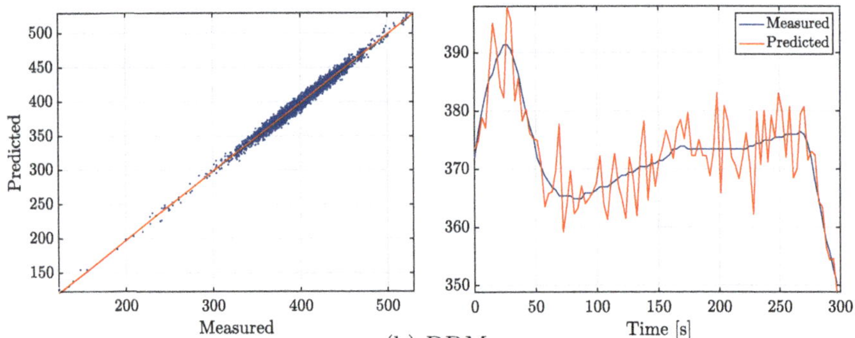

(b) DDM

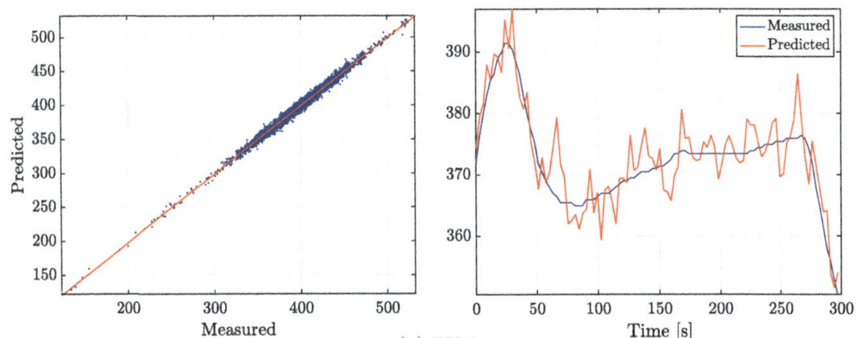

(c) HM

Fig. 6.8 Scatter plot (measured vs. predicted) and trend in time for $T_{t,\mathrm{out}}$ (°C) output feature using the different models (PMs, DDMs, and HMs) with $\Delta = 0$

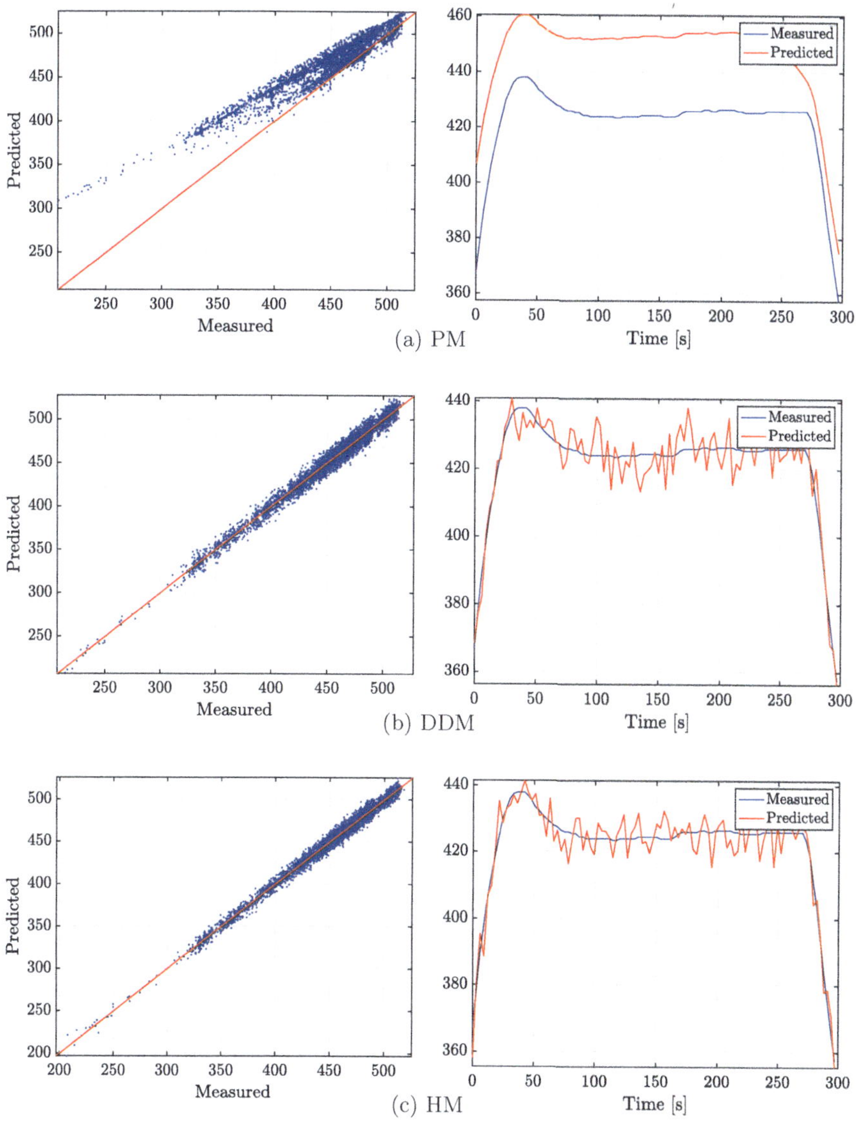

Fig. 6.9 Scatter plot (measured vs. predicted) and trend in time for T_{er} (°C) output feature using the different models (PMs, DDMs, and HMs) with $\Delta = 0$

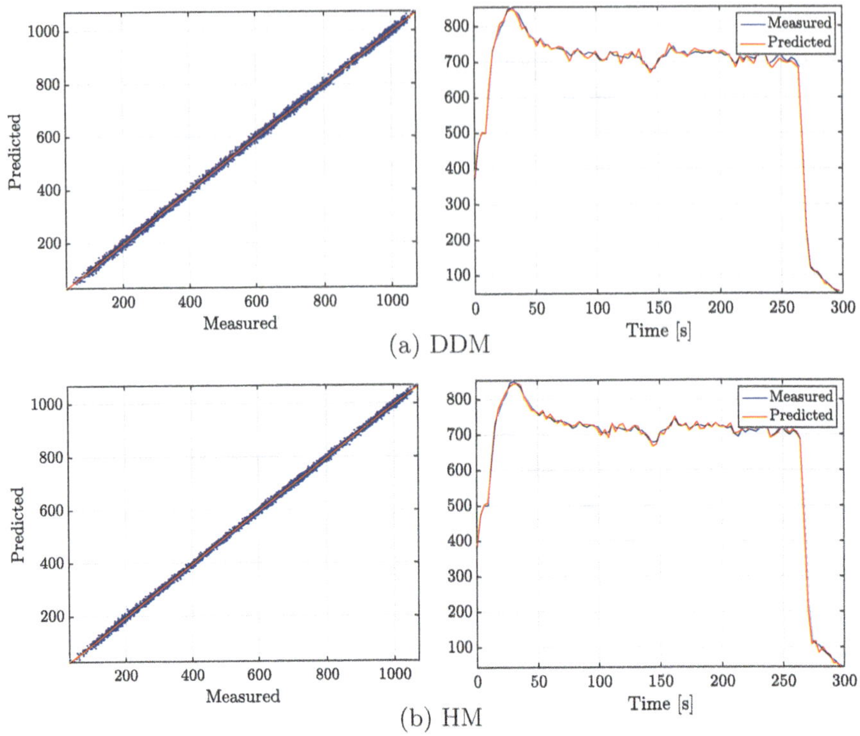

Fig. 6.10 Scatter plot (measured vs. predicted) and trend in time for $\dot{m}_f$ (kg/h) output feature using the different models (DDMs and HMs) with $\Delta = 20$

plot (measured and predicted values) and an example of trend in time (measured and predicted values) for the different targets using the different models (in this case PM, DDM, and HM). Moreover, Figs. 6.10, 6.11, 6.12 and 6.13 report for the value of Δ characterised by the best results in terms of accuracy ($\Delta = 20$) the scatter plots (measured and predicted values) and examples of trend in time (measured and predicted values) for DDM and HM.

Compared to the PM, the proposed DDMs are more accurate in predicting the four targets ($\dot{m}_f$, N_{tc}, $T_{\text{t,out}}$, and T_{er}), even without taking into account past information ($\Delta > 0$). A substantial decrease of the errors can be observed from Table 6.8 across all the targets. Considering $\dot{m}_f$, we can observe a MAPE decrease from 26.93% (PM) to 6.30% (DDM), to 4.89% (HM). The same general trend can be reported for N_{tc}, $T_{t,\text{out}}$, and T_{er}.

Moreover, when taking into account past information ($\Delta > 0$), from the error metrics reported in Table 6.8 we can observe:

- $\dot{m}_f$ – 90% MAPE reduction from 26.93% ±1.54% to 2.50% ±0.15%
- N_{tc} – 93% MAPE reduction from 15.39% ±0.75% to 0.94% ±0.05%
- $T_{t,\text{out}}$ – 80% MAPE reduction from 2.53% ±0.13% to 0.48% ±0.02%
- T_{er} – 88% MAPE reduction from 4.81% ±0.15% to 0.56% ±0.03%

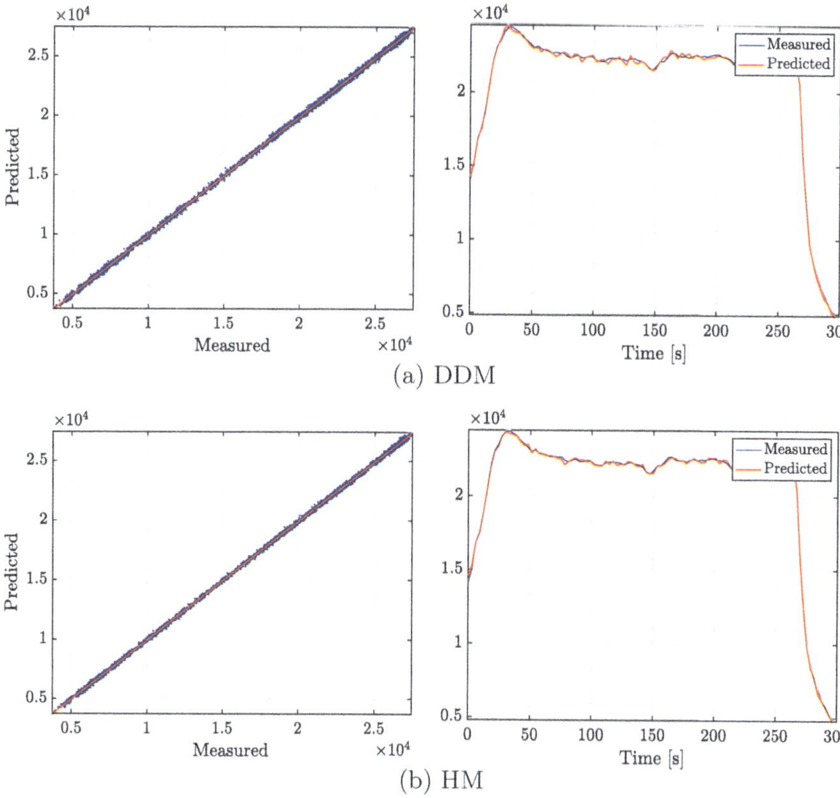

Fig. 6.11 Scatter plot (measured vs. predicted) and trend in time for N_{tc} (°C) output feature using the different models (DDMs and HMs) with $\Delta = 20$

From Figs. 6.6, 6.7, 6.8 and 6.9 it is possible to observe that DDMs are capable of fully capturing the transient behaviour of the fuel consumption (see Fig. 6.6b), the turbocharger rotational speed mechanical transient (see Fig. 6.7b), and the thermodynamic transients of both the turbine outlet gases (see Fig. 6.8b) and exhaust manifold (see Fig. 6.9b). Also from the results depicted in Figs. 6.6, 6.7, 6.8 and 6.9, it can be also observed that the DDMs are characterised by both lower bias and lower variance, with respect to the PM. The optimal time window (Δ) is found for a value equal to 20 seconds. For this value, minimal error metrics among all DDMs occur. According to Table 6.8, for this time window, the MAPE for $\dot{m}_f$ is as low as $1.79\% \pm 0.08\%$, for N_{tc} the MAPE is $0.83\% \pm 0.05\%$, for $T_{t,\text{out}}$ and T_{er} the same metric is identical and equal to $0.40\% \pm 0.01\%$. Furthermore, from the scatter plot of Figs. 6.10a, 6.11a, 6.12a, and 6.13a, it can be observed that minimum variance is achieved.

Table 6.8 Indexes of performances (MAE, MAPE, and PPMCC) of the different models (PMs, DDMs, and HMs) for different $\Delta \in \{0, 10, 20, 30\}$ for the different quantities to predict

Δ	Model	MAE [°C]	MAPE [%]	PPMCC
Fuel consumption $\dot{m}_f$ [kg/h]				
0	PM	76.62 ± 4.37	26.93 ± 1.54	0.98 ± 0.01
	DDM	24.11 ± 1.39	6.30 ± 0.38	0.99 ± 0.01
	HM	18.64 ± 0.98	4.89 ± 0.17	1.00 ± 0.01
10	DDM	9.55 ± 0.43	2.50 ± 0.15	1.00 ± 0.01
	HM	7.64 ± 0.48	2.01 ± 0.08	1.00 ± 0.01
20	DDM	6.83 ± 0.34	1.79 ± 0.08	1.00 ± 0.01
	HM	5.42 ± 0.25	1.43 ± 0.09	1.00 ± 0.01
30	DDM	11.36 ± 0.57	2.98 ± 0.15	1.00 ± 0.01
	HM	9.09 ± 0.60	2.39 ± 0.12	1.00 ± 0.01
TC rotational speed N_{tc} [rpm]				
0	PM	2090.10 ± 78.43	15.39 ± 0.75	0.97 ± 0.01
	DDM	302.62 ± 21.42	2.18 ± 0.15	1.00 ± 0.01
	HM	214.44 ± 9.54	1.53 ± 0.08	1.00 ± 0.01
10	DDM	130.12 ± 7.63	0.94 ± 0.05	1.00 ± 0.01
	HM	102.22 ± 4.02	0.74 ± 0.04	1.00 ± 0.01
20	DDM	114.57 ± 6.63	0.83 ± 0.05	1.00 ± 0.01
	HM	91.78 ± 4.59	0.66 ± 0.02	1.00 ± 0.01
30	DDM	157.90 ± 7.20	1.13 ± 0.06	1.00 ± 0.01
	HM	124.01 ± 7.58	0.90 ± 0.06	1.00 ± 0.01
Turbine outlet temperature $T_{t,\text{out}}$ [°C]				
0	PM	9.66 ± 0.57	2.53 ± 0.13	0.92 ± 0.01
	DDM	3.80 ± 0.20	0.97 ± 0.05	0.99 ± 0.01
	HM	3.18 ± 0.22	0.81 ± 0.05	0.99 ± 0.01
10	DDM	1.89 ± 0.12	0.48 ± 0.02	1.00 ± 0.01
	HM	1.54 ± 0.09	0.39 ± 0.02	1.00 ± 0.01
20	DDM	1.58 ± 0.11	0.40 ± 0.01	1.00 ± 0.01
	HM	1.27 ± 0.04	0.32 ± 0.02	1.00 ± 0.01
30	DDM	2.26 ± 0.11	0.57 ± 0.03	1.00 ± 0.01
	HM	1.76 ± 0.10	0.45 ± 0.02	1.00 ± 0.01
Exhaust manifold temperature T_{er} [°C]				
0	PM	19.92 ± 1.06	4.81 ± 0.15	0.96 ± 0.01
	DDM	5.02 ± 0.19	1.13 ± 0.04	0.99 ± 0.01
	HM	3.94 ± 0.24	0.88 ± 0.05	0.99 ± 0.01
10	DDM	2.51 ± 0.12	0.56 ± 0.03	1.00 ± 0.01
	HM	1.99 ± 0.07	0.45 ± 0.03	1.00 ± 0.01
20	DDM	1.78 ± 0.10	0.40 ± 0.01	1.00 ± 0.01
	HM	1.43 ± 0.05	0.32 ± 0.01	1.00 ± 0.01
30	DDM	3.23 ± 0.19	0.73 ± 0.05	1.00 ± 0.01
	HM	2.57 ± 0.11	0.58 ± 0.03	1.00 ± 0.01

Note that $\Delta = 0$ means that the authors do not exploit time series information from the past, for $\Delta > 0$ there is no PM result

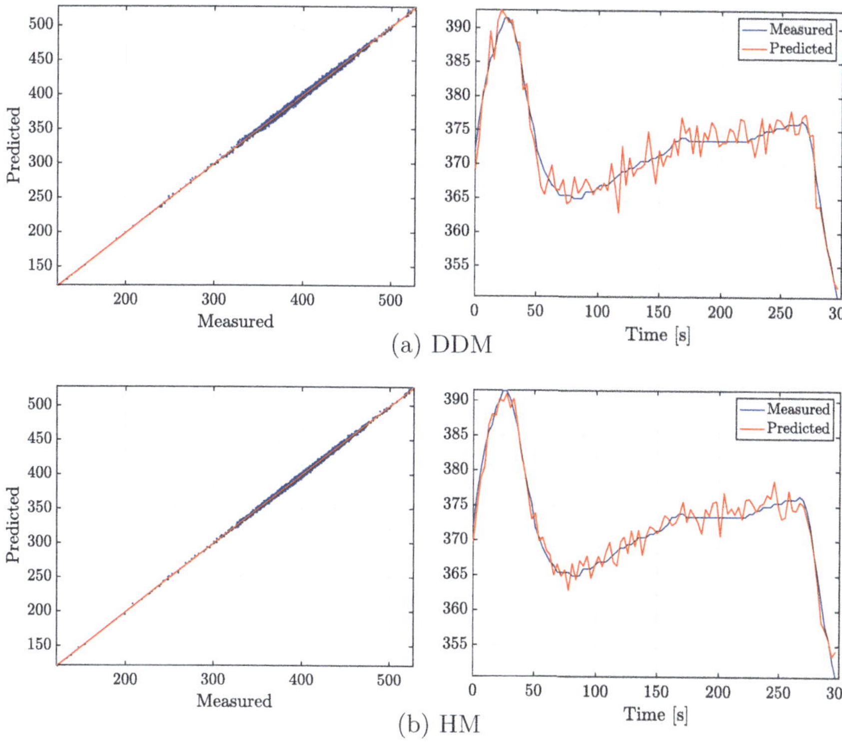

Fig. 6.12 Scatter plot (measured vs. predicted) and trend in time for $T_{t,\text{out}}$ (°C) output feature using the different models (DDMs and HMs) with $\Delta = 20$

It should be noted that, although DDMs are computationally demanding in the training phase, they are characterised by lower computational complexity in the feed-forward phase, as they just require matrix manipulation methods, in contrast with the solution of a system of first order differential equations that the PM requires (see Sect. 6.3.1). The combination of both accurate and fast predictions, makes DDMs an ideal candidate for real-time performance and condition estimation. However, the necessary data to reach this level of performance is rather high as reported in (Cipollini et al. 2018a, b), which makes this type of models applicable only after extensive measurement campaigns have been undertaken. In addition, another disadvantage of DDMs is the lack of interpretability as it is not supported by any physical interpretation (Shawe-Taylor and Cristianini 2004).

To overcome those limitations we proposed the use of HMs. These allow the exploitation of both the mechanistic knowledge of the underlying physical principles from the PM, and any available measurements taken during the operation of the vessel. An advantage of the HMs is their ability to exploit the coarse, but physically supported, predictions of the PM. Therefore, HMs have much smaller requirements regarding the use of actual measurements for the learning phase (Coraddu et al. 2017).

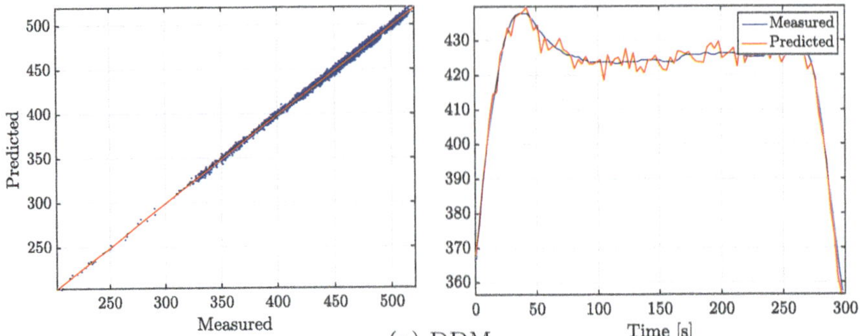

(a) DDM

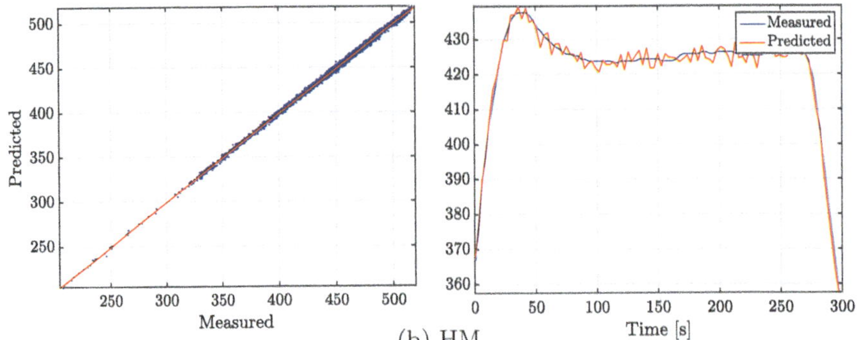

(b) HM

Fig. 6.13 Scatter plot (measured vs. predicted) and trend in time for T_{er} (°C) output feature using the different models (DDMs and HMs) with $\Delta = 20$

While they will still require a measurement campaign in order to be deployed, they can be reliably used already after a few months worth of measurements, in contrast with pure DDMs that would require at least half a year of available data, before they can be exploited.

The novelty introduced by the HMs led to more accurate predictions of the four targets compared to the rest of the models (PM and DDMs), regardless of the time window considered (Δ), as can be seen from Table 6.8. Furthermore, the same table shows that the optimal model is an HM with a time window of 20 seconds, which achieves MAPEs of $1.43\% \pm 0.09\%$ for $\dot{m}_f$, $0.66\% \pm 0.02\%$ for N_{tc}, $0.32\% \pm 0.02\%$ for $T_{t,\text{out}}$, and $0.32\% \pm 0.01\%$ for $T_{t,\text{out}}$. This is also supported by Figs. 6.10b, 6.11b, 6.12b, and 6.13b, which show representative time-series of the predictions of the four considered targets ($\dot{m}_f$, N_{tc}, $T_{\text{t,out}}$, and T_{er}) for time windows of 20 seconds. Finally, it can be noted that the variance has been completely eliminated, whereas the bias has been reduced to near-zero levels.

6.6 Conclusions and Future Perspectives

In this work the authors focused their attention on demonstrating a novel modelling framework for the hybridisation of physical and data driven models. The proposed framework is capable of delivering accurate, reliable, and computationally inexpensive models suitable for real-time performance assessment and condition monitoring applications. State-of-the-art data-driven methods have been presented, able to exploit the information provided by on-board measurements from one Holland Class Oceangoing Patrol Vessel, provided by the Royal Netherlands Navy and Damen Schelde Naval Shipbuilding. First, a 0-D physical model of a medium speed two-stroke diesel engine (MAN 12 V28-33D) was described in detail and validated against measured data. The results reported in Sect. 6.5.1 showed that the automatic calibration processes for stationary operations can provide suitable parameter values to adjust the model's response to the measured signals, capturing the stationary engine operation for a wide span of loads, ranging between 20–100%. The stationary relative errors are in general below 3.5% for the validation data. Nonetheless, the physical model proved to be not accurate enough to capture the engine behaviour in transient conditions. In this respect, the dynamic validation reported in Sect. 6.5.2 showed that the physical model model is capable of following the measured engine signals during transients, nonetheless, its response is not accurate. Moreover, its complexity depend upon computational requirements that are sometimes prohibitive, preventing the use of the physical model in real time applications.

Therefore, data-driven models have been discussed and proposed in Sect. 6.3.2, for predicting the behaviour of the engine, with a focus on four different targets: (i) fuel consumption, (ii) turbocharger rotational speed, (iii) turbine outlet temperature, and iv) exhaust receiver temperature. The models proved to be very accurate, with the enhanced capability of exploiting time series information from the past, achieving relative errors below 1% on the validation data, across all the considered output features. However, due to their nature, these data-driven models are hard to interpret.

To overcome the limitations of both the physical and the data-driven models, we proposed a hybrid approach that can take into consideration past information, capable of improving accuracy, easily interpreted, and have low computational time requirements. The hybridisation of physical and data driven models proved to be extremely accurate, achieving even lower errors when compared to the simple data-driven approach. These hybrid models can potentially also be used to improve accuracy of predictions for operation in other conditions than the measured ones, as purely data-driven models cannot be used for extrapolation, but the physical model contribution will improve hybrid model performance during extrapolation. While the hybrid approach will still require a measurement campaign in order to be deployed, this approach can be reliably used based on a significantly smaller dataset in comparison with the pure data-driven models, for the same average error, as shown in Sect. 6.5.2.

Acknowledgements This project is supported by the Royal Netherlands Navy supplying the operational measurement data from one Holland Class Oceangoing Patrol Vessel and Damen Schelde Naval Shipbuilding.

References

Ahmed R, El Sayed M, Gadsden SA, Tjong J, Habibi S (2015) Automotive internal-combustion-engine fault detection and classification using artificial neural network techniques. IEEE Trans Veh Technol 64(1):21–33

Argyriou A, Evgeniou T, Pontil M (2008) Convex multi-task feature learning. Mach Learn 73(3):243–272

Audet C, Dennis JE (2006) Mesh adaptive direct search algorithms for constrained optimization. SIAM J Optim 17(1):188–217

Audet C, Custódio AL, Dennis JE (2008) Erratum: mesh adaptive direct search algorithms for constrained optimization. SIAM J Optim 18(4):1501–1503

Bakker B, Heskes T (2003) Task clustering and gating for bayesian multitask learning. J Mach Learn Res 4:83–99

Baldi F, Johnson H, Gabrielii C, Andersson K (2014) Energy and exergy analysis of ship energy systems-the case study of a chemical tanker. In: ECOS, international conference on efficiency, cost, optimization, simulation and environmental impact of energy systems

Baldi F, Theotokatos G, Andersson K (2015) Development of a combined mean value-zero dimensional model and application for a large marine four-stroke diesel engine simulation. Appl Energy 154:402–415

Baxter J (2000) A model of inductive bias learning. J Artif Intel Res 12:149–198

Bidarvatan M, Thakkar V, Shahbakhti M, Bahri B, Aziz AA (2014) Grey-box modeling of hcci engines. Appl Therm Eng 70(1):397–409

Bouman EA, Lindstad E, Rialland AI, Strømman AH (2017) State-of-the-art technologies, measures, and potential for reducing ghg emissions from shipping-a review. Transp Res Part D: Transp Environ 52:408–421

Caruana R (1997) Multitask learning. Mach Learn 28(1):41–75

Chen S, Flynn P (1965) Development of a single cylinder compression ignition research engine. Technical report, SAE Technical Paper

Chiong MC, Kang HS, Shaharuddin N, Ma S et al (2021) Challenges and opportunities of marine propulsion with alternative fuels. Renew Sustain Energy Rev 149:111397

Cipollini F, Oneto L, Coraddu A, Murphy AJ, Anguita D (2018) Condition-based maintenance of naval propulsion systems with supervised data analysis. Ocean Eng 149:268–278

Cipollini F, Oneto L, Coraddu A, Murphy AJ, Anguita D (2018) Condition-based maintenance of naval propulsion systems: data analysis with minimal feedback. Reliab Eng Syst Safety 177:12–23

Commission European (2013a) Integrating maritime transport in the eu's greenhouse gas reduction policies: Communication from the commission to the european parliament, the council, the european economic and social committee and the committee of the regions. Technical report, European Union

Commission European (2013b) Proposal for a regulation of the european parliament and of the council on the monitoring, reporting and verification of carbon dioxide emissions from maritime transport and amending regulation (eu) no 525/2013. Technical report, European Union

Committee M.E.P. (2011) Resolution mepc. 203 (62); amendments to the annex of the protocol of (1997) to amend the international convention for the prevention of pollution from ships, 1973, as modified by the protocol of 1978 relating thereto. Technical report, International Maritime Organization

Conn AR, Scheinberg K, Vicente LN (2009) Introduction to derivative-free optimization. SIAM
Coraddu A, Oneto L, Ghio A, Savio S, Anguita D, Figari M (2016) Machine learning approaches for improving condition-based maintenance of naval propulsion plants. Proc Inst Mechan Eng Part M: J Eng Marit Environ 230(1):136–153
Coraddu A, Oneto L, Baldi F, Anguita D (2017) Vessels fuel consumption forecast and trim optimisation: a data analytics perspective. Ocean Eng 130:351–370
Coraddu A, Kalikatzarakis M, Oneto L, Meijn GJ, Godjevac M, Geertsmad RD (2018) Ship diesel engine performance modelling with combined physical and machine learning approach. In: International Naval engineering conference and exhibition
Coraddu A, Oneto L, Baldi F, Cipollini F, Atlar M, Savio S (2019a) Data-driven ship digital twin for estimating the speed loss caused by the marine fouling. Ocean Eng 186:106063
Coraddu A, Lim S, Oneto L, Pazouki K, Norman R, Murphy AJ (2019b) A novelty detection approach to diagnosing hull and propeller fouling. Ocean Eng 176:65–73
Coraddu A, Oneto L, de Maya BN, Kurt R (2020) Determining the most influential human factors in maritime accidents: a data-driven approach. Ocean Eng 211:107588
Coraddu A, Oneto L, Cipollini F, Kalikatzarakis M, Meijn GJ, Geertsma R (2021a) Physical, data-driven and hybrid approaches to model engine exhaust gas temperatures in operational conditions. Ships Offshore Struct 1–22
Coraddu A, Oneto L, Ilardi D, Stoumpos S, Theotokatos G (2021b) Marine dual fuel engines monitoring in the wild through weakly supervised data analytics. Eng Appl Artif Intel 100:104179
Cristianini N, Shawe-Taylor J (2000) An introduction to support vector machines and other Kernel-based learning methods. Cambridge University Press, Cambridge
Dahl J, Wassén H, Santin O, Herceg M, Lansky L, Pekar J, Pachner D (2018) Model predictive control of a diesel engine with turbo compound and exhaust after-treatment constraints. IFAC-PapersOnLine 51(31):349–354
Descieux D, Feidt M (2007) One zone thermodynamic model simulation of an ignition compression engine. Appl Therm Eng 27(8–9):1457–1466
Ding Y, Stapersma D, Knoll H, Grimmelius H, Netherland T (2010) Characterising heat release in a diesel engine: a comparison between seiliger process and vibe model. In: CIMAC world congress on combustion engine technology
Evgeniou T, Pontil M (2004) Regularized multi-task learning. In: ACM SIGKDD international conference on knowledge discovery and data mining
Fagerholt K, Psaraftis HN (2015) On two speed optimization problems for ships that sail in and out of emission control areas. Transp Res Part D: Transp Environ 39:56–64
Fernández-Delgado M, Cernadas E, Barro S, Amorim D (2014) Do we need hundreds of classifiers to solve real world classification problems? J Mach Learn Res 15(1):3133–3181
Floudas CA, Pardalos P (2008) Encyclopedia of optimization. Springer, Berlin
Galindo J, Climent H, Plá B, Jiménez VD (2011) Correlations for wiebe function parameters for combustion simulation in two-stroke small engines. Appl Therm Eng 31(6–7):1190–1199
Galinier P, Hamiez JP, Hao JK, Porumbel D (2013) Handbook of optimization. Springer, Berlin
García-Martos C, Rodríguez J, Sánchez MJ (2013) Modelling and forecasting fossil fuels, co2 and electricity prices and their volatilities. Appl Energy 101:363–375
Geertsma RD, Negenborn RR, Visser K, Loonstijn MA, Hopman JJ (2017) Pitch control for ships with diesel mechanical and hybrid propulsion: modelling, validation and performance quantification. Appl Energy 206:1609–1631
Geertsma RD, Visser K, Negenborn RR (2018) Adaptive pitch control for ships with diesel mechanical and hybrid propulsion. Appl Energy 228:2490–2509
Ghojel JI (2010) Review of the development and applications of the wiebe function: a tribute to the contribution of ivan wiebe to engine research. Int J Eng Res 11(4):297–312
Gogoi TK, Baruah DC (2010) A cycle simulation model for predicting the performance of a diesel engine fuelled by diesel and biodiesel blends. Energy 35(3):1317–1323
Goodfellow I, Bengio Y, Courville A (2016) Deep learning. MIT Press, Cambridge

Grimmelius HT (2003) Simulation models in marine engineering: from training to concept exploration. In: International EuroConference on Computer and IT Applications in the Maritime Industries

Grimmelius H, Boonen EJ, Nicolai H, Stapersma D (2010) The integration of mean value first principle diesel engine models in dynamic waste heat and cooling load analysis. In: CIMAC World Congress on Combustion Engine Technology

Grimmelius H, Mesbahi E, Schulten P, Stapersma D (2007) The use of diesel engine simulation models in ship propulsion plant design and operation. In: CIMAC international council on combustion engines

Guan C, Theotokatos G, Zhou P, Chen H (2014) Computational investigation of a large containership propulsion engine operation at slow steaming conditions. Appl Energy 130:370–383

Guan C, Theotokatos G, Chen H (2015) Analysis of two stroke marine diesel engine operation including turbocharger cut-out by using a zero-dimensional model. Energies 8(6):5738–5764

Gucwa M, Schäfer A (2013) The impact of scale on energy intensity in freight transportation. Transp Res Part D: Transp Environ 23:41–49

Guzzella L, Onder C (2009) Introduction to modeling and control of internal combustion engine systems. Springer, Berlin

Hamilton JD (2020) Time series analysis. Princeton University Press, Princeton

Hanson RK, Salimian S (1984) Survey of rate constants in the n/h/o system. Combust Chem 361–421

Hao C, Lu Z, Feng Y, Bai H, Wen M, Wang T (2021) Optimization of fuel/air mixing and combustion process in a heavy-duty diesel engine using fuel split device. Appl Therm Eng 186:116458

He Y, Lin C (2007) Development and validation of a mean value engine model for integrated engine and control system simulation. Technical report, SAE Technical Paper

Heywood JB (1988) Internal combustion engines fundamentals. McGraw-Hill, New York

Johnson K, Mollenhauer K, Tschoke H (2010) Handbook of diesel engines. Springer, Berlin

Kamal K, Hui C (2013) A semi-experimental modeling approach for a large two-stroke marine diesel engine simulation. In: CIMAC world congress on combustion engine technology

Keerthi SS, Lin CJ (2003) Asymptotic behaviors of support vector machines with gaussian kernel. Neural Comput 15(7):1667–1689

Kökkülünk G, Parlak A, Erdem H (2016) Determination of performance degradation of a marine diesel engine by using curve based approach. Appl Therm Eng 108:1136–1146

Larsen U, Pierobon L, Baldi F, Haglind F, Ivarsson A (2015) Development of a model for the prediction of the fuel consumption and nitrogen oxides emission trade-off for large ships. Energy 80:545–555

Lee B, Jung D, Kim Y, van Nieuwstadt M (2013) Thermodynamics-based mean value model for diesel combustion. J Eng Gas Turb Power **135**(9)

Lewis RM, Torczon V, Trosset MW (2000) Direct search methods: then and now. J Comput Appl Math 124(1–2):191–207

Lindstad H, Eskeland GS (2015) Low carbon maritime transport: how speed, size and slenderness amounts to substantial capital energy substitution. Transp Res Part D: Transp Environ 41:244–256

Lindstad H, Verbeek R, Blok M, van Zyl S, Hübscher A, Kramer H, Purwanto J, Ivanova O, Boonman H (2015) Ghg emission reduction potential of eu-related maritime transport and its impacts. Technical report, Van Mourik Broekmanweg

Lion S, Vlaskos I, Taccani R (2020) A review of emissions reduction technologies for low and medium speed marine diesel engines and their potential for waste heat recovery. Energy Convers Manage 207:112553

Liu Z, Zuo Q, Wu G, Li Y (2018) An artificial neural network developed for predicting of performance and emissions of a spark ignition engine fueled with butanol-gasoline blends. Adv Mechan Eng 10(1):1687814017748438

Livanos G, Papalambrou G, Kyrtatos NP, Christou A (2007) Electronic engine control for ice operation of tankers. In: CIMAC world congress on combustion engine technology

Llamas X, Eriksson L (2018) Control-oriented modeling of two-stroke diesel engines with exhaust gas recirculation for marine applications. J Eng Marit Environ, Proc Inst Mechan Eng Part M

Llamas X, Eriksson L (2019) Control-oriented modeling of two-stroke diesel engines with exhaust gas recirculation for marine applications. Proc Inst Mechan Eng Part M: J Eng Marit Environ 233(2):551–574
Malkhede DN, Seth B, Dhariwal HC (2005) Mean value model and control of a marine turbocharged diesel engine. Technical report, SAE Technical Paper
Maroteaux F, Saad C (2015) Combined mean value engine model and crank angle resolved in-cylinder modeling with nox emissions model for real-time diesel engine simulations at high engine speed. Energy 88:515–527
Martí R (2003) Multi-start methods. In: Handbook of metaheuristics
McBride BJ, Zehe MJ (2002) NASA Glenn coefficients for calculating thermodynamic properties of individual species. Natl Aeronaut Space Adm
Merker GP, Schwarz C, Stiesch G, Otto F (2005) Simulating combustion: simulation of combustion and pollutant formation for engine development. Springer, Berlin
Miglianti F, Cipollini F, Oneto L, Tani G, Viviani M (2019) Model scale cavitation noise spectra prediction: combining physical knowledge with data science. Ocean Eng 178:185–203
Miglianti L, Cipollini F, Oneto L, Tani G, Gaggero S, Coraddu A, Viviani M (2020) Predicting the cavitating marine propeller noise at design stage: A deep learning based approach. Ocean Eng 209:107481
Mishra C, Subbarao PMV (2021) A comparative study of physics based grey box and neural network trained black box dynamic models in an rcci engine control parameter prediction. Technical report, SAE Technical Paper
Miyamoto N, Chikahisa T, Murayama T, Sawyer R (1985) Description and analysis of diesel engine rate of combustion and performance using wiebe's functions. SAE Trans 622–633
Mohammadkhani F, Yari M, Ranjbar F (2019) A zero-dimensional model for simulation of a diesel engine and exergoeconomic analysis of waste heat recovery from its exhaust and coolant employing a high-temperature kalina cycle. Energy Convers Manage 198:111782
Namigtle-Jiménez A, Escobar-Jiménez RF, Gómez-Aguilar JF, García-Beltrán CD, Téllez-Anguiano AC (2020) Online ANN-based fault diagnosis implementation using an fpga: application in the EFI system of a vehicle. ISA Trans 100:358–372
Ni P, Wang X, Li H (2020) A review on regulations, current status, effects and reduction strategies of emissions for marine diesel engines. Fuel 279:118477
Nikzadfar K, Shamekhi AH (2014) Investigating the relative contribution of operational parameters on performance and emissions of a common-rail diesel engine using neural network. Fuel 125:116–128
Nikzadfar K, Shamekhi AH (2015) An extended mean value model (emvm) for control-oriented modeling of diesel engines transient performance and emissions. Fuel 154:275–292
Oberkampf WL, Trucano TG (2002) Verification and validation in computational fluid dynamics. Progr Aerosp Sci 38(3):209–272
Oberkampf WL, Trucano TG, Hirsch C (2004) Verification, validation, and predictive capability in computational engineering and physics. Appl Mechan Rev 57(5):345–384
Oneto L (2020) Model selection and error estimation in a nutshell. Springer, Berlin
Oneto L, Ghio A, Ridella S, Anguita D (2015) Support vector machines and strictly positive definite kernel: the regularization hyperparameter is more important than the kernel hyperparameters. In: IEEE international joint conference on neural networks (IJCNN)
Oneto L, Anguita D, Coraddu A, Cleophas T, Xepapa K (2016) Vessel monitoring and design in industry 4.0: a data driven perspective. In: 2016 IEEE 2nd international forum on research and technologies for society and industry leveraging a better tomorrow (RTSI), pp 1–6. IEEE
Özener O, Yüksek L, Özkan M (2013) Artificial neural network approach to predicting engine-out emissions and performance parameters of a turbo charged diesel engine. Therm Sci 17(1):153–166
Palmer KA, Bollas GM (2019) Active fault diagnosis for uncertain systems using optimal test designs and detection through classification. ISA Trans 93:354–369

Psaraftis HN, Kontovas CA (2014) Ship speed optimization: concepts, models and combined speed-routing scenarios. Transp Res Part C: Emerg Technol 44:52–69
Rakopoulos CD, Hountalas DT, Tzanos EI, Taklis GN (1994) A fast algorithm for calculating the composition of diesel combustion products using 11 species chemical equilibrium scheme. Adv Eng Softw 19(2):109–119
Rakopoulos CD, Rakopoulos DC, Mavropoulos GC, Giakoumis EG (2004) Experimental and theoretical study of the short term response temperature transients in the cylinder walls of a diesel engine at various operating conditions. Appl Therm Eng 24(5–6):679–702
Rosasco L, De Vito E, Caponnetto A, Piana M, Verri A (2004) Are loss functions all the same? Neural Comput 16(5):1063–1076
Sapra H, Godjevac M, Visser K, Stapersma D, Dijkstra C (2017) Experimental and simulation-based investigations of marine diesel engine performance against static back pressure. Appl Energy 204:78–92
Sapra H, Godjevac M, De Vos P, Van Sluijs W, Linden Y, Visser K (2020) Hydrogen-natural gas combustion in a marine lean-burn si engine: a comparitive analysis of seiliger and double wiebe function-based zero-dimensional modelling. Energy Convers Manage 207:112494
Scholkopf B (2001) The kernel trick for distances. In: Advances in neural information processing systems, pp 301–307
Schölkopf B, Herbrich R, Smola AJ (2001) A generalized representer theorem. In: Computat Learn Theor
Shalev-Shwartz S, Ben-David S (2014) Understanding machine learning: from theory to algorithms. Cambridge University Press, Cambridge
Shao L, Mahajan A, Schreck T, Lehmann DJ (2017) Interactive regression lens for exploring scatter plots. Comput Gr Forum
Shawe-Taylor J, Cristianini N (2004) Kernel methods for pattern analysis. Cambridge University Press, Cambridge
Shin S, Lee Y, Kim M, Park J, Lee S, Min K (2020) Deep neural network model with bayesian hyperparameter optimization for prediction of nox at transient conditions in a diesel engine. Eng Appl Artif Intel 94:103761
Sitkei G (1963) Über den dieselmotorischen zündverzug. MTZ 24(6):190–194
Stoer J, Bulirsch R (2013) Introduction to numerical analysis, vol 12. Springer, Berlin
Stoumpos S, Theotokatos G, Boulougouris E, Vassalos D, Lazakis I, Livanos G (2018) Marine dual fuel engine modelling and parametric investigation of engine settings effect on performance-emissions trade-offs. Ocean Eng 157:376–386
Stoumpos S, Theotokatos G, Mavrelos C, Boulougouris E (2020) Towards marine dual fuel engines digital twins-integrated modelling of thermodynamic processes and control system functions. J Marine Sci Eng 8(3):200
Sui C, Song E, Stapersma D, Ding Y (2017) Mean value modelling of diesel engine combustion based on parameterized finite stage cylinder process. Ocean Eng 136:218–232
Syed J, Baig RU, Algarni S, Murthy S, Masood M, Inamurrahman M (2017) Artificial neural network modeling of a hydrogen dual fueled diesel engine characteristics: an experiment approach. Int J Hydr Energy 42(21):14750–14774
Tang Y, Zhang J, Gan H, Jia B, Xia Y (2017) Development of a real-time two-stroke marine diesel engine model with in-cylinder pressure prediction capability. Appl Energy 194:55–70
Theotokatos G (2008) Ship propulsion plant transient response investigation using a mean value engine model. Int J Energy 2(4):66–74
Theotokatos G (2010) On the cycle mean value modelling of a large two-stroke marine diesel engine. Proc Inst Mechan Eng Part M: J Eng Marit Environ 224(3):193–205
Theotokatos G, Kyrtatos NP (2003) Investigation of a large high-speed diesel engine transient behavior including compressor surging and emergency shutdown. J Eng Gas Turb Power 125(2):580–589

Theotokatos G, Tzelepis V (2015) A computational study on the performance and emission parameters mapping of a ship propulsion system. Proc Instit Mechan Eng Part M: J Eng Marit Environ 229(1):58–76

Theotokatos G, Guan C, Chen H, Lazakis I (2018) Development of an extended mean value engine model for predicting the marine two-stroke engine operation at varying settings. Energy 143:533–545

Tikhonov AN, Arsenin VY (1979) Methods for solving III-posed problems. Nauka, Moscow

Tillig F, Mao W, Ringsberg J (2015) Systems modelling for energy-efficient shipping. Technical report, Chalmers University of Technology (2015)

Torczon V (1997) On the convergence of pattern search algorithms. SIAM J Optim 7(1):1–25

Vapnik VN (1998) Statistical learning theory. Wiley, New York

Vovk V (2013) Kernel ridge regression. In: Empirical inference

Wahlström J, Eriksson L (2011) Modelling diesel engines with a variable-geometry turbocharger and exhaust gas recirculation by optimization of model parameters for capturing non-linear system dynamics. Proc Inst Mechan Eng Part D: J Automob Eng 225(7):960–986

Wainberg M, Alipanahi B, Frey BJ (2016) Are random forests truly the best classifiers? J Mach Learn Res 17(1):3837–3841

Wang J, Wang Z, Stetsyuk V, Ma X, Gu F, Li W (2019) Exploiting bayesian networks for fault isolation: a diagnostic case study of diesel fuel injection system. ISA Trans 86:276–286

Wang H, Gan H, Theotokatos G (2020a) Parametric investigation of pre-injection on the combustion, knocking and emissions behaviour of a large marine four-stroke dual-fuel engine. Fuel 281:118744

Wang YS, Liu NN, Guo H, Wang XL (2020b) An engine-fault-diagnosis system based on sound intensity analysis and wavelet packet pre-processing neural network. Eng Appl Artif Intel 94:103765

Wang R, Chen H, Guan C (2021) Random convolutional neural network structure: an intelligent health monitoring scheme for diesel engines. Measurement 171:108786

Watson N, Janota M (1982) Turbocharging the internal combustion engine. Macmillan International Higher Education

Wen M, Pacino D, Kontovas CA, Psaraftis HN (2017) A multiple ship routing and speed optimization problem under time, cost and environmental objectives. Transp Res Part D: Transport Environ 52:303–321

Willmott CJ, Matsuura K (2005) Advantages of the mean absolute error (mae) over the root mean square error (rmse) in assessing average model performance. Clim Res 30(1):79–82

Woschni G (1968) A universally applicable equation for the instantaneous heat transfer coefficient in the internal combustion engine. SAE Trans 3065–3083

Woschni G, Anisits F (1973) Eine methode zur vorausberechnung der anderung des brennverlaufs mittelschnellaufender dieselmotoren bei geaenderten betriebsbedingungen. Motortech 34(4):106–115

Xing H, Spence S, Chen H (2020) A comprehensive review on countermeasures for CO2 emissions from ships. Renew Sustain Energy Rev 134:110222

Young DM (2003) Iterative solution of large linear systems. Dover Publications, Mineola

Yu H, Fang Z, Fu X, Liu J, Chen J (2021) Literature review on emission control-based ship voyage optimization. Transp Res Part D: Transp Environ 93:102768

Zheng A, Casari A (2018) Feature engineering for machine learning: principles and techniques for data scientists. O'Reilly Media, Inc

Zis T, Psaraftis HN (2018) Operational measures and logistical considerations for the decarbonisation of maritime transport. In: hEART 2018: 7th Symposium of the European Association for Research in Transportation

Zis T, Psaraftis HN, Ding L (2020) Ship weather routing: a taxonomy and survey. Ocean Eng 213:107697

Chapter 7
Soot and NO_x Modelling for Diesel Engines

Rahul Kumar Singh and Avinash Kumar Agarwal

Abstract Diesel engines find applications in many areas such as transportation of passengers/goods, farm machines, Generators, and construction equipment. The number of diesel-fueled engines is very large and is expected to increase further with time. However, diesel engines are responsible for anthropogenic NOx and soot emissions. To comply with the stringent future emission regulations, ways of inhibiting the formation of these emissions need to be explored further. Low-temperature combustion (LTC) strategies such as homogeneous charge compression ignition (HCCI), reactivity controlled compression ignition (RCCI), partially premixed charge compression ignition (PCCI) have the capabilities of ultra-low NOx and Soot emission formation. These LTC strategies improve the fuel–air mixing and reduce the peak combustion temperature, leading to low NOx and Soot emissions. These combustion strategies face challenges in controlling combustion, maintaining combustion stability, and operating the engine at a high load. Engine modelling could play a role in predicting Soot and NOx emissions from these complex combustion strategies. In this chapter, an effort has been made to understand the fundamentals of soot and NO_x modelling. It is essential to understand the chemical mechanisms of NOx and soot formation for accurate modelling. Empirical and phenomenological models for soot and NO_x formation mechanisms, factors affecting them, Zeldovich mechanism, and prompt NOx formation mechanisms are discussed. Hiroyasu-NSC model, Waseda model, Gokul model, and Dalian model to understand the soot formation and their capabilities are discussed in detail. The two most fundamental semi-empirical models for NO_x formation based on the Zeldovich mechanism and the prompt NO_x (Fenimore mechanism) are also discussed.

Keywords Soot formation · NOx modeling · Phenomenological models · Empirical models

---

R. K. Singh · A. K. Agarwal (✉)
Engine Research Laboratory, Department of Mechanical Engineering, Indian Institute of Technology Kanpur, Kanpur 208016, India
e-mail: akag@iitk.ac.in

A. K. Agarwal et al. (eds.), *Engine Modeling and Simulation*, Energy, Environment, and Sustainability, https://doi.org/10.1007/978-981-16-8618-4_7

7.1 Introduction

Complying with the stringent future emission norms set by regulatory bodies will be challenging for the automotive industry because of the complex combustion processes. The combustion phenomenon directly influences pollutant formation in the engine (Baskar and Senthilkumar 2016). Superior combustion results in fewer emissions, and inferior combustion leads to higher emissions and lower power output. Diesel engines are significant contributors to anthropogenic NOx and soot emissions. These pollutants cause severe environmental and human health problems such as cancer and cardiovascular diseases. Soot emissions mainly originating from the IC engine combustion are the second long-lived greenhouse gas (GHG) emissions (Ilkiliç and Aydin 2012). Fine and coarse soot particles consist of carbon which absorbs solar radiations of all wavelengths causing global warming (Adachi et al. 2010). Therefore, control of pollutants is a necessity for sustainable development. The automotive industry adopts exhaust gas after-treatment, low-temperature combustion, and exhaust gas recirculation (EGR) to meet the emission regulations. However, in the future, these emission regulations will be further tightened.

Therefore, further intensive research in combustion and emissions would be essential soon. Like combustion, Soot and NO_x formation are complex phenomenons. Experimental investigations of emissions from the engine exhaust have been explored for several decades and will continue in the future. However, these experimental approaches have limitations of cost and time. Therefore, computation approaches have been accepted widely in recent years to predict pollutant formation. With advancements in computational power and algorithm designs, numerical studies involving NOx and soot models can calculate emissions approximately. The predictions of such emissions help researchers explore and investigate newer combustion strategies. The computational approaches provide flexibility to optimize various parameters to achieve low emissions. Several models developed in the past two decades for NOx and soot formation quantification are discussed in the subsequent sections. These models are mainly classified into two categories (Kumar et al. 2018), phenomenological models and empirical models.

7.1.1 Empirical Models

In empirical models, a relationship between the model and input data is developed using the experimental data. Empirical models are fast in response, with relatively minor errors w.r.t. the experimentally measured data. The main limitation of these models is the lack of reliable far-field extrapolation. Empirical models are easy to implement with the CFD code. An example of an empirical model is the Hiroyasu-NSC model.

7.1.2 *Phenomenological Models*

In phenomenological models, the complex combustion phenomenon is modelled. This is quite challenging because a detailed mechanism of chemical reactions is required. These models make superior predictions but suffer from error propagation. Examples of such models include the Gokul model, Dalian model, Prompt model, and Waseda model.

Researchers have developed hybrid models that include both phenomenological and empirical phenomena. However, this chapter's primary focus is limited to the fundamental modelling of NOx and soot particles. NO_x formation in the combustion chamber can be predicted by single-zone, two-zone, multi-zone models, and detailed NO_x models (Savva and Hountalas 2012). The two-zone model approach is preferred over the multi-zone modelling because of lesser complexity and lower computational time. Single-zone models are not so suitable for NO_x prediction without the use of an empirical formula.

7.2 Overview of Diesel Engine Emissions

The fuel–air mixture in a diesel engine is non-uniform since fuel is injected just before the combustion starts. The exhaust constituent formation is mainly dependent on the fuel–air mixture quality. The engine-out emissions differ from the values calculated by assuming chemical equilibrium. Before modelling a system, it is essential to comprehensively understand the chemical reaction mechanisms and kinetics of pollutant formation.

7.2.1 *Soot Particles*

Soot particles form due to the incomplete combustion of fuel. Higher soot emissions reflect inadequate combustion and poor efficiency. Soot particles are largely carbonaceous, with some organic compounds adsorbed on their surfaces, such as aromatic compounds and unburned hydrocarbons (Daido et al. 2000). The soot emission level from the light-duty diesel engine ranges from 0.2 to 0.6 g/km, and for the heavy-duty diesel engines, it ranges from 0.5 to 1.5 g/kWh (Heywood 2018). Particulates are likely to be spherical, and their composition mainly depends on the exhaust conditions. The particle diameter ranges from ~ 15 to 30 nm at an exhaust temperature > 500 °C. Below 500 °C, the high molecular hydrocarbons get adsorbed on the particulates, causing an increase in their size (Heywood 2018). Soot particles are essentially necklace-like agglomerates. The shape of the basic unit of agglomerates, i.e., primary soot particles, is spherical or nearly spherical (Dobbin and Subramaniasivam 1994).

These spherules units are called "primary soot particles" (Dobbin and Subramaniasivam 1994), and the aggregates of spherules are called "secondary soot particles." The size of these spherules lies between 10 and 80 nm, but they are mostly are in size range of 15–50 nm (Dobbin and Subramaniasivam 1994; Walker et al. 1966; Xi and Zhong 2006; Glassman 1996). Hydrocarbons and inorganic substances adhere to the surface of these spherules. 10^5–10^6 carbon atoms are present in a primary particle/spherule (Walker et al. 1966). Secondary soot particles consist of 10–100 spherules. Soot particles observed by X-ray diffraction (XRD) reveal that these carbon atoms are arranged in hexagonal face-centred arrays. These arrays are referred to as platelets. Two to five platelets placed layer by layer form crystallites. The crystallites arranged in a turbostratic fashion (Xi and Zhong 2006), parallel to the surfaces, form a soot particle.

7.2.1.1 Fundamentals of Soot Formation

Soot formation in diesel engines occurs in an environment with a temperature ranging from 1000 to 2800 K and pressure of 50–100 atm with sufficient air to burn the entire fuel. It is affected by the temperature, pressure, and equivalence ratio. Soot formation is essentially a complex gaseous-to-solid phase transformation process that gets completed in few milliseconds. The initial process is the oxidation of hydrocarbon fuel leading to the formation of hydrocarbon radicals. These hydrocarbon radicals combine and create an aromatic ring. They make larger aromatic rings of polycyclic aromatic hydrocarbons (PAHs) via acetylene, known as the HACA (Hydrogen-abstraction-carbon-addition) mechanism. HACA consist of a repetitive sequence of two primary steps (a) abstraction of hydrogen from aromatic molecule for activation of an aromatic molecule, and (b) addition of carbon (acetylene) to the radical formed in the previous step. The coagulation of PAHs generates primary soot particles. Surface growth determines the final concentration of the soot particles in sooting flames, and coagulation determines the final size of these soot particles.

The processes involved in the formation of soot particles are as follows (Richter and Howard 2000):

i. Soot precursor formation
ii. Nucleation or inception of particles from heavy PAH molecules
iii. Mass growth of particles by addition of gas-phase molecules
iv. Coagulation via reactive particle–particle collisions
v. Carbonization of particulate material
vi. Oxidation of PAHs and soot particles.

7.2.1.2 Factors Affecting Soot Formation

I. **Fuel Composition**

Fuel composition affects the soot formed in the compression-ignition (CI) engine. Sulfur in the fuel increases the particulate mass. It oxidizes to SO_2, which is absorbed by neighbouring soot particles and reacts with hydrocarbons (Gülder 1993). An increase in the carbon to hydrogen ratio in the fuel leads to higher soot particles. The soot percentage reduces as the percentage of oxygen in the fuel increases (Du et al. 1991). Therefore, oxygenated fuels produce lower soot mass. Research confirmed that the soot formation is minimum when fuel oxygen is between 27 and 35% (w/w) (Tree and Svensson 2007). The oxygen required to avoid soot formation is independent of the fuel's molecular structure. However, the soot formation depends on the fuel structure (Olson et al. 1985).

II. **Engine Design Parameters**

Combustion chamber geometry is a factor that determines in-cylinder flow structures in the engine. The swirl generation increases turbulence in the chamber and accelerates the combustion rate. This results in the rapid burning of soot particles (Ogawa et al. 1996). However, too high a swirl causes one jet to curve and interact with an adjacent jet before reaching the cylinder walls. Therefore, a high swirl underutilizes fuel near the piston wall and liner and inhibits mixing rich fuel with the fresh charge. The combustion process and the oxidation of soot particles slow down; hence higher soot formation takes place (Ogawa et al. 1996).

In general, advanced injection timing reduces soot formation, and retarded injection timing increases soot formation (Zhang and Kook 2014). In diesel engines, higher intake temperature and pressure results in reduced soot formation. Interestingly, increasing the intake air temperature increases the soot formation rate. But soot oxidation dominates at high temperatures; hence the net effect of temperature rise is reduced soot formation. It was observed that a tremendous amount of soot particulates are emitted in the exhaust when there is a rapid change from idle/ low load to high-load and high-speed engine operation. This happens for two reasons: (a) turbocharging and (b) increased intake pressure. The multiple injection technique is effective in limiting soot formation (Han et al. 1996). The time gap between two injections allows fresh air to enter the spray jet, reducing the local equivalence ratio. The soot carried at the leading edge of the jet mixes with the entrained fresh air in-between the two injections. Combustion products of the first injection result in a high-temperature environment. Subsequent fuel injection in this environment causes rapid combustion of fuel, leading to reduced soot formation (Han et al. 1996).

7.2.2 NO_x Emissions

Combustion of fuels under oxygen-rich, high-temperature (> 1600 °C) conditions produce NO_x. The nitrogen in the air at temperatures > 1600 °C reacts with excess oxygen to form NOx. The amount of NOx formed is dependent on the maximum in-cylinder pressure, oxygen concentration, and residence time. Accurate measurement of NO_x from the engine depends on the chamber temperature, combustion mechanism, and O_2 concentration in the chamber. Nitric oxide (NO) and nitrogen dioxide (NO_2) together are referred to as NOx. NO comprises 85 to 95% of the total NOx, which gradually converts to NO_2 in the atmosphere. NOx emissions have severe environmental impacts, such as damage to the ozone layer, acid rain, poor ambient air quality, poor visibility due to smog, and global warming. NO_2 is five times more toxic than NO and is quite harmful to the human lungs. Therefore, It becomes essential to study the formation of NOx in detail and find solutions to minimize it. Researchers prescribed four NOx formation mechanisms: prompt mechanism, thermal mechanism, NOx formation via N_2O, and NOx formation from fuel Nitrogen. In diesel engines, NO formation is explained by the Zeldovich mechanism. This mechanism helps model the NOx formation in diesel engines and predicts the NOx formed during combustion.

7.2.2.1 NOx Formation

Three major mechanisms prescribed to explain the NOx formation during combustion are:

(i) Thermal NO_x (Zeldovich Mechanism)
(ii) Prompt NO_x (Fenimore Mechanism)
(iii) Fuel NO_x.

In diesel engines, NO_x is primarily formed by prompt NOx and Zeldovich mechanisms. Prompt NO_x accounts for a smaller contribution to the NO formation. The first two mechanisms are discussed in detail in the subsequent sections. Fuel NO_x is the NOx formation from the nitrogen bonded to the fuel in N–H and N–C bonds. N–H and N–C bonds lead to the formation of ammonia and cyanhydric acid, which dissociate through a chain reaction to form NO. In lean mixtures, up to two-thirds of nitrogen bonded to fuel gets converted to NO. In rich mixtures, lesser NO and higher ammonia and cyanhydric acid are formed, which convert to NO upon release into the atmosphere.

(a) **Nitric Oxide**

The two primary NO sources during diesel combustion are oxidation of atmospheric nitrogen and nitrogen-containing compounds present in the fuel. The oxidation of atmospheric nitrogen in the engine is found to be dominant over the other. The formation of NO from the atmospheric nitrogen during the combustion of lean and nearly stoichiometric fuel–air mixture follows the 'extended Zeldovich mechanism' (Bowman 1975).

$$O + N_2 = NO + N \tag{7.1}$$

$$N + O_2 = NO + O \tag{7.2}$$

$$N + OH = NO + H \tag{7.3}$$

The rate constants (cm^3/mol-s) for the above equations are

$$k_{1f} = 7.6 \times 10^{13} \exp\left[-\frac{38{,}000}{T}\right]$$

$$k_{1r} = 1.6 \times 10^{13} \exp\left[-\frac{425}{T}\right]$$

$$k_{2f} = 1.6 \times 10^{13} \exp\left[-\frac{3150}{T}\right]$$

$$k_{2r} = 6.4 \times 10^{9} \exp\left[-\frac{19,500}{T}\right]$$

$$k_{3f} = 4.1 \times 10^{13} \exp\left[-\frac{450}{T}\right]$$

$$k_{3r} = 2 \times 10^{14} \exp\left[-\frac{23,650}{T}\right]$$

Equation 7.3 contributes to NO formation in the near stoichiometric and rich fuel–air mixtures. The rate of formation of NO is given as:

$$\frac{d[NO]}{dt} = 2k_{1f}[O][N_2]\left\{\frac{1 - [NO]^2/K[O_2][N_2]}{1 + k_{1r}[NO]/\left(k_{2f}[O_2] + k_{3f}[OH]\right)}\right\} \tag{7.4}$$

The formation rate of NO is slower than the combustion; therefore, most NO is formed after fuel combustion. Hence, the formation of NO can be decoupled from the combustion, and the NO formation rate can be calculated by assuming equilibration of combustion. For lean mixture combustion, the formation of NO is given by

$$\frac{d[NO]}{dt} = 2k_1[O]_{eq}[N_2]_{eq}$$

The equilibrium concentration of O is given by

$$[O] = \frac{k_O}{(RT)^{1/2}}(O_2)^{1/2}$$

Using the value of k_{1f}, we get

$$K_O = 3.6 \times 10^3 \exp\left(-\frac{31090}{T}\right)$$

The formation rate of NO is given by

$$\frac{d[\mathrm{NO}]}{dt} = 6 \times 10^6 T_{eq}^{-1/2} \exp\left(-\frac{69090}{T_{eq}}\right) \times [\mathrm{O_2}]_{eq}^{1/2} [\mathrm{N_2}]_{eq} \ \mathrm{moles/cm^3\,s} \tag{7.5}$$

Equation 7.5 shows a strong dependence of the NO formation rate on the temperature of post-combustion gases. Near flame formation of NO increases as the equivalence ratio is increased.

(b) **Nitrogen dioxide**

NO formed reacts with oxygen and other oxygen-containing species to form nitrogen dioxide in the flame zone. This conversion is followed by the apparent conversion of NO_2 to NO in the near post flame zone, and this conversion occurs when NO_2 is quenched by mixing with the cooler fluid.

$$\mathrm{NO} + \mathrm{HO_2} = \mathrm{NO_2} + \mathrm{OH} \tag{7.6}$$

$$\mathrm{NO_2} + \mathrm{O} = \mathrm{NO} + \mathrm{O_2} \tag{7.7}$$

The above situation is consistent with the high NO_2/NO ratio for the light load diesel engine.

7.2.2.2 Factors Affecting NO_x Formation

Many factors affect NOx formation. A few of them are discussed in this section. Advanced injection timing increases, whereas retarded injection timing decreases the NO_x formation. Early start of ignition results in peak pressure when the piston reaches the TDC, causing a rapid increase in the in-cylinder temperature. This leads to an increased NOx formation by the Zeldovich mechanism. A higher fuel injection rate increases the mixing time or ignition delay during the injection period, generating hotter flames during combustion, leading to more NOx formation (Wang et al. 2017). The richer fuel–air mixtures form lower NO_x because additional fuel lowers the charge temperature. The reduction in the peak temperature in the combustion region reduces the NOx formation. However, the phenomenon is different for indirect injection compression ignition (IDCI). Prechamber operates in rich fuel conditions leading to lower NOx formation (Noor et al. 2014; Asad et al. 2013). Another factor is the compression ratio. As the compression ratio increases, the NO_x formation also increases because of the higher temperature of compressed air (Gnanamoorthi and Devaradjane 2015).

7.3 Soot Formation Models

Soot formation models are either phenomenological models and empirical models. An example of an empirical model is the Hiroyasu-NSC model. The examples of phenomenological models include the Gokul model, Dalian model, Prompt model, and Waseda model. These are discussed in the following sub-sections.

7.3.1 Empirical Model: Hiroyasu-NSC Model

In this model, two models (Hiroyasu and NSC) are coupled to simulate the soot formation. It is the most popular, simple, and easy model to implement in CFD applications. Hiroyasu provided a simple formulation to calculate the rate of soot mass formation, given by the difference between formation and oxidation rate of soot during combustion. The simple equations of this model are shown below (Hiroyasu and Kadota 1976):

$$\frac{dm_{soot}}{dt} = \left(\frac{dm_{soot}}{dt}\right)_{formation} - \left(\frac{dm_{soot}}{dt}\right)_{oxidation} \tag{7.8}$$

$$\left(\frac{dm_{soot}}{dt}\right)_{formation} = A_f m_{fuel} p^{0.5} \exp\left(-\frac{E_f}{RT}\right) \tag{7.9}$$

$$\left(\frac{dm_{soot}}{dt}\right)_{oxidation} = A_O m_{soot} X_{O_2} p^{1.8} \exp\left(-\frac{E_O}{RT}\right) \tag{7.10}$$

where m_{soot} is the net soot mass formed, m_{fuel} is the mass of vaporized fuel, X_{O_2} is the mole fraction of oxygen, and E_O and E_f are the activation energies of soot formation and oxidation, respectively. A_f and A_O are the Arrhenius pre-exponential factors based on the experimental data, and p is the cell pressure.

Nagle, Strickland, and Constable (NSC) extended the Hiroyasu model. Arrangement and location of carbon atoms over the surface of soot particles were considered in the soot oxidation. The sites were divided into more reactive site 'A' and less reactive site 'B.' As temperature increases, the less reactive sites B changes to more reactive sites A. Reactions of soot oxidation are given as follows (Nagle and Strickland-Constable 1962):

$$A_{site} + O_2(g) \overset{\frac{k_A}{k_Z}}{\rightarrow} A_{site} + 2CO(g)$$

$$B_{site} + O_2(g) \overset{k_B}{\rightarrow} +2CO(g)$$

$$A_{site} \overset{k_T}{\rightarrow} B_{site}$$

The rate of soot oxidation was given as:

$$\left(\frac{dm_{soot}}{dt}\right)_{oxidation} = \frac{6M_c}{\rho_{soot}d_{soot}} m_{soot} K_{O_2} \tag{7.11}$$

$$K_{O_2} = \left(\frac{k_A p_{O_2}}{1 + k_2 p_{O_2}}\right) x_A + k_B p_{O_2}(1 - x_A) \tag{7.12}$$

$$x_A = \left(1 + \frac{k_T}{k_B p_{O_2}}\right)^{-1} \tag{7.13}$$

$$k_A = 20 \times \exp\left(-\frac{30{,}000}{RT}\right)$$

$$k_B = 4.46 \times 10^{-3} \exp\left(-\frac{15{,}200}{RT}\right)$$

$$k_T = 1.51 \times 10^{5} \exp\left(-\frac{97{,}000}{RT}\right)$$

$$k_z = 21.3 \times \exp\left(-\frac{4100}{RT}\right)$$

where p_{O_2} is the partial pressure of oxygen in the atmosphere, x_A is the proportion of sites A on the soot particle surface. M_c, $\rho_{c,}$ and d_{soot} are the molecular mass of carbon atoms, the density of soot particles, and the diameter of soot particles, respectively. This two-step empirical model doesn't consider detailed information on the soot formation mechanism and is independent of fuel. The simulation using this model can predict flame and soot distribution in a diesel spray in the engine; however, the flame predicted by this model disagrees with the diesel flame structure in the near nozzle region (Tao et al. 2005). Therefore, detailed information of soot formation predicted is not accurate.

7.3.2 Phenomenological Model

7.3.2.1 Gokul Model

Gokul soot model (Vishwanathan and Reitz 2010) is based on a semi-detailed model (Leung et al. 1991). C_2H_2 is considered as inception for the Leung model. In contrast to the Leung model, pyrene is considered as the inception for the Gokul model. Here, reduced PAHs reaction mechanism is taken into account. PAHs and their inclusion growth beyond pyrene ($C_{16}H_{10}$) in the nucleation step lead to variations in the soot mass formed by a factor of 2. Therefore, PAHs reactions inception up to pyrene is considered. This model considers the NSC model for oxidation via O_2 and Neoh model for oxidation via OH (Leung et al. 1991). The pre-exponential factor and activation are considered as suggested by Ladommatos et al. (2002). The equations for the Gokul model for different steps are:

a. **Soot Inception Through Pyrene**

$$C_{16}H_{10}(A_4) \xrightarrow{R_1} 16C(s) + 5H_2$$
$$R_1 = k_1 \times [A_4]$$
$$k_1 = 2000\{s^{-1}\}$$

b. **C_2H_2-Assisted Surface Growth**

$$C(s) + C_2H_2 \xrightarrow{R_2} 3C(s) + H_2$$
$$R_2 = k_2 \times [C_2H_2]$$
$$k_2 = 9 \times 10^4 \exp\left(\frac{-12,100}{T}\right) \times \sqrt{S}\,\{s^{-1}\}$$
$$S = \pi d_p^2 N\,\{cm^{-1}\}$$
$$d_p = \left(\frac{6Y_{c(s)}\rho}{\pi \rho_{c(s)} N}\right)\{cm\}$$

Here S is the surface area of soot per unit volume, calculated from the diameter of the soot particle d_p, $Y_{c(s)}$ is the mass fraction of soot, ρ is ambient density, $\rho_{c(s)}$ is the density of soot, N is the soot number density in particles (cm^{-3}). The pre-exponential factors for this model are increased by 1.5, as suggested by Leung et al. (1991), to extract more accurate results.

c. **Soot Coagulation**

$$nC(s) \xrightarrow{R_3} C(s)_n$$
$$R_3 = K(T) \times \left[\frac{\rho Y_{c(s)}}{M_{c(s)}}\right]^{1/6} \times [N]^{1/6}$$
$$K(T) = 2C_a\left(\frac{6M_{c(s)}}{\pi \rho_{c(s)}}\right)^{1/6} \times \left(\frac{6K_{bc}T}{\rho_{c(s)}}\right)^{1/2}\{particles\ cm^{-3}s^{-1}\}$$

Here $M_{c(s)} = 12$ g/mol, T is the temperature of the CFD cell, K_{bc} is Boltzmann constant, C_a is agglomeration constant. C_a value for laminar flames can be 0, 3, and 9. For diffusion flame, its value is 9.

d. **Soot Oxidation**

$$\text{Oxidation through } O_2\text{: } C(s) + \frac{1}{2}O_2 \rightarrow CO$$
$$\text{Oxidation through OH: } C(s) + OH \rightarrow CO + \frac{1}{2}H_2$$
$$R_3 = R_{O_2} + R_{OH}$$

7.3.2.2 Dalian Model

This phenomenological model describes the soot formation and oxidation for homogeneous charge compression ignition (HCCI) combustion (Jia et al. 2009). This model's chemical soot formation mechanism involves a reduced reaction mechanism of n-heptane with 29 species and 52 reactions (Patel et al. 2004).

Additional 13 reactions are included due to the lack of C_2H_2 species in the n-heptane mechanism. These reactions form C_2H_2 from C_2H_3 and C_2H_4. Further, it leads to oxidation of C_2H_2 by O and OH (Park and Reitz 2007), as shown in Table 7.1.

$$K = AT^N \exp\left(-\frac{E}{RT}\right)$$

R_1, R_2, R_3, and R_4, are the reaction rates of soot inception, soot surface, soot oxidation, and soot coagulation as shown in Fig. 7.1. The rates for these reactions are assumed to be in global Arrhenius expressions. Another assumption in this model is that the soot precursor and soot particles are composed of the carbon atom, and their thermochemical properties are similar to that of graphite. The suggested number of carbon atoms in the soot precursor and the initial soot nucleus are 50 and 100,

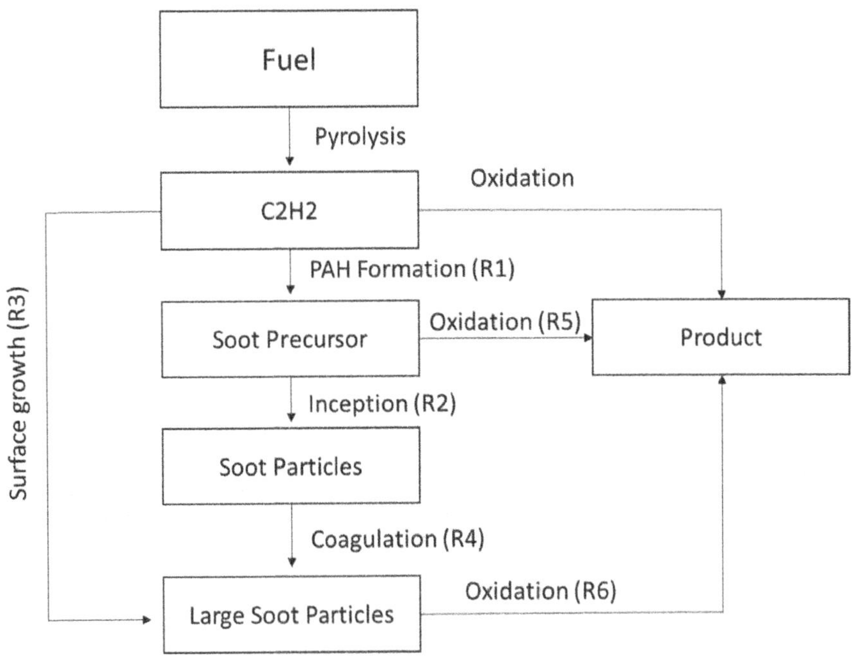

Fig. 7.1 Block diagram for improved phenomenological soot model

Table 7.1 C_2H_2 oxidation and formation (Jia et al. 2009)

S. No.	Reactions	A (cm^3/mol s)	N	E (kcal/mol)
1	$CH_2 + CH_2 = C_2H_2 + H_2$	1.2×10^{13}	0	800
2	$CH_2 + CH_2 = C_2H_2 + H + H$	1.2×10^{14}	0	800
3	$C_2H_4 + M = C_2H_2 + H_2 + M$	1.5×10^{15}	0	55,800
4	$C_2H_2 + O_2 = HCO + HCO$	4×10^{12}	0	28,000
5	$C_2H_2 + O = CH_2 + CO$	1.02×10^{7}	2	1900
6	$C_2H_2 + H + M = C_2H_3 + M$	5.54×10^{12}	0	2410
7	$C_2H_3 + H = C_2H_3 + H_2$	4×10^{13}	0	0
8	$C_2H_3 + OH = C_2H_2 + H_2O$	3×10^{13}	0	0
9	$C_2H_3 + CH_2 = C_2H_2 + CH_3$	3×10^{13}	0	0
10	$C_2H_3 + C_2H_3 = C_2H_2 + C_2H_4$	1.45×10^{13}	0	0
11	$C_2H_3 + O = C_2H_2 + OH$	1×10^{13}	0	0
12	$C_2H_2 + OH = CH_3 + CO$	4.830×10^{-4}	4	– 2000
13	$C_2H_3 = C_2H_2 + H$	4.6×10^{e}	– 8.8	46,200

respectively, resulting in a 2 nm size of incipient soot particle with the density of 2 g/cm^2 (Kazakov and Foster 1998; Tao et al. 2004).

a. **Soot Inception**

$$C_{50} \rightarrow nC(s)_m$$

$$R_1 = [C_{50}]8 \times 10^{10} \exp\left(-\frac{61{,}597}{RT}\right)$$

b. **Soot Surface Oxidation**

$$C(s)_m + C_2H_2 \rightarrow C(s)_{m+2} + H_2$$

$$R_2 = 1.05 \times 10^4\left(-\frac{6159.7}{RT}\right)[C_2H_2]\sqrt{\pi d_p^2 N}$$

c. **Soot Oxidation**

$$\text{Oxidation through } O_2\text{: } C(s) + \frac{1}{2}O_2 \rightarrow CO$$

$$\text{Oxidation through OH: } C(s) + OH \rightarrow CO + \frac{1}{2}H_2$$

$$R_3 = R_{O_2} + R_{OH}$$

d. **Soot Coagulation**

$$nC(s)_m \rightarrow C(s)_{m*n}$$

$$R_4 = \frac{1}{2}\beta N^2$$

Here β is the collision frequency of particles in the free collision regime and continuum regime, estimated by the Kazakov-Foster model (Kazakov and Foster 1998). Based on the above model definition, the net rate of soot density Y_s and soot number density N are given as:

$$\frac{dY_s}{dt} = M_C(50R_1 + R_2 - R_3) \tag{7.14}$$

$$\frac{dN}{dt} = \frac{N_A}{2}R_2 - R_4 \tag{7.15}$$

where N_A is the Avogadro number, and M_C is the molecular weight of the carbon atom.

This model was validated by the shock tube experiment of rich n-heptane for a wide range of temperatures and pressures (Jia et al. 2009). The particle diameter, particle number density, and soot yield values predicted by the model followed the same trend as the experiment (Kellerer and et al. 1996). Here, the soot yield was the total carbon available as per the assumption made earlier. The trends show that a large number of particles formed initially; however, the particle diameter increased rapidly due to coagulation, and the particle number density decreased. After 1.5 ms, the soot yield became almost constant.

The experimental values are taken from the experimental study performed for soot formation with 20, 40, and 80 bar pressure and temperature variations from 1600 to 2100 K (Patel et al. 2004).

The predicted trends agreed very well with the experimental values. The soot yield increased with an increasing cylinder pressure; hence, soot formation dominated over soot oxidation. Soot yield decreased with increased temperature after the critical point and showed good agreement with the model. The model was simple and could be applied to a three-dimensional computational domain with acceptable computational time.

7.3.2.3 Waseda Model

A gas-phase PAH model was coupled with the detailed phenomenological soot formation model. The soot precursor is acenaphthylene radical (A_2R5) with two aromatic rings, which form the initial soot particle after nucleation. The process involved in the soot formation for model development is described as follows:

a. **Nucleation**

Gas-phase gets converted to solid carbon. Assuming that three-dimensional structure formation takes place via five-membered rings. The equations for this process are as follows:

$$A_2R5 \rightarrow 12C(s) + 4H_2$$
$$R_1 = k_{NC}[A_2R5] \quad (7.16)$$

$$k_{NC} = 10^3 \times \exp\left(\frac{-5000}{RT}\right) \quad (7.17)$$

$$\frac{dN_{n,NC}}{dt} = \frac{N_A}{C_{min}} \times 12R_1 \quad (7.18)$$

The rate of change of particle number density is calculated by Eq. 7.18. $C_{min} = 100$.

b. **Surface Growth**

The assumption is that soot particle surface is covered by radical sites, where acetylene can attack, and H is bonded to inactive sites. The below reactions represents the surface growth of soot:

$$C_{soot}H + H \leftrightharpoons C_{soot^*} + H_2$$

$$C_{soot}H + OH \leftrightharpoons C_{soot^*} + H_2$$

$$C_{soot^*}H + H \leftrightharpoons C_{soot}H$$

$$C_{soot^*}H + C_2H_2 \leftrightharpoons C_{soot+2}H + H$$

The change in the mass of carbon (s) can be modelled as

$$C(s) + C_2H_2 \rightarrow 3C(s) + H_2$$

$$R_2 = k_2 P_{C_2H_2}\left(0.5\left[\tanh\left(\frac{8168}{T} - 4.57\right) + 1\right]\frac{\chi^*{}_{SOOT}}{N_A} A_{soot} f_n\right) \quad (7.19)$$

Here k_s is the coefficient of reaction rate, $P_{C_2H_2}$ is the partial pressure of $P_{C_2H_2}$, $\chi^*{}_{SOOT}$ is the active surface site density of soot particles, A_{soot} is the surface area function of soot particles per unit volume, and f_n is the soot density.

c. **Surface Oxidation**

The oxidants in this model are O_2, OH, and NO for the soot particles. Oxidation by O_2 follows the model proposed by NSC. Soot oxidation by OH and NO is according to kinetic molecular theory with a certain probability of reaction (Gersum and Roth

1992). The oxidation probability of soot by NO is lesser among NO, O_2, and OH, but it is involved in the modelling to get more accurate results.

$$\text{Oxidation by } O_2\text{: } C(s) + \frac{1}{2}O_2 \rightarrow CO$$

$$\text{Oxidation by OH: } C(s) + OH \rightarrow CO + \frac{1}{2}H_2$$

$$\text{Oxidation by NO: } C(s) + NO \rightarrow CO + \frac{1}{2}N_2$$

The net reaction rate of oxidation is given as

$$R_3 = R_{O_2} + R_{OH} + R_{NO}$$

d. **Coagulation**

The soot particle coagulation phenomenon starts before the exhaust stroke. Therefore, it is considered in this model.

$$nC(s)_m \rightarrow C(s)_{m*n}$$

$$R_4 = -\frac{5}{6}k_{coag}g_v^{1/6}g_n^{11/6} \quad (7.20)$$

Here k_{coag} is the coagulation coefficient, g_v is the soot volume fraction, g_n is the soot number density. The rate of soot mass formed is given as:

$$\text{For species density: } \dot{S}_M = (R_1 + R_2 - R_3) \times M_C$$

$$\text{For number density: } \dot{S}_M = 0.12N_A R_1 - R_4$$

This model was applied on an in-line four-cylinder, 3L turbocharged engine. Quantitative soot predictions by this model were acceptable when compared with the empirical results. The study was conducted to predict soot emission with different exhaust gas recirculation (EGR) rates. It was found that with an increase in the EGR rate, the soot mass increases rapidly. This was due to a decrease in soot oxidation (Gnanamoorthi and Devaradjane 2015).

7.4 NO_x Formation Models

7.4.1 Semi-Empirical Model (Thermal NO_x)

Purely empirical models are based on the correlation between engine operating parameters such as temperature, pressure, and NOx emissions. Empirical NO_x

models are valid only for a narrow range of operating conditions, and they cannot predict the values outside the calibration range. Semi-empirical models were developed to overcome this issue. In the semi-empirical models, the NO_x formation chemical mechanism is also taken into account. This model predicts NO_x over a wider operating range for various engine types/configurations. The semi-empirical model is a pseudo-multi-zone model and is faster than the phenomenological model. In these models, the in-cylinder charge from the start of compression to the ignition start is treated as a single zone. Assuming a uniform fuel composition, temperature, and pressure in the cylinder, the temperature history is calculated from measured pressure using the ideal gas equation. The temperature calculated also accounts for the heat losses from the cylinder, which is important for the NO_x calculations. The computational time is the time taken for each crank angle degree rotation. The burnt fuel is discretized in the control volume with required air for each crank angle degree rotation, known as zone. Each zone is treated as one thermodynamic system with a particular temperature and fuel composition. For the first estimation of temperature, combustion is assumed to be perfect without any dissociation. O_2, N_2, CO_2, H_2O, H, H_2, N, NO, O, OH, CO species are considered in this reaction mechanism. The species concentration is calculated using a detailed chemical mechanism (Rakopoulos et al. 1994; Vickland et al. 1962; Way 1976). NO formation is calculated via chemical kinetics for each zone using the concentration of species. As discussed in the previous section, this model uses the extended Zeldovich mechanism for NO formation. The NO formation is dominant over other oxides of nitrogen; therefore, the NO concentration is equivalent to the NO_x. The concentration of NO in each zone is calculated parallelly. Integrating NO values for each zone provides overall NO formation. The total NO concentration per stroke is the total value of NO at the time of the exhaust valve opening.

Calculations

The volume of burning zone ‘j’ at CA ‘i’ using the ideal gas equation is given as (Çengel and Boles 2006):

$$V_{j,i} = m_j \times \frac{\overline{R}}{\sum_l^n x_{j,i} \cdot M_l} \cdot \frac{T_{j,i}}{p_i} \tag{7.21}$$

Here

m_j is the zone mass,

$\overline{R}$ is the universal gas constant,

$x_{j,i}$ is the mole fraction and M_l is the molecular weights of lth species in the ‘j’ zone at examined CA ‘i.’

Zone $T_{j,i}$ temperature is given by:

$$T_{j,i} = T_{j,i-1}\left(\frac{p_i}{p_{i-1}}\right)^{\frac{\Upsilon_p - 1}{\Upsilon_p}} \tag{7.22}$$

The rate of formation of NO is given by Eq. 7.4

$$\frac{d[\mathrm{NO}]}{dt} = 2k_{1f}[\mathrm{O}][\mathrm{N}_2]\left\{\frac{1 - [\mathrm{NO}]^2/K[\mathrm{O}_2][\mathrm{N}_2]}{1 + k_{1r}[\mathrm{NO}]/(k_{2f}[\mathrm{O}_2] + k_{3f}[\mathrm{OH}])}\right\}$$

The formation of NO per change of CA for examined volume $V_{\mathrm{j,I}}$ is given by Eq. 7.23, which is further simplified to Eq. 7.24

$$\left(\frac{d\mathrm{NO}}{dCA}\right)_{j,i} = \left(\frac{d[\mathrm{NO}]}{dt}\right)_{j,i} \cdot V_{j,i} \cdot \frac{dt}{dCA} \quad (7.23)$$

$$\left(\frac{d\mathrm{NO}}{dCA}\right)_{j,i} = 2k_{1f}[\mathrm{O}][\mathrm{N}_2]\left\{\frac{1 - [\mathrm{NO}]^2/K[\mathrm{O}_2][\mathrm{N}_2]}{1 + k_{1r}[\mathrm{NO}]/(k_{2f}[\mathrm{O}_2] + k_{3f}[\mathrm{OH}])}\right\} \cdot \frac{V_{j,i}}{6 \cdot N} \quad (7.24)$$

The instantaneous amount of NO inside each zone can be obtained by integrating Eq. 7.24 using Euler method. The total NO amount for each CA interval is the sum of NO in the zones for the crank angle step under investigation.

This model was applied on a heavy (truck) and a light-duty (passenger car) engine to check its ability to predict NOx trends (Provataris et al. 2017). The NOx formation was predicted for variations in the start of injection (SOI), exhaust gas recirculation (EGR) rate, injection pressure, and boost pressure. As shown in Fig. 7.2, results for the heavy-duty truck engine predicted by the model followed almost the same trends as the measured data. The NO formed in the tailpipe shows exponential reduction with injection retard and linear reduction with the EGR rates. NO formation rate increased linearly with the injection pressure and boost pressure. The model approaches absolute values of NO quite well without the use of any correction or scaling factor. Thus, it minimizes the calibration efforts for this model and makes this model suitable for research and practical applications.

However, on a percentages basis, the relative error was high, attributed to low NO values as shown in Fig. 7.3. The average percentage error for the test cases was ~ 18% for heavy-duty engines. Further development of this model in the future can minimize the uncertainty and errors.

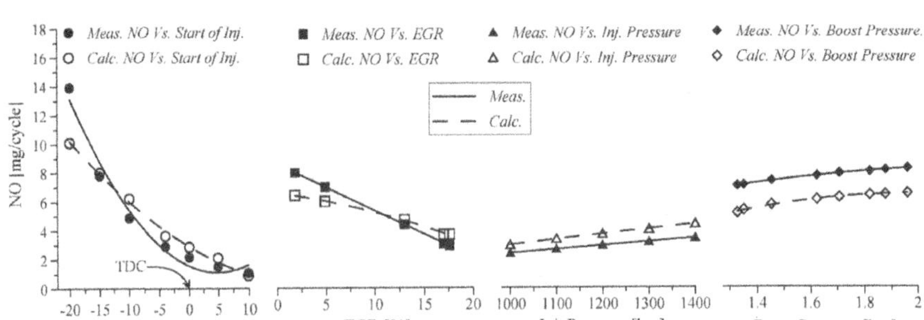

Fig. 7.2 NO variations with different parameters (Provataris et al. 2017)

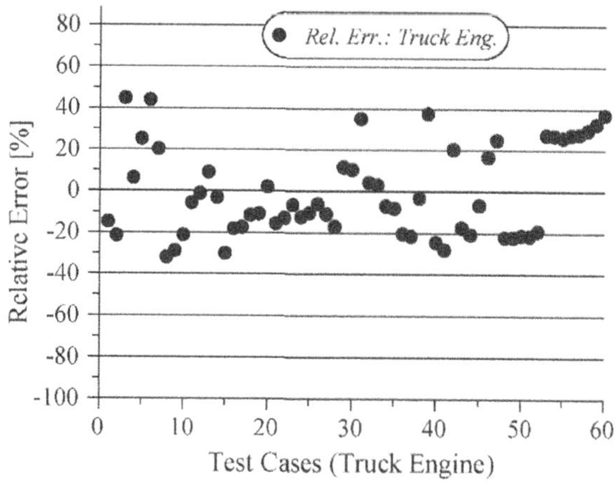

Fig. 7.3 Relative errors between the measured and calculated tailpipe NO emission (Provataris et al. 2017)

7.4.2 *Prompt NO_x Model*

The rapid NO formation in fuel-rich and low-temperature conditions is termed 'prompt NO_x.' Prompt NOx makes only a small contribution to the NO formation. The mechanism involved is known as the Fenimore mechanism (Fenimore 1971). NO is formed in the flame front, where a rich fuel–air ratio is present due to the unavailability of oxygen. CH radicals react with nitrogen and create HCN and NO at the termination of the reaction. The basic concept behind prompt NO is that the nitrogen reacts with the hydrocarbon radicals to form amines or cyano compounds. These get converted to an intermediate compound to form NO.

$$\mathrm{CH} + \mathrm{N_2} \rightarrow \mathrm{HCN} + \mathrm{N}$$
$$\mathrm{C} + \mathrm{N_2} \rightarrow \mathrm{NH} + \mathrm{CO}$$

The second reaction becomes the primary path in the entire mechanism and limits the rate of the mechanism. The conversion of HCN to NO is as follows:

$$\mathrm{HCN} + \mathrm{O} \rightarrow \mathrm{NCO} + \mathrm{H}$$
$$\mathrm{NCO} + \mathrm{H} \rightarrow \mathrm{NH} + \mathrm{CO}$$
$$\mathrm{NH} + \mathrm{H} \rightarrow \mathrm{N} + \mathrm{H_2}$$
$$\mathrm{N} + \mathrm{OH} \rightarrow \mathrm{NO} + \mathrm{H}$$

The rate of NO formation given by De Soete (1975) is:

$$\frac{d[\mathrm{NO}]}{dt} = \dot{m}_{\mathrm{NO_x}} - \dot{m}_{\mathrm{N_2}} \tag{7.25}$$

Here, $\dot{m}_{NO_x}$ and$\dot{m}_{N_2}$ are the formation rate of overall prompt NOx and nitrogen, respectively.

When the oxygen concentration is high in the fuel-rich region, NOx formation takes place instead of nitrogen. Therefore, the rate of NO_x formation is equal to the overall rate of prompt NO_x formation. The expression for NO is given by:

$$\frac{d[\mathrm{NO}]}{dt} = k_{pr}[\mathrm{O_2}]^c[\mathrm{N_2}][fuelspecies]e^{-E_a/RT} \tag{7.26}$$

Applying a correction factor f to De Soete model for higher hydrocarbon-containing fuel and rich fuel conditions to improve the performance of the model given by Eq. 7.26. The equation then becomes:

$$\frac{d[\mathrm{NO}]}{dt} = fk'_{pr}[\mathrm{O_2}]^c[\mathrm{N_2}][fuelspecies]e^{-E'_a/RT} \tag{7.27}$$

The correction factor f is

$$f = 4.75 + 0.081a - 23.2\varphi + 32\varphi^2 - 12.2\varphi^3$$

Here φ is the equivalence ratio, and a is the number of carbon atoms present in one molecule of hydrocarbon fuel. f is valid for φ value between 0.6 and 1.6 for aliphatic hydrocarbon fuels only (Dupont et al. 1993). The prompt model is valid only for single-component hydrocarbon fuel; otherwise, it gives incorrect values. Order of oxygen in the Eq. 7.27 depends on the mole fraction of oxygen in each cell.

$$c = \begin{cases} 1.0 & X_{O_2} \le 4.1e-3 \\ -3.95 - 0.9\ln X_{O_2} & 4.1e-3 \le X_{O_2} \le 1.11e-2 \\ -0.35 - 0.1\ln X_{O_2} & 1.11e-3 \le X_{O_2} < 0.03 \\ 0 & X_{O_2} \ge 0.03 \end{cases}$$

7.5 Future Research Directions

Accurate modelling of physical and chemical processes for NOx and soot formation remains challenging, opaque, and complex, even with the availability of data for detailed chemical reaction mechanisms. Another challenging task that needs research effort in the future is validating these models for engine warm-up conditions. This is challenging due to errors in measuring heat losses to the wall and the lubricating oil. Multi-step models are more accurate and reliable. Therefore, two-step empirical soot models can be improved by multi-step soot models. In the Gokul model, aromatics formation is via propargyl radicals. Hence another pathway of formation of aromatics should be considered to improve the model. Individual vaporization of diesel components such as paraffin, aromatics, etc., could be explored in the future for accurate modelling. Improvement in soot precursor formation and oxidation mechanism will make the soot model more accurate and in agreement with the shock tube experiments.

7.6 Summary

In the current chapter, a comprehensive review of Soot and NOx models and their classification is presented. Soot and NOx models are classified into two categories: (a) Phenomenological and (b) Empirical models. It is essential to consider the detailed formation mechanisms for accurate and reliable model development. Processes such as Soot inception, nucleation, growth, oxidation, and coagulation are involved in the detailed soot formation mechanisms. Phenomenological NOx models are generally developed based on extended Zeldovich and Fenimore mechanisms. Some developed models are described in detail, along with their capabilities and limitations. Hiroyasu-NSC model is simple to use in CFD simulations. Dalian model agrees accurately with the shock tube experiments and can be implemented in the 3D computational domains. Waseda model is developed for medium-duty diesel engines. The models discussed in this chapter provide an insight into the fundamentals of emission modelling. The accuracy and reliability of the models depend on the model parameters, chemical mechanism, and assumptions made.

References

Adachi K, Chung SH, Buseck PR (2010) Shapes of soot aerosol particles and implications for their effects on climate. J Geophys Res Atmos 115(D15)

Asad U, Divekar P, Zheng M, Tjong J (2013) Low-temperature combustion strategies for compression ignition engines: operability limits and challenges. SAE Technical paper no 2013-01-0283. https://doi.org/10.4271/2013-01-0283

Baskar P, Senthilkumar A (2016) Effects of oxygen-enriched combustion on pollution and performance characteristics of a diesel engine. Eng Sci Technol Int J 19(1):438–443

Bowman CT (1975) Kinetics of pollutant formation and destruction in combustion. Prog Energy Combust Sci 1(1):33–45. https://doi.org/10.1016/0360-1285(75)90005-2

Çengel YA, Boles MA (2006) Mass and energy analysis of control volumes. Thermodyn Eng Approach:216–246

Daido S, Kodama Y, Inohara T, Ohyama N, Sugiyama T (2000) Analysis of soot accumulation inside diesel engines. JSAE Rev 21(3):303–308. https://doi.org/10.1016/S0389-4304(00)00048-5

De Soete GG (1975) Overall reaction rates of NO and N_2 formation from fuel nitrogen. Symp (Int) Combust 15(1):1093–1102. https://doi.org/10.1016/S0082-0784(75)80374-2

Dobbin RA, Subramaniasivam H (1994) Soot formation in combustion. Chem Phys 5

Du DX, Axelbaum RL, Law CK (1991) The influence of carbon dioxide and oxygen as additives on soot formation in diffusion flames. Symp (Int) Combustion 23(1):1501–1507. https://doi.org/10.1016/S0082-0784(06)80419-4

Dupont V, Pourkashanian M, Williams A, Woolley R (1993) The reduction of NOx formation in natural gas burner flames. Fuel 72(4):497–503. https://doi.org/10.1016/0016-2361(93)90108-E
Fenimore CP (1971) Formation of nitric oxide in premixed hydrocarbon flames. Symp (Int) Combust 13(1):373–380. https://doi.org/10.1016/S0082-0784(71)80040-1
Glassman I (1996) Combustion, 3rd edn. Academic Press, Inc., San Diego
Gnanamoorthi V, Devaradjane G (2015) Effect of compression ratio on the performance, combustion, and emission of DI diesel engine fueled with ethanol—diesel blend. J Energy Inst 88(1):19–26. https://doi.org/10.1016/j.joei.2014.06.001
Gülder ÖL (1993) Influence of sulfur dioxide on soot formation in diffusion flames. Combust Flame 92(4):410–418. https://doi.org/10.1016/0010-2180(93)90152-S
Han Z, Uludogan A, Hampson GJ, Reitz RD (1996) Mechanism of soot and NOx emission reduction using multiple-injection in a diesel engine. SAE Trans:837–852
Heywood JB (2018) Internal combustion engine fundamentals. McGraw-Hill
Hiroyasu H, Kadota T (1976) Models for combustion and formation of nitric oxide and soot in direct injection diesel engines. SAE Trans 513–526. https://doi.org/10.4271/760129
İlkiliç C, Aydin H (2012) The harmful effects of diesel engine exhaust emissions. Energy Sour Part A Recov Utilization Environ Effects 34(10):899–905
Jia M, Peng ZJ, Xie MZ (2009) Numerical investigation of soot reduction potentials with diesel homogeneous charge compression ignition combustion by an improved phenomenological soot model. Proc Inst Mech Eng Part D J Autom Eng 223(3):395–412. https://doi.org/10.1243/09544070JAUTO993
Kazakov A, Foster DE (1998) Modeling of soot formation during DI diesel combustion using a multi-step phenomenological model. SAE Trans:1016–1028
Kellerer H, Müller A, Bauer HJ, Wittig S (1996) Soot formation in a shock tube under elevated pressure conditions. Combust Sci Technol 113(1):67–80.https://doi.org/10.1080/00102209608935488
Kumar M, Tsujimura T, Suzuki Y (2018) NOx model development and validation with diesel and hydrogen/diesel dual-fuel system on diesel engine. Energy 145:496–506. https://doi.org/10.1016/j.energy.2017.12.148
Ladommatos N, Song H, Zhao H (2002) Measurements and predictions of diesel soot oxidation rates. Proc Inst Mech Eng Part D J Autom Eng 216(8):677–689. https://doi.org/10.1177/095440700221600806
Leung KM, Lindstedt RP, Jones WP (1991) A simplified reaction mechanism for soot formation in non-premixed flames. Combust Flame 87(3–4):289–305. https://doi.org/10.1016/0010-2180(91)90114-Q
Nagle J, Strickland-Constable RF (1962) Oxidation of carbon between 1000 and 2000 °C. In: Proceedings of the fifth carbon conference. Pergamon Press, London, p 154
Noor MM, Wandel AP, Yusaf T (2014) Effect of air-fuel ratio on temperature distribution and pollutants for biogas MILD combustion. Int J Autom Mech Eng 10(1):1980–1992. https://doi.org/10.15282/ijame.10.2014.1%205.0166
Ogawa H, Matsui Y, Kimura S, Kawashima J (1996) Three-dimensional computation of the effects of the swirl ratio in direct-injection diesel engines on NOx and soot emissions (no. 961125). SAE technical paper. https://doi.org/10.4271/961125
Olson DB, Pickens JC, Gill RJ (1985) The effects of molecular structure on soot formation II. Diffusion flames. Combustion Flame 62(1):43–60. https://doi.org/10.1016/0010-2180(85)90092-6
Park SW, Reitz RD (2007) Numerical study on the low emission window of homogeneous charge compression ignition diesel combustion. Combust Sci Technol 179(11):2279–2307. https://doi.org/10.1080/00102200701484142
Patel A, Kong SC, Reitz RD (2004) Development and validation of a reduced reaction mechanism for HCCI engine simulations. SAE technical paper no. 2004-01-0558. https://doi.org/10.4271/2004-01-0558

Provataris SA, Savva NS, Chountalas TD, Hountalas DT (2017) Prediction of NOx emissions for high-speed DI Diesel engines using a semi-empirical, two-zone model. Energy Convers Manage 153:659–670. https://doi.org/10.1016/j.enconman.2017.10.007

Rakopoulos CD, Hountalas DT, Tzanos EI, Taklis GN (1994) A fast algorithm for calculating the composition of diesel combustion products using 11 species chemical equilibrium scheme. Adv Eng Softw 19(2):109–119. https://doi.org/10.1016/0965-9978(94)90064-7

Richter H, Howard JB (2000) Formation of polycyclic aromatic hydrocarbons and their growth to soot—a review of chemical reaction pathways. Prog Energy Combust Sci 26(4–6):565–608. https://doi.org/10.1016/S0360-1285(00)00009-5

Savva N, Hountalas D (2012) Detailed evaluation of a new semi-empirical multi-zone NOx model by application on various diesel engine configurations. SAE technical paper no. 2012-01-1156. https://doi.org/10.4271/2012-01-1156

Tao F, Golovitchev VI, Chomiak J (2004) A phenomenological model for the prediction of soot formation in diesel spray combustion. Combust Flame 136(3):270–282. https://doi.org/10.1016/j.combustflame.2003.11.001

Tao F, Srinivas S, Reitz RD, Foster DE (2005) Comparison of three soot models applied to multi-dimensional diesel combustion simulations. JSME Int J Ser B 48(4):671–678. https://doi.org/10.1299/jsmeb.48.671

Tree DR, Svensson KI (2007) Soot processes in compression ignition engines. Prog Energy Combust Sci 33(3):272–309. https://doi.org/10.1016/j.pecs.2006.03.002

Vickland CW, Strange FM, Bell RA, Starkman ES (1962) A consideration of the high-temperature thermodynamics of internal combustion engines. SAE Trans 70:785–795. https://doi.org/10.4271/620564

Vishwanathan G, Reitz RD (2010) Development of a practical soot modelling approach and its application to low-temperature diesel combustion. Combust Sci Technol 182(8):1050–1082. https://doi.org/10.1080/00102200903548124

Von Gersum S, Roth P (1992) Soot oxidation in high-temperature N_2O/Ar and NO/Ar mixtures. Symp (Int) Combust 24(1):999–1006. https://doi.org/10.1016/S0082-0784(06)80118-9

Walker PL, Austin LG, Nandi SP (1966) In chemistry and physics of carbon. In: Walker PL (ed)

Wang Q, Shao C, Liu Q, Zhang Z, He Z (2017) Effects of injection rate on combustion and emissions of a pilot ignited direct injection natural gas engine. J Mech Sci Technol 31(4):1969–1978. https://doi.org/10.1007/s12206-017-0346-3

Way RJB (1976) Methods for determination of composition and thermodynamic properties of combustion products for internal combustion engine calculations. Proc Inst Mech Eng 190(1):687–697. https://doi.org/10.1243/PIME_PROC_1976_190_073_02

Xi J, Zhong BJ (2006) Soot in diesel combustion systems. Chem Eng Technol Indust Chem-Plant Equip-Process Eng-Biotechnol 29(6):665–673. https://doi.org/10.1002/ceat.200600016

Zhang R, Kook S (2014) Influence of fuel injection timing and pressure on in-flame soot particles in an automotive-size diesel engine. Environ Sci Technol 48(14):8243–8250. https://doi.org/10.1021/es500661w

Chapter 8
NO_x Emission Prediction for DI Diesel Engine: Numerical Modelling

Rajesh Kumar Prasad

Abstract Strict emission norms, fewer complex emissions prediction models forces to develop a numerical model which actively control both combustion phase as well as after-treatment systems. This study is focused on the simulation and experimental investigation of simplified model for prediction of NO_x emissions along with unburned zone, burned zone, adiabatic flame temperatures and species concentration especially atomic oxygen [O], Nitrogen [N_2] and nitric oxide [NO]. Burned flame temperature and thermal NO concentrations were simulated by enthalpy balance and Zeldovich mechanism respectively. The simulated results were validated with experimental result of Turbocharged direct injected Diesel engine at steady-state operating conditions. The maximum temperature (T_{max}) simulated within burned zone at 2200 rpm and 100% load is 2917 K while at 75% load, and 50% load it reduces to 2853 K, and 2776 K respectively. It was also observed that the equilibrium concentrations of [O], [NO] and [N_2] were directly proportional to burned zone temperature. The accuracy of proposed model was tested at 2200 rpm rated speed and also at 1400 rpm with full load, 75% load, and 50% load. NOx reduces with speed for identical operating conditions.

Keywords Zero-dimensional two zone combustion model · Burned zone temperature · Adiabatic flame temperature · NO_x emissions · Zeldovich mechanism · DI diesel engine

8.1 Introduction

Diesel engines are the most fuel efficient engines (Agarwal et al. 2016) for off road and on road applications nowadays; nevertheless emissions from these either regulated or unregulated still creates imbalances on environment. For active control measures of diesel combustion and optimization of after-treatment requires an adequate and reliable pollutants prediction model which became alternative to gas sensors (Arrègle

R. K. Prasad (✉)
Department of Mechanical Engineering, ICFAI University Jharkhand, Ranchi 835222, India
e-mail: rajeshkumar.p@iujharkhand.edu.in

A. K. Agarwal et al. (eds.), *Engine Modeling and Simulation*, Energy, Environment, and Sustainability, https://doi.org/10.1007/978-981-16-8618-4_8

et al. 2010; Moos 2005). Oxides of nitrogen (NO, NO_2 and N_2O etc.), and particulate matter (PM) are main pollutants in diesel engines while NO_x is prime emission in gasoline engine. NO_x formation in gasoline engine is mainly temperature dependent and reduced by leaning the mixture (Prasad and Agarwal 2021) while diesel combustion could be due to either thermal route or fuel bound or prompt formation route or via N_2O route. In this study, only thermal NO_x prediction model was considered using Zeldovich mechanism (Egnell 1998; Timoney et al. 2005; Cipolat 2007; Hernández et al. 2008) which is a temperature dependent phenomena (Prasad et al. 2020). In the past, few researchers proposed model for NO_x prediction which were based on mathematical correlations with different operation variables (Re et al. 2010), while others considered cyclic NO_x evolution using the in-cylinder pressure (P) Vs crank angle (Θ) data as an input parameter (Egnell 1998; Andersson et al. 2006; Arrègle et al. 2008). The in-cylinder pressure signal provides details about combustion phenomena, air flow rate (Desantes et al. 2010), online combustion parameter detection (Luján et al. 2010), misfire detection (Leonhardt et al. 1999), control strategy of exhaust gas recirculation (Hasegawa et al. 2006), engine torque estimation (Shimasaki et al. 2004) and combustion noise control (Payri et al. 2005). In this study, engine operating parameters along with engine specifications were noted to predict in-cylinder pressure, temperature, ignition delay, fuel mass fraction burned, heat release, adiabatic flame temperature, and steady-state NO_x production. Instantaneous NO_x simulation models were developed with real time calculations (Andersson et al. 2006; Arrègle et al. 2008; Hountals et al. 2010). All these proposed models combined with closed-loop control of few input parameters have a big potential on NO_x reduction during combustion as well as in after-treatment (Devarakonda et al. 2009; Katare et al. 2007). In this way, optimizing agent flow into the catalytic converter and thus extending its lifetime. The main challenge with NO_x prediction models is to maintain equilibrium between accuracy and speed of calculation. This could be overcome by predicting NO_x with varying load and speed with physical modeling approach. At the beginning, a physical representation of each minute detail has been provided when the engine is in off-design operation. The zero-dimensional thermodynamic model is the simplest model which is based on empirical heat release model and predict NO_x using time based ordinary differential equation while multi-dimensional CFD model is most complex which uses conservation of mass, energy and momentum and NO_x prediction by solving partial differential equation which is the function of space and time (Stone 1999).

Johansson et al. (2009) developed vehicle on-board NO_x prediction model. This study focused on various ways to handle temperature prediction within the different combustion zones which avoid iterative computing of energy balance. A two-zone model was considered for this study, NOx formed in burned zone while unburned zone composed of fuel–air mixture. With the help of number of moles and global temperature during combustion burned zone temperature was predicted. After predicting burned zone temperature, NO was computed using the well-known 'Zeldovich mechanism'. Arsie et al. (2004) predicted NOx and Soot emissions of diesel engine using phenomenological models. A single-zone model was used for 1.9 L engine to prediction of ignition delay for stages of fuel injection such as pilot, pre and main injection.

The heat release rate was simulated using Watson model. NO and soot emissions were simulated using Zeldovich mechanism and approach proposed by Hiroyasu respectively. Three sub-models were used to simulate mixing, ignition delay, combustion for pilot, pre and main injection. Their predicted results for NO and soot showed good agreement with experimental results.

Andersson et al. (2006) developed a fast NO_x simulation model for optimization of engine operating parameters, and after treatment system control or virtual mapping. For this cylinder pressure was used as input data interpolation table for high computation speed of equilibrium burned zone and global temperatures and species concentrations. They proposed zero-dimensional, multi-zone model in which products of combustion mixed with the surrounding cylinder air. The thermal NO_x was simulated with the help of two reactions of the original Zeldovich mechanism. They validated the proposed model in various circumstances such as varying compression ratios, EGR rates, injection timing, inlet pressures etc.

Ericson et al. (2006) proposed a two-zone zero dimensional combustion model which can simulate engine-out NO_x and various operating parameters such as in-cylinder pressure, combustion temperature and suction and exhaust gas flows. They used a quasi steady gas exchange model for exhaust gas recirculation and variable geometry turbocharger and a zero dimensional two-zone combustion model was used for NO_x prediction. Heat release was predicted by fuel flow parameters and pressure traces, simplified combustion used for species concentrations and burned zone temperature and for NO_x concentration was predicted by Zeldovich mechanism. They validated their proposed model by both ways: steady state condition and transient condition.

Mellor et al. (1998) developed a two-zone combustion model for direct injection diesel engine NOx prediction. For this, they formed skeletal mechanism which consists of seven elementary reactions. They reported low NO concentration in Zone-1 (stoichiometric region; $\Phi = 1$) formed only by forward reactions. Zone-2 consists of burned gases at an overall equivalence ratio. NO was formed in zone-2 by the decomposition or reverse N_2O mechanism. Chmela et al. (1999) proposed a non-dimensional combustion model that relies on the concept of mixing controlled combustion (MCC) which is independent of fuel -air mixture formation and fuel oxidation processes. They reported that rate of heat release (ROHR) was depend upon instantaneous fuel mass injected and at that instant, local density of turbulent kinetic energy. This study was performed without knowing combustion chamber geometry, spray characteristics, fuel–air mixture formation and evaporation. Authors used simple physical law for the simulation which is limited to the heat release rate of diffusion combustion.

The main objective of this study is to develop mathematical modeling to simulate the engine NO_x emission for DI Diesel engine which includes;

- Predicting and investigating different control parameters that affect the NO_x formation during combustion,
- Experimental investigation to study the different models and select the appropriate one,

- Exhaust emission characteristics at different operating conditions.

8.2 Experimental Setup and Methodology

Figure 8.1 shows experimental setup of a turbocharged direct injection diesel engine which is coupled with eddy current dynamometer to simulate road like conditions in the laboratory. The experimental diesel engine specification is provided in Table 8.1.

The engine suction side is fitted with air-conditioning system, air measuring system, and fuel measuring unit while exhaust side is partially connected to emission analyzer. AVL Indiset, was used as data acquisition system. Un-cooled miniature pressure transducer (GM12D) was used to acquire the in-cylinder combustion and motoring pressure data w.r.t. crank angle. AVL 365C angle encoder was used to convert the analogue angle pulses into digital output signals and it determines the

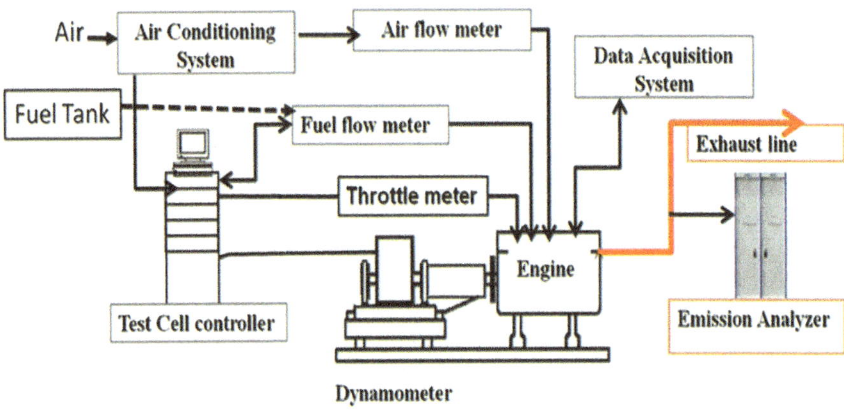

Fig. 8.1 Engine experimental setup to collect data for validation of simulated results

Table 8.1 Engine specification

Parameter	Specifications
Engine model	4- cylinder turbocharged DI diesel
Bore (mm)/stroke (mm)	102/110
Connecting rod length (mm)	220
IVC (CA)/EVO (CA)	210/540
Rated speed (rpm)	2200
Compression ratio/swept volume (liter)	17.4/0.9 per cylinder
Type of injection system	Common rail

position of the shaft w.r.t. TDC. The parameters noted from the engine experiment were Brake Power (kW), Engine Speed (rpm), Crank angle Measurement, Air Flow Rate (kg/h), In-cylinder gas pressure measurement (bar), Fuel consumption (kg/h), and NO_x concentration (ppm). To measure diesel flow rate accurately, "FEV Fuel rate (FI-15)" was used which is based on principle of coriolis theory. It measures flow rate accurately whatever may variations in pressure, density, viscosity and temperature. The measuring range of this instrument is up to 100 kg/h with 0.01 kg/h resolution; the accuracy of instrument is ± 0.5 kg/h at full scale. Air flow measurement is also an important parameter for diesel combustion, performance calculation and emission characteristics. ABB sensy flow meter which works on principle of Hot Wire Anemometer was used to measure air flow into the engine cylinders. This flow meter can measure maximum up to 400 kg/h. Yantrashilpa's coolant condition unit YS4025 installed to maintain the engine water temperature within band of ± 0.2 °C under specific test conditions. Major pollutants emitted from the diesel engine exhaust are oxides of nitrogen (NO_x), Smoke, Unburned Hydrocarbon (UHC) and Carbon-monoxides (CO). Exhaust gas analysis was done for all the species. For measurement of NO_x, UHC and CO emissions, PEUS R1D1 Gas Analyzer has been used. The R1D1 equipped with CLD700LEV, NDIR and FID14 for measurement of NO_x, CO and UHC emissions. AVL439 was used to measure smoke in terms of percentage opacity and AVL415 used to measure smoke in terms of Filter Smoke Number (FSN).

The chemiluminescence technique has been used to measure nitrogen oxides worldwide. Sample gas from engine exhaust was drawn into the analyzer and mixed with internally produced ozone. The following reaction takes place,

$$NO + O_3 \rightarrow NO_2^* + O_2 \tag{8.1}$$

Only about 20% of the NO_2 goes into the excited state NO_2* in reaction (Rakopoulos et al. 2009). This exited NO_2* comes back to the ground state NO_2 while emitting photon energy or say electromagnetic radiation:

$$NO_2^* \rightarrow NO_2 + h_\vartheta \tag{8.2}$$

The radiation wavelength lies in the range of 600–3000 nm with an intensity maximum to ~ 1200 nm. When O_3 is present in excess, the output signal is directly proportional to the [NO] of the sample gas.

Figure 8.2 shows burned and unburned zones of diesel combustion. The burned zone in diesel combustion referred to that region in the combustion chamber which is already burned. This zone filled with air and small amount of residual unburnt organic part of fuel, and CO_2 and water vapor. Ahead of it, combustion front zone exist where oxidation of fuel occurs. Remaining portion of the combustion chamber is called unburned zone which contain mixture of fuel and air.

A zero dimensional two zone model was used for prediction of unburned zone temperature, burned zone temperature and adiabatic flame temperature. NO_x formation takes place only inside the burned zone and unburned zone consists of air.

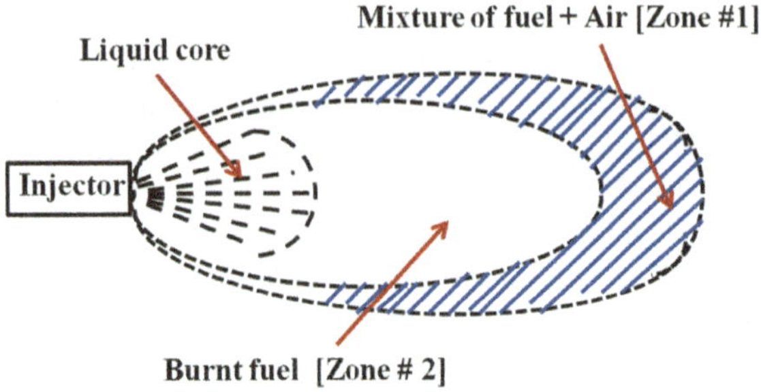

Fig. 8.2 Two zone concept (Jung and Assanis 2001)

Assumptions made to develop simulation model are listed below:

- Zero dimensional two zone model.
- Each zone constituents behaves as an ideal gas.
- Compression and expansion takes place isentropic manner and at local stoichiometric conditions.
- Complete combustion and energy released originates from injected fuel only.
- There is no heat interaction in between the zones.

Tables 8.2 and 8.3 show the steps needed for NO emission simulation and experimental validation for 4-cylinder turbocharged DI diesel engine. Unburned zone temperature can be calculated using the polytropic relation between temperature and pressure (Stone 1999; Heywood 1988):

$$T_1 = T_0\left(\frac{P_1}{P_0}\right)^{\frac{\gamma-1}{\gamma}} \tag{8.3}$$

Table 8.2 Steps for NO simulation

S. No.	Prediction model
1	Input data from HR and combustion pressure predictions
2	Unburned zone temperatures
3	Flame temperatures calculations
4	Burned zone temperatures and mass calculations
5	NO emission predictions

Table 8.3 Steps for NO emission experimental validation

S. No.	Experimental validation
1	Adaptation for emission measurement probe
2	Setup preparation on engine dynamometer with emission measurement system
3	Emission measurements at the same operating points
4	Comparison prediction results with measured values

Burned zone temperature before considering flame temperature can be calculated using isentropic relationship (Stone 1999):

$$T_{exp,1} = T_{burn,0} \times \left(\frac{P_1}{P_0}\right)^{\frac{\gamma-1}{\gamma}} \tag{8.4}$$

Burned zone temperature after considering flame temperature is computed as:

$$T_{burn,1} = \left\{T_{exp,1} \times m_{burn,0} + T_{flame,1} \times m_{flame,1}\right\} / \left\{m_{burn,1}\right\} \tag{8.5}$$

NO_x is the prime emission produced by diesel combustion which comprises nitric oxide (NO) and nitrogen dioxide NO_2 with traces of di nitrogen trioxide (N_2O_3) and di-nitrogen pentoxide (N_2O_5). NO produced in premixed combustion phase on the weak side of the reaction zone in the post-flame gases. At elevated temperature, dissociated molecular nitrogen and oxygen combine together to form NO. This reaction occurs in a slightly lean or near stoichiometric fuel–air mixtures. All these Eqs. (8.6), (8.7), and (8.8) are commonly known as extended Zeldovich mechanism (Stone 1999; Rakopoulos and Giakoumis 2009; Pundir 2007; Heywood 1988).

$$\mathrm{O + N_2 = NO + N} \tag{8.6}$$

$$\mathrm{N + O_2 = NO + O} \tag{8.7}$$

$$\mathrm{N + OH = NO + H} \tag{8.8}$$

The third Eq. (8.8) is considered only when the mixture is near Stoichiometric.

The rate constants used for these equations are chosen from Ragland et al. (2011) in units of cm^3/mol-s and temperature in K given in Table 8.4. The NO formation rate using the three reactions (8.6)–(8.8) can be expressed by the following equation;

$$\begin{aligned}\frac{d}{dt}[\mathrm{NO}] = {} & k_1[\mathrm{O}][\mathrm{N_2}] - k_{-1}[\mathrm{NO}][\mathrm{N}] \\ & + k_2[\mathrm{N}][\mathrm{O_2}] - k_{-2}[\mathrm{NO}][\mathrm{O}] \\ & + k_3[\mathrm{N}][\mathrm{OH}] - k_{-3}[\mathrm{NO}][\mathrm{H}]\end{aligned} \tag{8.9}$$

Table 8.4 Reaction rates for no formation mechanism, cm^3/mol-s, T in k

Reaction	Forward	Reverse
$O + N_2 = NO + N$	$k_1 = 1.8 \times 10^{14} \times \exp(-38{,}370/T)$	$k_{-1} = 3.8 \times 10^{13} \times \exp(-425/T)$
$N + OH = NO + H$	$k_2 = 1.8 \times 10^{10} \times \exp(-4680/T)$	$k_{-2} = 3.8 \times 10^{9} \times \exp(-20{,}820/T)$
$N + O_2 = NO + O$	$k_3 = 7.1 \times 10^{13} \times \exp(-450/T)$	$k_{-3} = 1.7 \times 10^{14} \times \exp(24{,}560/T)$

where k_1, k_2 and k_3 are forward reaction rate constants for Eqs. (8.6), (8.7), and (8.8) respectively while negative subscript used with k_1, k_2 and k_3 denotes the backward reaction rate constants. The symbol [] denotes the species concentration in moles/cm^3.

By solving the Eq. (8.9), concentration of O_2, N_2, and radicals O, N, H, OH are to be estimated. For this calculation, it was assumed that, concentration of N_2 remains nearly constant. Further, two more assumptions are as follows:

(1) The concentrations of O, OH, O_2 and O are equilibrated.

Steady state assumption of [N] leads to,

$$\begin{aligned}\frac{d}{dt}[N] &= +k_1[O][N_2] - k_{-1}[NO][N] \\ &\quad - k_2[N][O_2] + k_{-2}[NO][O] \\ &\quad - k_3[N][OH] + k_{-3}[NO][H] = 0\end{aligned} \tag{8.10}$$

Equations (8.6) and (8.7) yield the rate of NO formation,

$$\frac{d}{dt}[NO] = 2\{k_1[O][N_2] - k_{-1}[NO][N]\} \tag{8.11}$$

Using Eq. (8.7), steady state concentration of N is achieved,

$$[N]_{SS} = \frac{k_1[O][N] + k_{-2}[NO][O] + k_{-3}[NO][H]}{k_{-1}[NO] + k_2[O_2] + k_3[OH]} \tag{8.12}$$

From the assumption of O, OH, H, and O_2,

$$\frac{[O_2][H]}{[O][OH]} = \frac{k_{-2} \cdot k_3}{k_2 \cdot k_{-3}}$$

Eliminating [N] and [H] using Eqs. (8.11) and (8.12), the Eq. (8.10) gives,

$$\frac{d}{dt}[NO] = 2k_1[O][N_2]\frac{1 - \frac{[NO]^2}{k[O_2][N_2]}}{1 + \frac{k_{-1}[NO]}{k_2[O_2] + k_3[OH]}} \tag{8.13}$$

where $K = (k_1/k_{-1})(k_2/k_{-2})$ is equilibrium constant for the reaction for the reaction.

$$N_2 + O_2 \leftrightarrow 2NO$$

The NO formation rate may be calculated by Eq. (8.13) using equilibrium concentrations of O, O_2, OH and N_2. The rate of formation NO is much slower compared to combustion rates. NO formation takes place behind the flame front in the burned gases.

$$R_1 = k_1[O]_e[N_2]_e = k_{-1}[NO]_e[N]_e$$

where, R_1 is the reaction rate using equilibrium concentrations for the reaction (8.4). Similarly

$$R_2 = k_2[N]_e[O_2]_e = k_{-2}[NO]_e[O]_e; \text{ and}$$
$$R_3 = k_3[N]_e[OH]_e = k_{-3}[NO]_e[H]_e$$

Using the above notations, the Eq. (8.13) is simplified as;

$$\frac{d[NO]}{dt} = \{2 \times R_1\{1 - ([NO]/[NO]_e)^2\}/\{1 + w[NO]/[NO]_e\}\} \quad (8.14)$$

where, $w = R_1/\{R_2 + R_3\}$.

8.3 Results and Discussions

Zero-dimensional two zone combustion model was used for calculation of various combustion parameters those were needed for NO calculation. For this simulation, a C-code was developed and run it for different operating parameters (engine speeds and loads). During the experimentation, every measurement was noted on average data of 30 s just after 5 min stabilization. The results obtained are discussed in the following paragraph.

8.3.1 Zone Temperature

A two-zone burning model approach was used for combustion phenomena, assuming uniform pressure as input for each zone separately.

Figure 8.3 shows comparison of unburned zone, burned zone and bulk temperature profile at 2200 rpm and 1400 rpm respectively. The maximum unburned temperature (T_{max}) simulated at 100% load was 1218 K at 368° CA while T_{max} at 75% load, 50% load and 10% load are 1145 K, 1135 K, 1110 K and their θ_{Tmax} were 368° CA, 365.5° CA and 362.5° CA respectively. The unburned zone temperature profile reflects the compression process, reaching its peak pressure (P_{max}) close to TDC, and then reduces as the piston moves downward. At high load, T_{max} was higher compared to low load due to presence of higher pressure and temperature at inlet. If engine speed was reduced from 2200 to 1400 rpm then the simulated unburned T_{max} at 100% load, 75% load and 50% load were reduced to 1147 K, 1084 K, 1033 K and simulated θ_{Tmax} were 370° CA, 369° CA and 366.5° CA respectively [refer Fig. 8.3(iv)]. Inlet

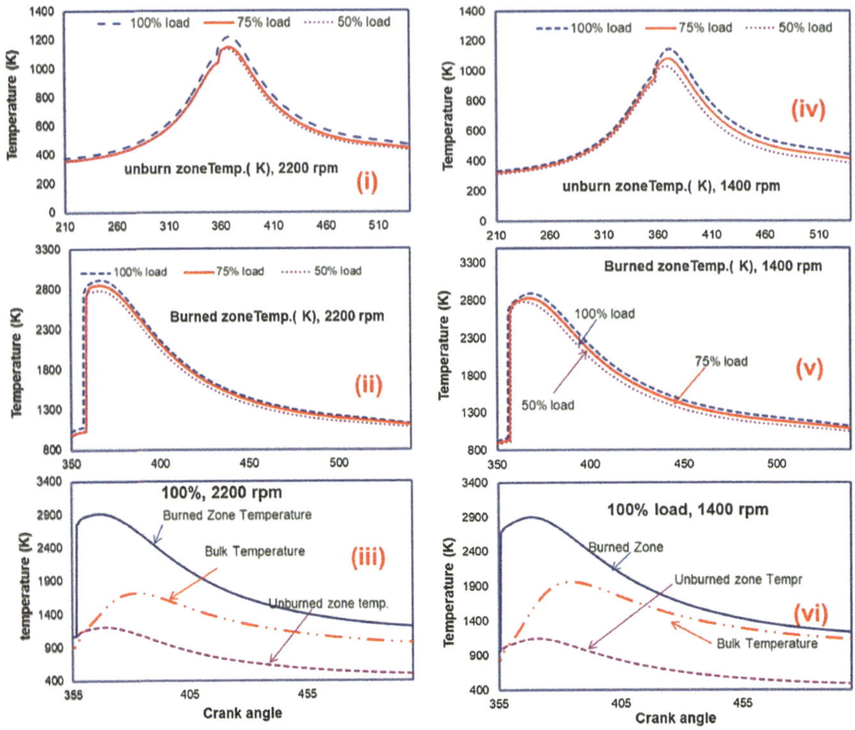

Fig. 8.3 Comparison of unburned zone and burned zone temperatures with varying speed and loads

pressure and temperature condition at inlet was responsible for this reduction in temperature and retarted θ_{Tmax}.

Burned zone temperature before considering flame temperature can be calculated using isentropic relationship as per Eq. (8.4) and during combustion according to Eq. (8.5). Figure 8.3(ii) shows the burned zone temperature profile at 2200 rpm. The T_{max} simulated at 100% load was 2917 K at 366.5° CA while at 75% load and 50% load were 2853 K and 2776 K and their θ_{Tmax} was 366° CA, 366° CA and 362° CA respectively. It was observed that burned zone temperature was very high near the TDC just after start of combustion (SOC) and reduces as the expansion progresses.

If engine speed was lowered to 1400 rpm then the simulated burned zone Tmax at 100% load, 75% load and 50% load were 2896, 2833, 2781 K and θ_{Tmax} were retarded to 369° CA, 367.5° CA and 365° CA. At higher load, presence of more mass of fuel along with higher pressure and temperature at SOC results higher burned zone temperature. It was observed that the global or bulk cylinder temperature lies between the temperatures of burned zones and the unburned zone. Figures 8.3(iii) and 8.4(vi) shows the comparison of temperature profile between unburned, burned and global. At 2200 rpm under full load, bulk temperature simulated was 1728 K and corresponding θ_{Tmax} was 382° CA [refer Fig. 8.3(iii)] while at 1400 rpm with

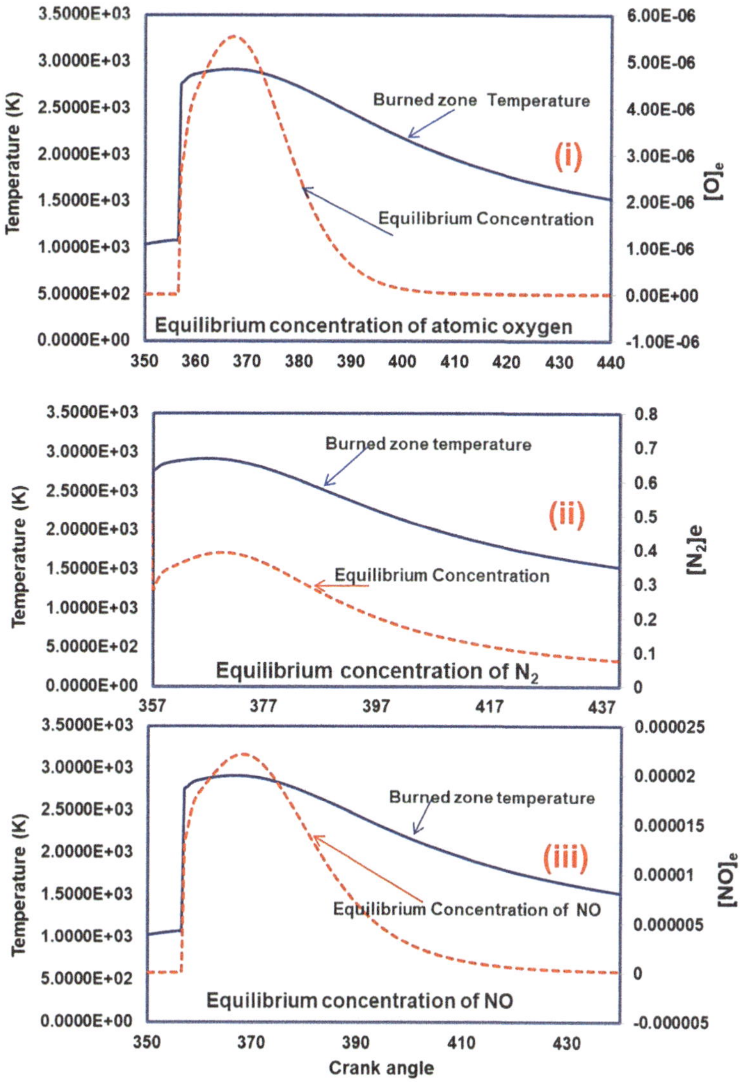

Fig. 8.4 Equilibrium concentration of O, N_2, and NO at 2200 rpm and 100% load

identical operating condition simulated bulk temperature was increased to 1964 K at 384.5° CA.

8.3.2 Equilibrium Concentration of Species

The equilibrium concentrations of O, N_2 and NO in the burned zone are functions of temperature (Rakopoulos et al. 2009) (refer Fig. 8.4) and it was observed that these concentrations were varied with crank position. The maximum equilibrium concentration noted to be 0.00000551 mol/cm^3 at 367° CA for Oxygen, 0.000022124 mol/cm^3 at 368° CA for NO and 0.3914 mol/cm^3 at 369.5° CA for N_2 at 2200 rpm under full load. It is observed from Fig. 8.4(iii) that [NO] concentration exponentially changes with crank angle due to variation in burn zone temperature. As combustion proceeds, burn zone temperature increased which results dissociation of N_2. The atomic nitrogen combines with O_2 produces NO. As combustion progresses, the [N_2] and [O_2] reduces while [NO] increased with CAD but later in expansion stroke [NO] decreases with combustion temperature as shown in Fig. 8.4.

8.3.3 Effect of Engine Load and rpm on Engine NOx

Exhaustive study has been carried to analyze the effect of load on engine on NO_x emission. A series of experiments were performed at different operating modes at 2200 and 1400 rpm. Figure 8.5 shows comparison between simulated and experimental NOx emission at different loads.

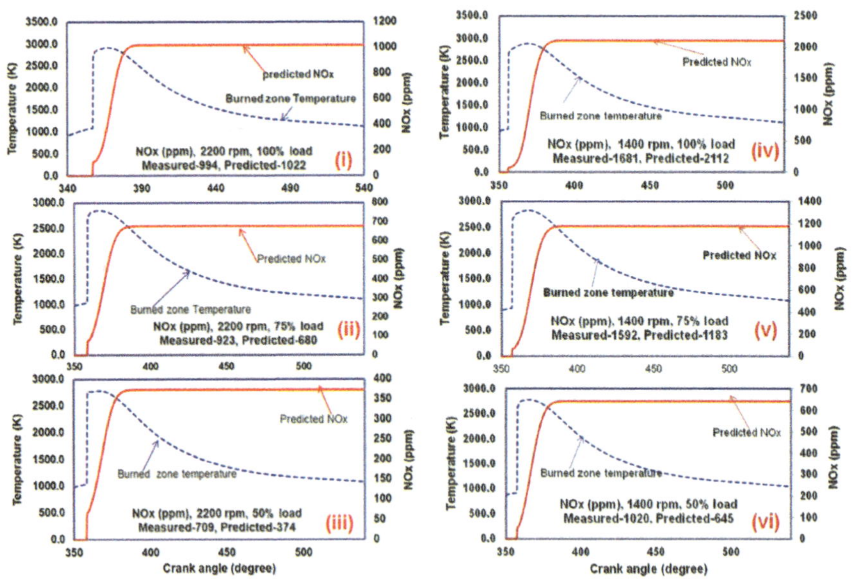

Fig. 8.5 Comparison of simulated versus measured NO_X at 2200 rpm and 100% load

Figure 8.5(i)–(iii) shows the simulated NO_x versus burned zone temperature at 2200 rpm but with decreasing load i.e., 100%, 75% and 50% respectively. At 100% operating mode the mass of fuel consumption was 50.91 mg at an equivalence ratio of 0.51. The simulated NO_x was 1022 ppm whereas measured was 994 ppm. The NO_x formation freezes at crank angle 387.5° CA. At 75% load, the simulated NO_x was reduced to 680 ppm whereas measured was 923 ppm [refer Fig. 8.5(ii)]. The NO_x formation freezes at crank angle 388° CA. At this operating mode the mass of fuel consumption was 39.24 mg at an equivalence ratio of 0.44. Similarly at 50% load, the mass of fuel consumption was 28.11 mg at an equivalence ratio of 0.35. The simulated NO_x was exponentially reduces to 374 ppm whereas measured was 709 ppm. The NO_x formation freezes at crank angle 388.5° CA. At higher load condition, higher combustion temperature results faster NO formation and higher [NO] results compared to 75% load and 50% load condition as shown in Fig. 8.5.

Figure 8.5(iv)–(vi) shows the simulated NO_x versus burned zone temperature at 1400 rpm but with decreasing load i.e., 100%, 75% and 50% respectively. At 100% load, the mass of fuel consumption was 56.85 mg at an equivalence ratio of 0.73. The simulated NO_x was 2112 ppm whereas measured was 1681 ppm [refer Fig. 8.5(iv)]. The NO_x formation freezes at 390.5° CA, fuel mass requirement increased by 5.94 mg and equivalence ratio increased to 0.73 instead of 0.51. At 75% load and 1400 rpm, the simulated NO_x was 1183 ppm whereas measured was 1592 ppm as shown in Fig. 8.5(v). The NO_x formation freezes at crank angle 390° CA. At this operating mode the mass of fuel consumption was 42.44 mg at an equivalence ratio of 0.58. But at 50% load and 1400 rpm, the simulated NO_x was reduced to 645 ppm whereas measured was 1020 ppm as shown in Fig. 8.5(vi). The NO_x formation freezes at crank angle 388.5° CA. At this operating mode the mass of fuel consumption was 28.87 mg at an equivalence ratio of 0.42. At lower 1400 rpm engine produces a higher concentration of NO for a given power due to the longer time available for the reactions.

It is concluded that generation of NO_x is highly depending on burned-zone temperature. Maximum NO_x concentrations was formed between the SOC and shortly after maximum cylinder pressure at which burned gas temperature reached maximum. Mixture which burn early in the premixed combustion phase is especially important since it is compressed to a higher temperature, increasing the thermal NO formation route. After θ_{Pmax}, burned gas temperature reduced as the cylinder gas expands. The temperature reduces due to expansion and mixing of high temperature gas with air or cooled burned gas freezes the NO chemistry. It was also observed that the NO_x formation rate was very high above 2500 °C.

Figure 8.6 shows different operating points where the engine-out NO_x emission. This variation indicated that only Zeldovich mechanism is not enough to predict the NO_x even though other mechanism are also play a vital role at low temperatures.

However, curve-fit of simulated NO_x results a simple mathematical equation as given below (refer Fig. 8.7). This equation incorporated in the NO_x prediction model to get précised accurate results as an empirical correction algorithm. The data point those are deviated from curve fit as shown in Fig. 8.7 is because of NO concentration was predicted amongst oxides of nitrogen. At low load operating condition, the

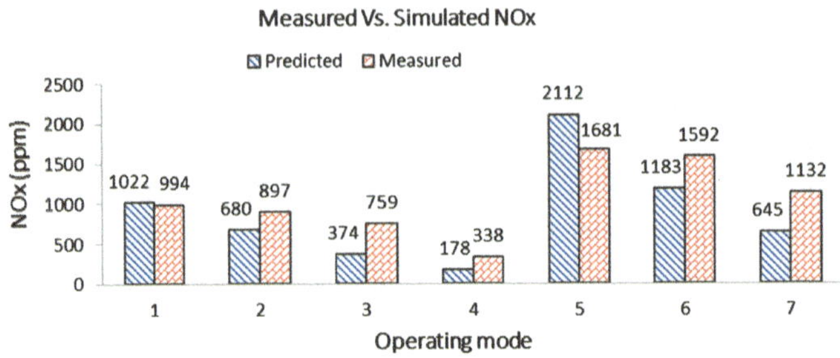

Fig. 8.6 Comparison between measured versus simulated NO_x

Fig. 8.7 Curve-fit of NO_x data

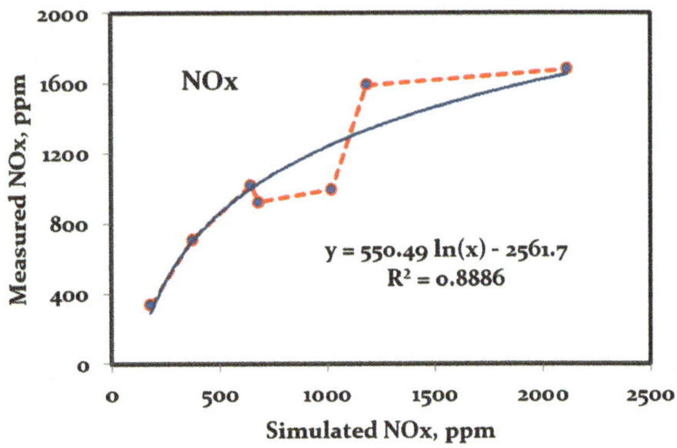

influence of backward reactions consuming NO is greater than that of higher load. For this, the third reaction plays an important contributor to the backward reaction.

8.4 Summary

The NO_x predicting numerical model is developed with for 4-stroke turbocharged DI diesel engine. In this work, zero-dimensional two zone combustion model was used to predict unburned zone, burned zone and adiabatic flame temperatures along with NO_x. The main findings using model is validated with experimentation and findings are summarized below:

- The simulated T_{max} at 2200 rpm under full load was 1218 K at 368° CA while at 75% load, and 50% load were 1145 K, 1135 K, and their θ_{Tmax} were 368° CA, and 365.5° CA respectively but at lower speed of 1400 rpm T_{max} reduced to 1147 K, 1084 K, and 1033 K and simulated θ_{Tmax} were 370° CA, 369° CA and 366.5° CA respectively at identical load conditions.
- Maximum burned zone temperature at 2200 rpm under 100% load was 2917 K at 366.5° CA while at 75% load and 50% load were 2853 K and 2776 K and

their θ_{Tmax} was 366° CA, and 366° CA respectively but if speed was lowered to 1400 rpm under 100% load, 75% load and 50% load then the simulated maximum burned zone temperature showed reverse trend that of unburned i.e., maximum temperature increased to 2896 K, 2833 K, 2781 K.

- Both $[O]_e$ and $[NO]_e$ exponentially varies with crank angle due to variation in burn zone temperature while variation in $[N_2]_e$ is directly proportional to burned zone temperature.
- At 50% load, simulated NO_x exponentially reduced to 374 ppm whereas experimental result showed 709 ppm.
- The simulated NO_x at 2200 rpm under full load showed 1022 ppm whereas experimentally measured was 994 ppm. Reducing engine load from full load to 75% load and 50% load, the simulated NO_x was reduced to 680 ppm whereas experimentally measured was 923 ppm and 374 ppm measured was 709 ppm.
- Further reduction to 50% load, the simulated NO_x was reduced to 374 ppm whereas experimentally measured was 709 ppm.
- If the engine speed was reduced to 1400 rpm and load were varied identically then the simulated NO_x was observed to be 2112, 1183, and 645 ppm while experimentally measured NO_x were 1681, 1592, and 1020 ppm.
- After curve fit the simulated NO_x data with experimental data, the deviations observed at various operating modes were reduced. The prediction accuracy is higher at full load operating points.

References

Agarwal AK, Shukla PC, Patel C, Gupta JG, Sharma N, Prasad RK, Agarwal RA (2016) Unregulated emissions and health risk potential from biodiesel (KB5, KB20) and methanol blend (M5) fuelled transportation diesel engines. Renew Energy 98:283–291

Andersson M, Johansson B, Hultqvist A, Noehre C (2006) A predictive real time NOx model for conventional and partially premixed diesel combustion. SAE Trans:863–872

Arrègle J, López JJ, Guardiola C, Monin C (2008) Sensitivity study of a NOx estimation model for on-board applications (no. 2008-01-0640). SAE technical paper

Arrègle J, López JJ, Guardiola C, Monin C (2010) On board NOx prediction in diesel engines: a physical approach. In: Automotive model predictive control. Springer, London, pp 25–36

Arsie I, Di Genova F, Pianese C, Sorrentino M, Rizzo G, Caraceni A, Cioffi P, Flauti G (2004) Development and identification of phenomenological models for combustion and emissions of common-rail multi-jet diesel engines (no. 2004-01-1877). SAE technical paper

Chmela FG, Orthaber GC (1999) Rate of heat release prediction for direct injection diesel engines based on purely mixing controlled combustion. SAE Trans:152–160

Cipolat D (2007) Analysis of energy release and NOx emissions of a CI engine fuelled on diesel and DME. Appl Therm Eng 27(11–12):2095–2103

Del Re L, Allgöwer F, Glielmo L, Guardiola C, Kolmanovsky I (eds) (2010) Automotive model predictive control: models, methods and applications, vol 402. Springer

Desantes JM, Galindo J, Guardiola C, Dolz V (2010) Air mass flow estimation in turbocharged diesel engines from in-cylinder pressure measurement. Exp Thermal Fluid Sci 34(1):37–47

Devarakonda M, Parker G, Johnson JH, Strots V (2009) Model-based control system design in a urea-SCR aftertreatment system based on NH 3 sensor feedback. Int J Automot Technol 10(6):653

Egnell R (1998) Combustion diagnostics by means of multizone heat release analysis and NO calculation. SAE Trans:691–710
Ericson C, Westerberg B, Andersson M, Egnell R (2006) Modelling diesel engine combustion and NOx formation for model based control and simulation of engine and exhaust aftertreatment systems. In: SAE international, pp 01–0687
Hasegawa M, Shimasaki Y, Yamaguchi S, Kobayashi M, Sakamoto H, Kitayama N, Kanda T (2006) Study on ignition timing control for diesel engines using in-cylinder pressure sensor (no. 2006-01-0180). SAE technical paper
Hernández JJ, Pérez-Collado J, Sanz-Argent J (2008) Role of the chemical kinetics on modeling NOx emissions in diesel engines. Energy Fuels 22(1):262–272
Heywood JB (1988) Internal combustion engine fundamentals. McGraw Hill Inc., New Delhi
Hountals DT, Savva N, Papagiannakis R (2010) Development of a new physically based semi-empirical NOx model using the measured cylinder pressure. In: Proceedings of the THIESEL conference on thermo-and fluid dynamic processes in diesel engines. CMT
Johansson B, Wilhelmsson C, Tunestål P, Johansson R, Widd A (2009) A physical two-zone NOx model intended for embedded implementation. In: SAE world congress. SAE
Jung D, Assanis DN (2001) Multi-zone DI diesel spray combustion model for cycle simulation studies of engine performance and emissions. SAE Trans:1510–1532
Katare SR, Patterson JE, Laing PM (2007) Diesel aftertreatment modeling: a systems approach to NOx control. Ind Eng Chem Res 46(8):2445–2454
Leonhardt S, Muller N, Isermann R (1999) Methods for engine supervision and control based on cylinder pressure information. IEEE/ASME Trans Mechatron 4(3):235–245
Luján JM, Bermúdez V, Guardiola C, Abbad A (2010) A methodology for combustion detection in diesel engines through in-cylinder pressure derivative signal. Mech Syst Signal Process 24(7):2261–2275
Mellor AM, Mello JP, Duffy KP, Easley WL, Faulkner JC (1998) Skeletal mechanism for NOx chemistry in diesel engines. SAE Trans:786–801
Moos R (2005) A brief overview on automotive exhaust gas sensors based on electroceramics. Int J Appl Ceram Technol 2(5):401–413
Payri F, Broatch A, Tormos B, Marant V (2005) New methodology for in-cylinder pressure analysis in direct injection diesel engines—application to combustion noise. Meas Sci Technol 16(2):540
Prasad RK, Agarwal AK (2021) Experimental evaluation of laser ignited hydrogen enriched compressed natural gas fueled supercharged engine. Fuel 289:119788
Prasad RK, Mustafi N, Agarwal AK (2020) Effect of spark timing on laser ignition and spark ignition modes in a hydrogen enriched compressed natural gas fuelled engine. Fuel 276:118071
Pundir BP (2007) I.C. engine (cumbustion and emissions). Narosa Publishing House Pvt. Ltd., New Delhi
Ragland KW, Bryden KM (2011) Comustion engineering. CRC Press, Taylor & Francis Group
Rakopoulos CD, Dimaratos AM, Giakoumis EG, Rakopoulos DC (2009) Exhaust emissions estimation during transient turbocharged diesel engine operation using a two-zone combustion model. Int J Veh Des 49(1–3):125–149
Rakopoulos CD, Giakoumis EG (2009) Diesel engine transient operation. Springer
Shimasaki Y, Kobayashi M, Sakamoto H, Ueno M, Hasegawa M, Yamaguchi S, Suzuki T (2004) Study on engine management system using in-cylinder pressure sensor integrated with spark plug (no. 2004-01-0519). SAE technical paper
Stone R (1999) Introduction to internal combustion engines. Macmillan, London
Timoney DJ, Desantes JM, Hernández L, Lyons CM (2005) The development of a semi-empirical model for rapid NOx concentration evaluation using measured in-cylinder pressure in diesel engines. Proc Inst Mech Eng Part D J Autom Eng 219(5):621–631

Part IV
SI Engine Modeling

Chapter 9
Fundamentals, Evolution, and Modeling of Ignition Systems for Spark Ignition Engines

Dhananjay Kumar and Avinash Kumar Agarwal

Abstract The advancement of technologies has led researchers to explore new ways to comply with stringent emission norms globally and fulfil the energy requirements. The trends in engine development favour computational studies for initial investigations due to lesser time demand and economy. In a spark ignition (SI) engine, ignition of the fuel–air mixture is achieved by the spark discharge across the spark plug electrodes. The discharge is of very high intensity for a very short interval, providing sufficient energy in the form of plasma kernel to initiate chemical reactions necessary to generate a self-sustaining flame. Direct injection SI combustion system is considered an upcoming next-generation technology capable of meeting stringent emission norms with improved engine performance. Conventional spark plug system undergoes various issues such as erosion of spark plug, heat losses at the electrodes, hindrance in working at high in-cylinder pressures, and fixed spark location. Therefore, the researchers explore alternate ignition concepts/ systems that provide greater flexibility than conventional ignition systems. These alternate ignition concepts/ systems include laser ignition, turbulent jet ignition, corona ignition, and microwave ignition. These all are also referred to as advanced ignition systems. Advanced ignition has emerged as an alternative way to ignite leaner fuel–air mixture owing to higher engine performance and lower emissions. These systems offer significant advantages; however, they are still under research, and many challenges need to be overcome before they are commercialized. In this chapter, the evolution of the spark ignition systems has been discussed. Modelling aspects of spark ignition engines using 1D and 3D simulation tools have been summarised. The working of these advanced ignition systems has been discussed in detail, and their challenges are also summarised.

Keywords Advance ignition system · Modelling · Spark ignition engine · Laser ignition · Turbulent jet ignition

---

D. Kumar · A. K. Agarwal (✉)
Engine Research Laboratory, Department of Mechanical Engineering, Indian Institute of Technology Kanpur, Kanpur 208016, India
e-mail: akag@iitk.ac.in

D. Kumar
e-mail: dhanu@iitk.ac.in

A. K. Agarwal et al. (eds.), *Engine Modeling and Simulation*, Energy, Environment, and Sustainability, https://doi.org/10.1007/978-981-16-8618-4_9

Abbreviations

1D	One Dimensional
2D	Two Dimensional
3D	Three Dimensional
bTDC	Before Top Dead Centre
BSFC	Brake Specific Fuel Consumption
CAD	Crank Angle Degree
CFD	Computational Fluid Dynamics
CI	Compression Ignition
CNG	Compressed Natural Gas
CO	Carbon Monoxide
CPOA	Cylinder Pressure Only Analysis
CW	Continuous Wave
EGR	Exhaust Gas Recirculation
ECU	Electronic Control Unit
IC	Internal Combustion
SI	Spark Ignition
GDI	Gasoline Direct Injection
H_2	Hydrogen
HCNG	Hydrogen Enriched Compressed Natural Gas
HC	Hydrocarbons
HCCI	Homogeneous Charge Compression Ignition
LI	Laser Ignition
MFB	Mass Fraction Burned
NOx	Oxides of Nitrogen
TPA	Three Pressure Analysis
TKE	Turbulent Kinetic Energy
TJI	Turbulent Jet Ignition
R-L-C	Resistance-Inductance-Capacitance

9.1 Introduction

Future generation spark-ignition (SI) engines would have higher power output with lesser emissions to meet the emission norms adopted by various legislative bodies. The compression ratio of the SI engine is significantly lower than the compression ignition (CI) engine, resulting in lower temperature and pressure conditions in the combustion chamber at the end of the compression stroke. The self-ignition temperature of high-octane/ low-cetane fuel is high, which requires an ignition source to ignite the fuel–air mixture inside the combustion chamber. Close to the end of the compression stroke, electrical discharge produced between the two spark plug electrodes is responsible for initiating the SI engine's combustion process. Spark plug creates a high-temperature plasma kernel in the thin reaction sheet (between the central and ground electrodes). This plasma kernel develops into a self-sustaining flame front. The spark needs to be produced repeatedly over a range of speeds and loads at an appropriate crank angle of the engine cycle to ensure smooth engine running. The ignition process is a small-scale local phenomenon in a small zone inside the engine combustion chamber.

9.1.1 Importance and Challenges of Ignition System

In the SI engine, ignition of the fuel–air mixture is achieved by spark discharge across the spark plug electrodes. The discharge is of very high intensity for a very short interval, providing sufficient energy in the form of heat to initiate chemical reactions necessary to generate a self-sustaining flame. The heat released from the chemical reactions must be greater than the heat losses across the cylinder walls, spark plug electrodes, and unburnt gases. There is a need for higher power output with the existing engines with the same smoothness level at idle and part loads. With an ideal thermodynamic cycle for SI engines, three primary ways of improving engine efficiency are: (i) increasing the engine's compression ratio or (ii) burning the leaner mixture by adapting EGR or (iii) employing a superior ignition system. Both stratified and homogenous approaches are used to adapt to the burning of the leaner mixture. Moreover, they differ by the needs of engine demand for the burning processes. For the stratified mode of operation, spray formation provides favourable conditions at the location of the ignition.

Furthermore, spatial and temporal variations in charge distribution exist near the spark plug's tip (Scarcelli et al. 2016). Therefore, an appropriate ignition system preferably covers considerable space and multiple bursts to average the fluctuations due to each cycle's spatial and temporal charge variations. For the homogeneous mode operation, fuel injection starts early, and mixture gradients are quite small. Therefore lesser spatial variations of charge are present, resulting in a longer delay period. To have smoother and faster combustion, higher levels of turbulence and charge motion are generally desirable.

Moreover, higher turbulence also increases the heat loss near the cylinder wall. For the ignition process, the ability to adapt to turbulence is essential for efficient combustion. In tumble dominated SI engines, the tumble breaks down to reduce the TKE at the TDC. Resolving this breakdown (through finer meshes and correct turbulence models) is important for accurate spark modelling. In other words, the ignition system should cope with both leaner and richer mixtures with different turbulence levels. Lean combustion offers a considerable reduction in emissions and fuel consumption at part load operation. However, igniting the extended leaner mixture imposes several challenges to the conventional ignition systems. Another key challenge of critical consideration is the capability to adapt to the transient behaviour of the engine at different loads. The particle formation during combustion and probability of ignition components fouling is of significant importance under the

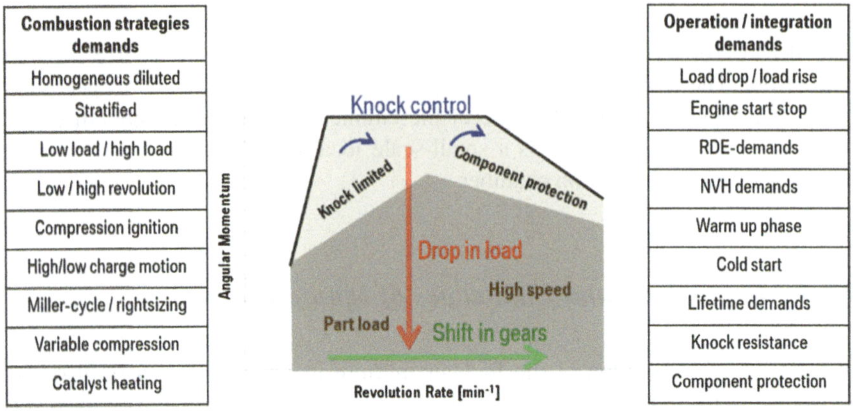

Fig. 9.1 Challenges and demands associated with modern SI combustion systems (Schenk et al. 2017)

dynamic engine operating conditions of the engine. Figure 9.1 summarises major challenges associated with the ignition systems at different engine loads (Schenk et al. 2017).

9.1.2 *Evolution of Ignition System*

An ignition system involves electrical/ electronic components to ignite the fuel–air mixtures in the engine cylinder. In 1860, Etienne Lenoir demonstrated the first internal combustion (IC) engine (gas engine) integrated with an electrical spark plug (History of the Ignition System 2021). Tesla (1898) patented the first spark plug. Bosch's (1902) invented the spark plug capable of producing high voltage and commercialized it as a part of the magneto ignition system for IC engine applications. After that, the basic magneto system was improved and integrated with a replaceable battery. Then the battery ignition systems came into the picture ~ 1910. The battery-operated coil, capacitors, contactor points, and distributer were integrated to achieve the breakdown voltage necessary to achieve electrical plasma ignition.

Nowadays, modern ignition systems are used in vehicles that employ electronic ignition instead of mechanical devices like contactor points. These systems are becoming more advanced with time and providing flexibility in controlling the ignition timing. A modern ignition system can be coupled to a computer interface for real-time observation, and also it can be programmed for optimizing the engine power/ torque output. With tightening emission legislations and higher power requirements, the ignition of extended lean burning of fuel–air mixture poses increasing challenges to these modern ignition systems. Conventional spark plug ignition systems face several challenges such as erosion of spark plug, heat losses at the electrodes, hindrance in working at high cylinder pressures, fixed spark location (Soldera et al.

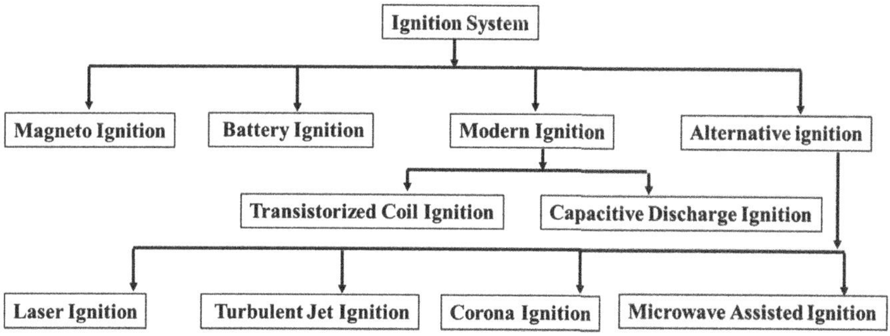

Fig. 9.2 Evolution of ignition system for IC engine applications

2004; Pischinger and Heywood 1990). Therefore, several researchers have explored alternate ignition systems that provide greater flexibility over conventional ignition systems. These alternate ignition systems include laser ignition (LI) (Vasile and Pavel 2021; Singh et al. 2020; Kumar and Agarwal 2020; Prasad and Agarwal 2021; Wermer et al. 2021; Prasad and Agarwal 2021), turbulent jet ignition (Attard et al. 2010; Gholamisheeri et al. 2017; Taskiran 2020; Distaso et al. 2020; Biswas and Ekoto 2019), corona ignition (Cimarello et al. 2017; Cruccolini et al. 2020; Vinogradov et al. 2008), and microwave ignition (Hwang et al. 2017, 2021; Chen et al. 2020). These are also referred to as advanced ignition systems. These systems offer significant advantages; however, these are still under development, and several technical challenges need to be addressed before being commercialized in practical vehicles. These advanced ignition systems have been discussed in detail in the following sections. Figure 9.2 categorizes different advanced ignition systems for IC engine applications and their evolution pathways.

9.2 Ignition Process

Four phases of spark generation were suggested by Maly (1984). These phases consisted of pre-breakdown, breakdown, arc, and glow discharge (Maly 1984). Pre-breakdown refers to the initial phase when the electric field between the electrodes increases continuously for few microseconds. The breakdown phase (~nanosecond range) occurs in the electrode gap when the electric field reaches the conductivity threshold of the air–fuel mixture present in the nearby volume. The required maximum voltage developed in the breakdown phase is referred to as the 'breakdown voltage.' Various parameters contribute to the breakdown voltage, including the fuel–air mixture strength, fuel type, engine load, electrode temperature, electrode orientation, etc. For instance, gaseous fuels such as H_2, CNG, HCNG, etc., require higher breakdown voltage than liquid fuels such as gasoline, methanol, etc. Also, for the leaner mixtures, the required breakdown voltage is much higher than the richer mixture since the leaner mixtures possess more air, which reduces the electrical conductivity in the electrode gaps.

The temperature in this phase reaches in the range of (50–60 × 10^3) K (Fernandes et al. 2016). This phase of spark generation is particularly important for combustion. After the breakdown for a particular cycle, a sharp drop in temperature (5000–6000 K) happens with the arc discharge phase lasting for a few microseconds. The arc discharge phase is responsible for damaging the electrode surfaces by erosion. Therefore it is necessary to understand the arc discharge phase. The transition from arc discharge to glow discharge is considered the fourth phase of spark generation. The glow discharge phase generally lasts for several milliseconds.

9.3 Modelling of Spark Ignition Systems for IC Engines

The trend in engine development favours computational study for initial investigations owing to lesser time and economy (Valera et al. 2020). Non-uniform fuel–air mixture distribution in the vicinity of spark electrodes can lead to unfavourable conditions for efficient combustion and cyclic fluctuations for port injection SI combustion and direct injection SI combustion (Drake et al. 2005). It is important to understand the underlying processes and mixture behaviour in the vicinity of the spark plug. Experimental studies require efforts to develop an experimental setup, which is time-consuming and expensive. On the other hand, computational studies offer flexibility in predicting the engine parameter variations with time economically and efficiently. Computational Fluid Dynamics (CFD) studies of engines use various assumptions for modelling the in-cylinder phenomenon. Computational studies have shown significant advantages in engine performance improvement and emission reduction in the past few decades. Computational studies use various techniques to solve complex combustion phenomena in the engine.

9.3.1 One Dimensional (1-D) Simulations

The prediction of a performance parameter of an engine uses a deterministic approach. The deterministic modelling approach can be of 0D, 1D, 2D, or 3D. Different approaches refer to the number of independent space variables and time variables. A model can be transient or steady. 0D modelling uses ordinary differential equations to model the variable parameters. It does not consider the spatial variations of the parameters and only solves for time variations. However, 1D, 2D, and 3D modelling approaches consider spatial and time variations together. 1D modelling is a simple approach for understanding an engine's performance, combustion, and emission characteristics. Many commercial codes, such as GT-Power, Ricardo Wave, AVL BOOST, etc., are available to perform a 1-D simulation. 1D modelling has a handicap in predicting emissions, especially HC and CO. The 1D model cannot predict spatial and temporal emissions variations, which can be done in 3-D simulations.

9.3.2 3-D CFD Simulations

The in-cylinder charge motion greatly influences the fuel–air mixing and combustion process in the engine. Studying turbulence flows characteristics and bulk flow patterns are essential since they significantly impact heat transfer. The intake process sets up the initial in-cylinder flow pattern and is subsequently modified during the compression and expansion processes. In the context of 3D, it felicitates understanding of the temporal and spatial variations of the in-cylinder phenomenon, which is not possible with 1D modelling (Kumar et al. 2021). It allows understanding the basic phenomenon occurring within the cylinder, such as fuel–air mixing, heat transfer, etc., and its effects on the global engine parameters. Many commercial codes, such as CONVERGE, KIVA, AVL-Fire, are available for simulating engine-like conditions.

9.3.3 1D Model Preparation Using GT Power

There are several commercial software capable of solving complex engine processes with fair accuracy. Commercially available software such as AVL BOOST, Ricardo Wave, GT Power are popular tools for 1D simulations. Figure 9.3 shows a simple setup for a 1D model of SI engine developed using GT Power. The flow model comprises Navier Stoke, continuity, and energy equations. In 1D modelling, this equation is solved in a single direction; variables in the other directions are averaged only in the flow direction. This method of solving the flow process reduces the simulation time. However, spatial variations are not possible with a 1D model, but this model is best suited for predicting overall performance with reduced simulation efforts. In Fig. 9.3, the intake system is shown using a blue box, and the red

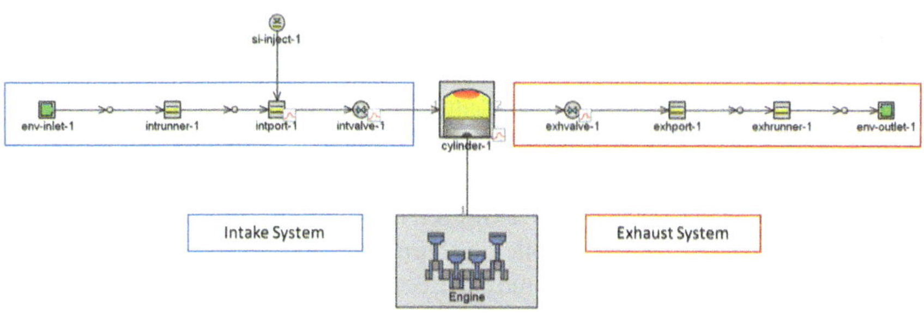

Fig. 9.3 A general outline of a 1D model for a single-cylinder spark-ignition engine

Table 9.1 Typical parameters required for 1D modelling of different components

Sub-system	Component name	Parameters required
Intake and exhaust system	Environment conditions	Temperature, pressure, air composition, humidity
	Runner pipe and inlet and outlet port dimensions	Diameter at inlet and outlet, Discretization length
	Intake and exhaust valve	Cam timing, valve lift with crank angle, the flow discharge coefficient
Injection system	Injector	Injector delivery rate, fuel ratio, fuel name, injection timing
Cylinder	Engine cylinder	Initial condition, wall temperature (head, piston, cylinder), heat transfer model, combustion timing, duration
Engine	Engine crank train	Engine type (4-stroke or 2-stroke), engine speed, cylinder geometry, firing order, inertia, the start of a cycle

box represents the exhaust system. The intake system comprises inlet environment conditions, runner pipe, port dimensions, and intake valve. While modelling, each component needs to be considered and, some critical parameters are required as input for a decent accuracy while predicting. Table 9.1 shows the required general parameters for 1D modelling of a SI engine.

9.3.3.1 Combustion Model Selection

There are three main combustion models: non-predictive combustion model, predictive combustion model, and semi-Predictive combustion model (Kumar 2021).

Non-Predictive Combustion Model

This type of combustion model requires an imposed burn rate as a function of crank angle degrees. Combustion would be controlled by the imposed burn rate, regardless of any other in-cylinder conditions. This model assumes that sufficient fuel is available in the combustion chamber to follow the imposed burn rate. Hence, the burn rate remains unaffected by other variables such as residual mass fraction, trapped mass, injection timing, etc. This model is helpful to study the effect of variable parameters, which do not affect the fuel burn rate. However, this model is not suitable for studying engine variables, which directly implies the burn rate. A predictive or semi-predictive combustion model is preferred to study these variables that directly correlate with the burn rate. For example, suppose someone wants to study the effect of injection profile and injection timing in a diesel engine. In that case, the predictive capability must have meaningful results since these parameters strongly affect the burn rate.

Semi-Predictive Combustion Model

The semi-predictive combustion model is weakly sensitive to the engine variables that substantially affect the burn rate. This model responds appropriately to these variables. However, this does not use any physical model to predict their implications. This model may be an appropriate choice for predictive modelling in some cases.

Predictive Combustion Model

This type of combustion model is preferable for different engine simulations. Predictive combustion models are an appropriate choice for all simulations. There are practical factors that make non-predictive combustion models preferable in certain situations. It requires calibration with measured data to provide accurate results. The model imposes a burn rate to predict the in-cylinder pressure. However, if the burn rate is not available, it can be obtained using experimental data generated under known test conditions.

Burn-rate prediction from the known in-cylinder pressure history is obtained using the reverse run method. In this method, inputs and outputs of the calculations are reversed from the typical engine combustion calculations. In the forward run, in-cylinder pressure comes out as an output, and the burn rate serves as an input, and the reverse run is vice-versa. Hence two-zone combustion approach has been followed to predict the burn rate. Two-zone generally considered are the burnt zone and the unburned zone. The reverse run uses the same set of equations as the forward run, including thermodynamics and combustion. In the reverse run method, the amount of fuel being transferred from the unburned zone to the burned zone is iterated for each time step until its in-cylinder pressure matches the measured in-cylinder pressure. Two approaches can be used to predict the apparent burn rate from the experimentally measured in-cylinder pressure trace. The first approach is based on closed volume analysis, also known as 'cylinder pressure only analysis' (CPOA), and the second approach is 'three pressure analysis' (TPA). Both methods follow the same basic functions to calculate the burn rate and produce similar results. However, these two approaches differ in the way they require necessary additional input data such as trapped air and fuel mass, heat transfer in addition to the measured in-cylinder pressure-crank angle history. GT-power uses different heat transfer models as per data availability, such as the WoshiniGT model is used to predict the heat transfer when swirl parameters are not available. There are several other heat transfer models, such as WoschiniHuber, WoshiniSwirl, where more accurate prediction is possible with the availability of required initial conditions. The details of these models have been explained in our previous publication (Valera et al. 2020). If spatial and temporal variations within the cylinder are required with high accuracy, using more complex 3D CFD modelling becomes essential. 3D computational modelling requires high computational resources.

9.3.4 3D Model Preparation Using CONVERGE CFD

In this section, 3D model preparation steps for the spark-ignition engine are discussed. For the model preparation, the computational boundary needs to be prepared. CONVERGE has two platforms, namely 'CONVERGE Studio' and 'Solver,' to set up the model and run simulations, respectively. CONVERGE studio is designed for case Preparation and setting up different boundaries. CAD models of different components such as cylinders, valves, ports, pistons need to be prepared and assembled. Initial geometry diagnosis may be required and taken up in CONVERGE studio, and boundary conditions need to be assigned, as mentioned in the 1D modelling section. Moving boundary conditions can also be specified by enabling transient boundary conditions.

Figure 9.4 shows an example model prepared for 3D simulations. Unlike 1D, 3D simulations solve the governing equations (mass, momentum, and energy) in all three directions. Thus, the spatial and temporal variations within the computational domain can be analyzed. It requires a complete reaction and transport mechanism for a particular fuel for solving involved transport phenomena. CONVERGE can generate run time mesh structure and assign the embedding region where more fine meshing is required with appropriate grid size. This is one of the essential features of this software since mesh generation was always challenging and time-consuming for complex geometry, especially near the wall region.

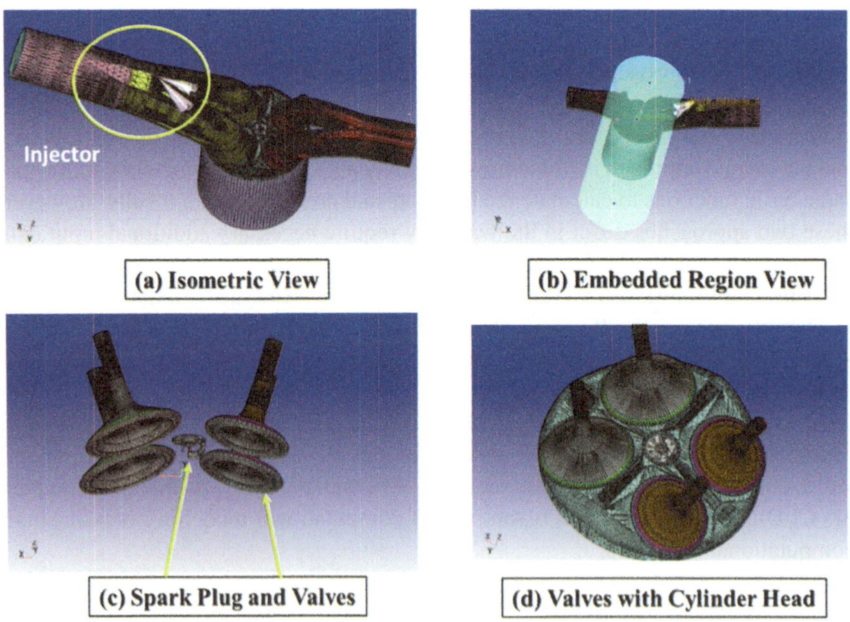

Fig. 9.4 Different views of port fuel injected (PFI) spark ignition engine

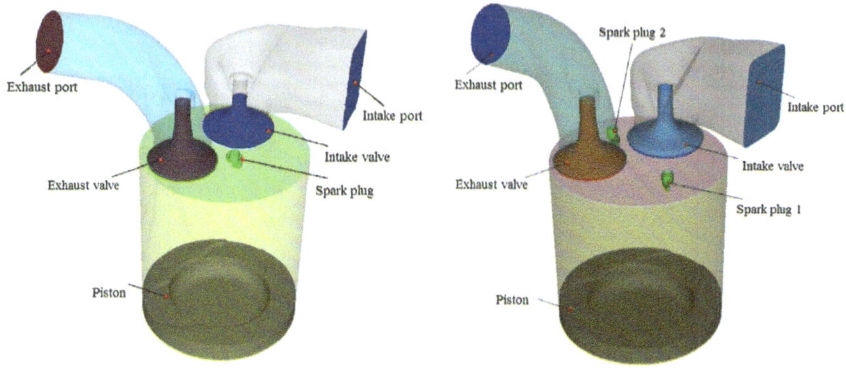

(a) Centrally located single-plug (original combustion chamber)
(b) Symmetrically twin-spark plugs

Fig. 9.5 Computational domain of the model, showing single and twin spark plug configurations (Duan et al. 2020)

Case Study: Effect of twin Spark Plug on Engine Performance and Emissions

Duan et al. analyzed the effect of the twin-spark plugs on the combustion and performance for the lean operation of a natural gas engine (Duan et al. 2020). They first validated the model with the experimental results using one centrally located spark plug for carrying out the investigations, as shown in Fig. 9.5a. Other boundaries such as exhaust and intake port, intake and exhaust valves, piston liner can also be observed in Fig. 9.5.

The location of the spark plasma has a strong impact on the initiation of combustion as it starts the initial flame kernel. After that, flame development occurs in the combustion chamber. Any deviation in the spark timing, spark location, and the number of sparks will severely affect the engine performance, combustion, and emission characteristics. The main advantage of multiple sparks is that they can extend the lean limit of the fuel–air mixture and contribute to lesser NOx emissions. In the present case, a physical model was developed using 3D CONVERGE CFD, which offers run time mesh generation and can solve complex chemical kinetics. The model validation using the experimental data was done for a single spark plug.

Further, the twin spark plug location was chosen to be symmetrically opposite, as shown in Fig. 9.5b. The different percentages of EGR (5, 10, 15%) and their effects were compared for single and twin spark plug configurations. Figure 9.6 shows a comparative plot of the in-cylinder pressure and heat release rates for different conditions. Figure 9.6a shows that the peak in-cylinder pressure for the twin spark plug was significantly higher than the single spark plug with no EGR case. By introducing EGR (5 and 10%), the peak in-cylinder pressure was somewhat lower than a twin spark plug with no EGR case. However, the cylinder pressure was still significantly higher than a single spark plug engine case. Moreover, with 15% EGR and twin spark plug, the in-cylinder pressure was lowest among all cases. This trend was also supported by the heat release rate curve, as shown in Fig. 9.6b. The heat release rate

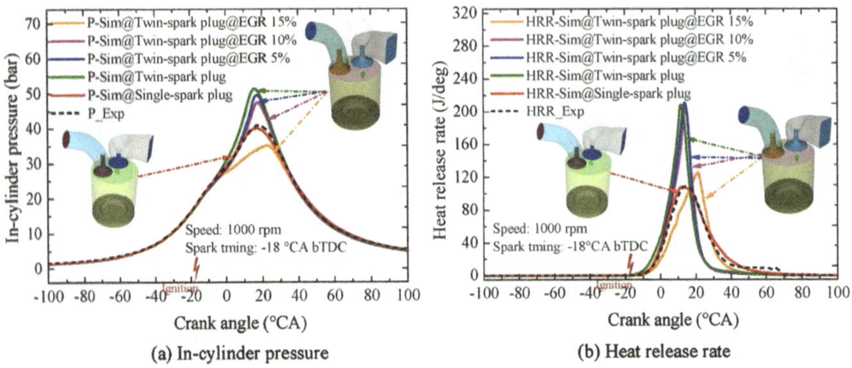

Fig. 9.6 In-cylinder pressure and heat release rate comparison of simulated and experimental results for different EGR % obtained for single and twin spark plug configurations (Duan et al. 2020)

of the twin spark plug (with and without EGR) was much higher than the single spark plug, except for a twin spark plug with a 15% EGR case. It can be concluded that an EGR of up to 10% was favourable with a twin spark plug configuration. The result obtained is from numerical analysis needs experimental validation. The authors also investigated spatial parameters such as flame development within the combustion chamber and used the 'SAGE' combustion model to predict the combustion. The total flame kernel volume of the initial spark was observed to be larger than a single spark, which was obvious. Flame propagation speed was faster for the twin spark plug.

9.4 Advance Ignition System for SI Engines

Conventional ignition systems have several limitations and shortcomings. The required voltage and energy are relatively higher for initiating dilute stoichiometric combustion with a conventional spark plug. Furthermore, higher voltage imposes reliability problems such as spark plug erosion, degradation of electrodes, and recurrent replacement of spark plugs (Bisetto et al. 2006). Some advanced ignition systems in the research phase can downsize the engines and ignite the leaner mixture. Several new spark plug systems include laser-induced ignition, corona ignition, and turbulent jet ignition systems. In the following sub-section, some of these alternative ignition systems have been discussed.

9.4.1 Laser-Induced Ignition

Laser-induced ignition is an innovative concept to ignite the fuel–air mixtures in an internal combustion engine. It uses a couple of optical elements to focus the laser beam at any location within the combustion chamber. It has the potential to overcome various problems common to conventional ignition systems. LI can be used for different engines such as IC engines, rocket engines (ramjets, scramjet), and gas turbines (Kumar and Agarwal 2020; Lorenz et al. 2016; Azarmanesh and Targhi 2021; Morsy 2012). A LI system for IC engine applications needs to overcome critical challenges faced by SI engines such as inappropriate combustion, lesser lean limit, higher NOx emissions, and lower efficiency.

9.4.1.1 Fundamentals of Laser Ignition

An intense laser beam is focused in the combustion chamber, where a subsequent breakdown of gaseous molecules initiates combustion. Figure 9.7 shows the basic principle of operation. An intense laser beam of a short pulse is focused into the combustion chamber with the help of suitable optical components, and a bright plasma is generated at the focal spot of the laser focusing lens (Azarmanesh and Targhi 2021).

There are four well-known modes by which laser beam can interact with the fuel–air mixture to initiate the combustion. These modes are (i) photochemical ignition, (ii) thermal ignition, (iii) resonant ignition, and (iv) non-resonant ignition (Kumar and Agarwal 2020; Morsy 2012).

In the photochemical mode of interaction, a laser source of resonant wavelength with absorbing medium is used. Target molecules get dissociated into atoms, and the chain branching mechanism generates highly reactive radical species. Also, there is no electrical breakdown. The local temperature rises due to the collision of produced radicals with nearby molecules, leading to ignition. However, due to collision, produced radicals may recombine with themselves or nearby molecules. Studies (Chou and Zukowski 1991; Lavid and Stevens 1985) suggest that full-scale

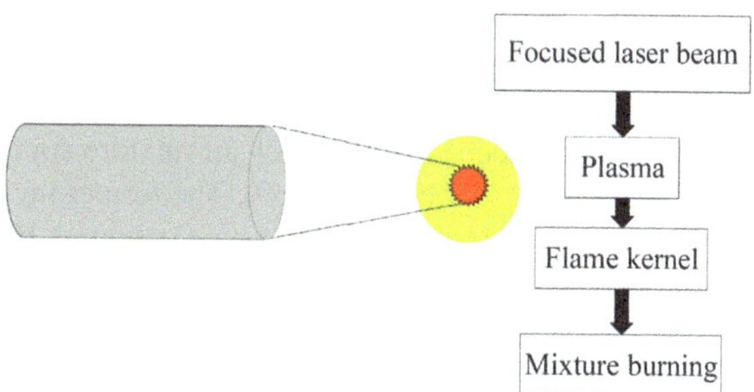

Fig. 9.7 Different stages of laser ignition of fuel–air mixture (Azarmanesh and Targhi 2021)

combustion starts when the production rate of radicals is higher than their recombination rate in photochemical ignition. Reported studies also concluded that radical concentration to initiate the combustion process should be in the range of 10^{17}–10^{18} atoms/cm^3. This is not favourable for IC engine applications due to disadvantages such as close match requirement of laser wavelength and absorbing molecules in the medium. However, fuel concentration varies from rich to lean in the engine, depending on the engine load and test conditions. In the thermal mode of interactions, ignition occurs without the electrical breakdown. The laser source is used to excite the gaseous molecule in translational, vibrational, or rotational modes. This leads to the generation of radicals of gaseous molecules, and indirect heating leads to the start of chemical reactions, consequently starting the ignition. A continuous-wave (CW) laser can achieve this ignition mode. However, this mode of LI is generally not preferred in IC engine applications since it takes longer to ignite the fuel mixture, requires a solid target to start the chemical reactions, and is difficult to ignite the leaner mixtures (Tanoff et al. 1995).

In the resonant mode of laser-induced ignition, the ignition process starts with non-resonant photodissociation of the molecules of the medium, followed by resonant photoionization of atoms created during the dissociation processes (Forch and Miziolek 1991). Both these events occur in a sequence, leading to the formation of seed electrons in the vicinity of the incident laser beam. After that, the seed's electrons gain energy by absorbing photons or from generated energy because of collisions. This phenomenon is also referred to as an 'inverse bremsstrahlung process. When these seed's electrons gain enough energy, they collide with nearby molecules, generating more free electrons. This way, a chain reaction starts, which increases the electron density in the vicinity of the focal region of the beam. This further leads to electron avalanche and the electrical breakdown of gaseous molecules. This mode of laser-induced ignition is exceptionally efficient over the others. However, it involves rigorous complexity. This mechanism requires tuning the laser to produce a laser beam of wavelength equivalent to the molecular length of the combustible mixture. Otherwise, two different lasers, one for the photodissociation process and the other for the ionization process, are used. Therefore, the resonant mode is not suitable for practical engine applications. In the non-resonant mode of laser-induced ignition, the process is somewhat similar to the resonant mode of ignition. However, it involves non-resonant multiphoton ionization of molecules. This laser-induced ignition does not require a close match of molecules and laser wavelengths, making it feasible for IC engine applications. Two different processes, multiphoton ionization breakdown and electron cascade ionization dominate the non-resonant laser-induced ignition, depending on fuel–air mixture composition and the laser parameters (Phuoc 2006; DeMichelis 1969). The former involves simultaneous absorption of photons required to achieve ionization potential. Electrons generated gain energy via the bremsstrahlung process. However, the cascade ionization is the reverse of the bremsstrahlung process. Electrons lose their energy to nearby molecules through collisions. Here ignition starts when the rate of energy absorption from photons overcomes the rate of losses. Table 9.2 summarizes the significant characteristics of different laser-induced ignitions.

Table 9.2 Summary of characteristics associated with different types of laser-induced ignitions

Ignition types	Characteristics
Photo chemical ignition	Minimum radical concentration to initiate the combustion process should be in the range of 10^{17}–10^{18} atoms/cm^3 Require close match between laser wavelength and absorbing medium molecules Not favourable for IC engine applications
Thermal ignition	A laser source is used to excite the gaseous molecules A continuous-wave laser can achieve it It takes a long time to ignite the fuel–air mixtures Require solid target to start the chemical reactions Difficult to ignite lean mixtures Not Suitable for IC engine applications
Resonant Ignition	Extremely efficient ignition Require tunable laser to achieve ignition Complex and expensive Not favourable for IC engine applications
Non-resonant ignition	Different types of lasers can be used A tunable laser is not required Can ignite different types of fuels (gaseous or liquid) Any particular wavelength is not necessary to initiate the combustion Can ignite leaner fuel–air mixtures Favourable for practical engine applications

9.4.1.2 Laser-Induced Multi-point Ignition

One of the primary advantages offered by the LI system is to direct the laser beam to several spots in the combustion chamber that eventually allow the ignition of the fuel–air mixture at several locations simultaneously. This technique of achieving laser-induced ignition, called multi-point ignition, increases the engine's combustion efficiency and overall performance. Laser-induced multi-point ignition offers many advantages over single-point ignition. It improves the in-cylinder combustion characteristics by shortening the flame propagation distance, resulting in a shorter combustion duration (Lavid and Stevens 1985). Some researchers (Dale et al. 1997; Saito et al. 2017; Kuang et al. 2017; Grzeszik 2017) explored various aspects of the multi-point technique and highlighted its advantages over the conventional spark plug and single-point LI techniques. Evaluation of multi-point ignition is still in its infancy and needs further experimental investigations and development. Grzeszik et al. experimented with three-point ignition for gasoline engines and found an increase in operational lean limit, lower fuel consumption, and lower NOx emissions (Grzeszik 2017). Vasile and Pavel demonstrated that the multi-point ignition of the methane-air mixture using a fibre-coupled laser diode-pumped, passively Q-switched Nd: YAG/Cr^{4+}:YAG compact laser with four-output laser beams in a constant volume combustion chamber (CVCC) (Vasile and Pavel 2021). Four laser beam outputs were used with beam pulse energy of 3.8 mJ at 0.9 ns pulse duration. Sapphire windows were installed in the periphery of the CVCC for the incoming laser beam, and several inlets and outlets port were made for facilitating the gas exchange process. They used four fibres of 600 μm diameter with a numerical aperture, NA = 0.22, to pump the laser beam. They performed several primary investigations to ensure the four-beam output of the same energy. Beam quality factor (M^2) of 4.8 was reported for all four-beam output from the laser. The schematic of the laser and optical breakdown of the air is shown in Fig. 9.8.

Figure 9.9 shows the excess pressure variations with time for different λ under single and multi-point ignition in a CVCC. They concluded that the multi-point LI showed increased excess pressure than the baseline single-point ignition. It was reported that the multi-point technique widened the flammability limit of ignition up to λ = 0.91–1.77 for the rich and lean mixtures. For all values of λ, the excess pressure was somewhat higher for the multi-point ignition technique than for the single-point ignition.

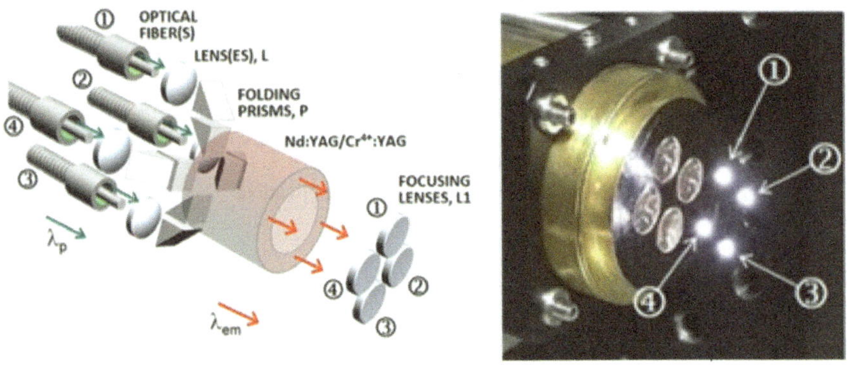

(a) (b)

Fig. 9.8 **a** Schematic of multi-point laser ignition with four focused beams, and **b** multi-point air breakdown by each beam (Vasile and Pavel 2021)

Fig. 9.9 Variations of the excess pressure with time at 1 bar initial CVCC pressure for different methane-air mixture proportions (Vasile and Pavel 2021)

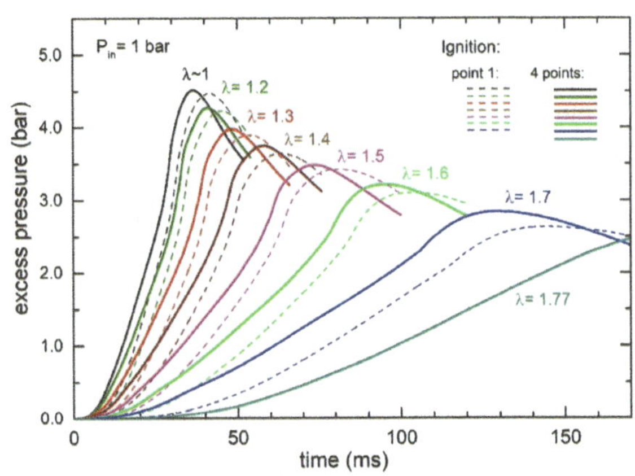

9.4.1.3 Laser-Induced Engine Characteristics

IC engines with conventional spark ignition can be retrofitted with a laser-induced ignition system. An engine's performance and combustion characteristics with LI are superior to conventional ignition. Azarmanesh and Targi compared the laser and conventional spark plugs by modelling the thermodynamic combustion of methane-air mixtures (Azarmanesh and Targhi 2021 Feb). They reported an 8% improvement in thermal efficiency for the lean methane-air mixture ($\lambda = 1.42$) and 10% higher maximum pressure for LI. Also, after extending the lean limit about ($\lambda = 1.66$), the unburnt mixture proportion was relatively higher within the cylinder, resulting in inferior efficiency when ignited using the conventional spark plug; however, after implementing a LI system, the combustion quality improved, and efficiency increased by ~ 21%. Several other studies reported an increased peak pressure, rate of pressure rise, and HRR with LI (Pal and Agarwal 2015; Srivastava and Agarwal 2014).

Birtas et al. (2019) developed a LI system mounted directly to the cylinder head, shown in Fig. 9.10. A similar approach was also reported (Pavel et al. 2017; Birtas et al. 2017) for analyzing the potential of the LI system. They used optical fibres (0.6 mm diameter, the numerical aperture NA = 0.22) to transfer the pump light to the active medium. This system delivered energy of ~ 4 mJ with 0.8 ns pulse duration. The engine used for the study was a 4-cylinder Renault engine. The laser system was triggered by ECU installed with the engine. By optimizing the spark timing for the air–fuel mixtures, they concluded that only minor engine performance improvements could be achieved using the LI system for the stoichiometric mixture. However, for lean mixture operation, LI improved the engine performance significantly. With a leaner mixture, combustion duration was lower, and combustion stability was better.

Lied et al. investigated the effect of plasma location using different converging lenses in the GDI engine cylinder and concluded that the optimum location of the laser-plasma was the center of the combustion chamber. At this location, the lowest BSFC was observed (Liedl et al. 2005).

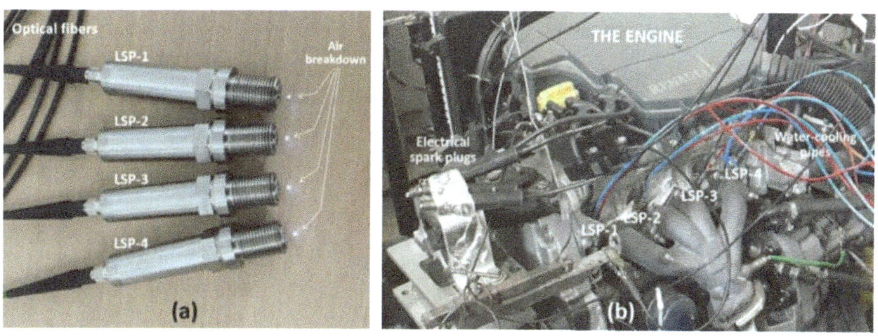

Fig. 9.10 **a** Laser ignition system with four laser spark plugs showing plasma formation in the air, **b** experimental setup installed with the laser ignition system (Birtas et al. 2019)

Prasad et al. (2021) experimented with the laser ignited hydrogen-enriched compressed natural gas (HCNG) fueled boosted single-cylinder engine. They concluded that brake-specific fuel consumption (BSFC) in LI mode was lower than conventional spark ignition (SI) mode. A recent study adopted a computational approach to predict the LI for methane-air mixture in IC engines (Azarmanesh and Targhi 2021). They modelled the combustion in four different stages, as shown in Fig. 9.11a. This multi-zone methodology (Mahabadipour et al. 2019) validated the in-cylinder pressure, as shown in Fig. 9.11b. For simulating LI conditions, it was assumed that the mixture would absorb the energy ~ 150 mJ.

Figure 9.12a shows that peak in-cylinder pressure for LI was higher than the conventional spark plug for stoichiometric conditions. Also, combustion duration was reduced (~3 ms) for LI. After diluting the fuel–air mixture, LI became more promising and showed a significant increase in the in-cylinder pressure at $\lambda = 1.66$,

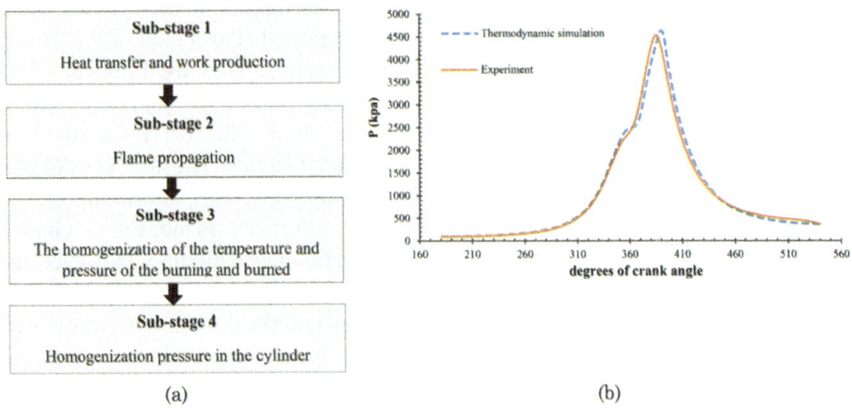

Fig. 9.11 **a** Flow chart of stages involved in combustion, **b** validation of experimental in-cylinder pressure (Azarmanesh and Targhi 2021)

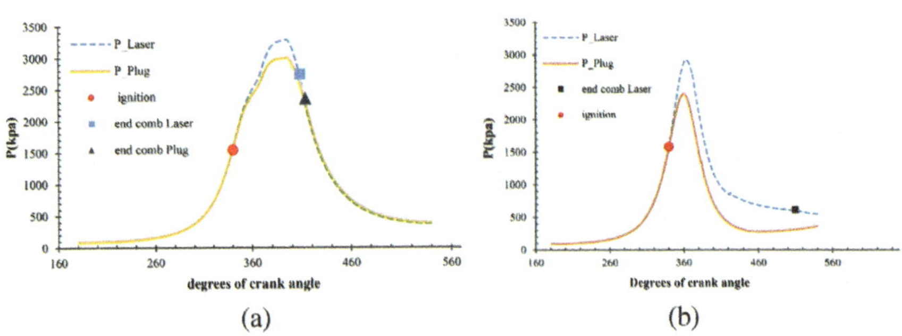

Fig. 9.12 Comparison of the in-cylinder pressure vs crank angle degree curves for the laser and spark plug ignitions at different stoichiometric conditions **a** $\lambda = 1$, and **b** $\lambda = 1.66$ (Azarmanesh and Targhi 2021)

as shown in Fig. 9.12b. They also reported that the rate of burning of the methane-air mixture was higher for the LI.

9.4.1.4 Advantages, Limitations and Future Prospects of Laser Ignition

From the literature discussed in the previous section, it can be concluded that the LI system has a significant advantage over the conventional ignition systems. It improves the combustion process by allowing the ignition of extended lean fuel–air mixtures. LI system provides the flexibility to choose an ignition location within the combustion chamber. Location change of ignition point can be changed by employing converging lenses of different focal lengths or by translating the converging lens in the plug along the direction of the laser beam (Grzeszik 2017). There is no intrusion, unlike conventional spark plug, and hence there is no quenching effect to the flame kernel in the combustion chamber. Several challenges, including the delivery of laser beam to the combustion chamber, powering the laser from the small-sized vehicle battery and making compact laser need to be addressed for realizing the LI system adaptation for practical engine applications. The open beam path approach is suitable only for laboratory-scale experiments and is unsuitable for practical engine applications. The open beam path approach uses several optical components and suffers from alignment issues and high-power transmissivity losses in the optics (lenses and mirrors). For multi-cylinder engine applications, closed path fibre delivers the laser beam, providing flexibility and resolving the alignment issues. The fibre delivery approach can also be used with a multiplexing system to ignite the multi-cylinder engine using a single laser source. However, the fibre used for transmitting the laser beams involves higher losses. A multiplexing system is a desirable alternative by which one laser may be used to ignite multiple cylinders. For successful implementation, engines should be optimized for laser beam multiplexing. This approach may overcome high-cost issues associated with a single laser per cylinder for multi-cylinder engine applications. Some other technical aspects must be considered and resolved to realize the LI system for production-grade engines. Optical windows for passing the laser beams should be more robust to sustain harsh conditions at different engine loads with minimum transmission losses. The size of the laser spark plug should be optimized to fit in the existing engine, and mass production of laser ignition systems should be undertaken to make them economical. A close-loop ignition control coupled with ECU may help trigger the laser at the right time for different cylinders. Multi-location plasma would help reduce the combustion duration with faster flame propagation to improve gaseous and liquid fuel combustion. Underlying physics at optimizing the plasma locations is still in the research phase. A computational study may help to understand the molecular characteristics at the ignition location. LI for the GDI engine application seems promising for different engine operation modes, particularly in the stratified mode.

9.4.2 *Corona Ignition*

This ignition system utilizes a series of circuits with a coil and capacitor. Series of circuits produce a high voltage at the tip of the igniter to generate corona. When a very high electrical field is enforced to a sharply curved and blunt surface, the voltage reaches a state that can break down the gas molecule in the vicinity. This local breakdown is in the form of a glow discharge, usually referred to as corona or electric wind, and hence the term corona ignition has been conceptualized. The magnitude of local voltage required here is lesser than the voltage generated in the conventional spark plug ignition for the same gap length. Corona ignition does not contain any ground electrode. Corona ignition covers a large combustion chamber volume and can ignite leaner mixtures that may not be ignitable by the conventional spark ignition.

9.4.2.1 Fundamentals of Corona Ignition

Corona ignition system uses series of resonant resistance-inductance-capacitance (R-L-C) circuits. The basic layout of the circuit is shown in Fig. 9.13a (Cruccolini et al. 2020). This electronic system is integrated with an igniter assembly, which is responsible for delivering high-frequency power to the igniter as per the feedback received from the ECU. The combination of resistance and inductor is connected in series, and this system amplifies the low to high voltage. Capacitors are installed at the firing end of the igniter plug installed in the cylinder head, as shown in Fig. 9.13b.

When a high voltage is applied to the igniter, the free electrons are accelerated because of the developed electromotive force and collide with the available gas molecules in the vicinity of the igniter plug. The ionization of gas molecules starts when the kinetic energy of electrons reaches the threshold limit of the ionization energy of the fuel–air mixture (gas molecules). This results in electron avalanche,

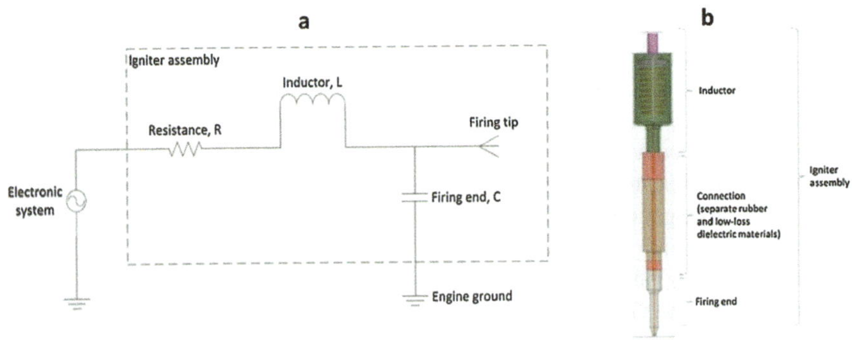

Fig. 9.13 **a** Electric circuit for corona ignition, and **b** schematic of corona igniter assembly (Cruccolini et al. 2020)

Fig. 9.14 **a** Corona igniter, **b** corona igniter installed in the combustion chamber (Discepoli et al. 2018)

and branch chain reaction leads to the breakdown of gas molecules. Discepoli et al. analyzed the energy required for the primary circuit of single and multi-point spark ignition and then compared it with the corona ignition (Fig. 9.14) (Discepoli et al. 2018). They reported that with increasing chamber pressure, the required energy for a primary circuit remains approximately constant, while for multiple spark cases, it decreases slightly. The required energy for corona ignition is independent of the chamber pressure.

9.4.2.2 Corona Ignition Engine Characteristics

Cimarello et al. investigated the effect of corona ignition using an optical engine for gasoline-air mixture and compared it with the conventional spark-ignition engine (Cimarello et al. 2017). Their study focused on flame development, stability, and lean limit operations. ECU was used to trigger the circuit at the right timing. They concluded that corona ignition showed four times faster flame propagation compared to conventional spark ignition. They claimed ~ 88% NOx reduction by adopting the corona ignition system. NOx reduction was observed primarily because corona could ignite leaner mixture at multi-spots simultaneously. At low load and speeds, corona ignition allowed the extension of lean operation by 0.25 λ. Cruccolini et al. also experimented with the corona ignition in optical engines using gaseous fuels (H_2 and Methane) (Cruccolini et al. 2020). They reported that when corona ignition was employed for methane fuelled engine, there was a guaranteed extension of 0.15 λ for the lean working limit at different engine loads. By hydrogen enrichment of the mixture, a further extension of 0.25 λ was achieved in the lean limits.

A sharp transition between stable and unstable conditions was observed for conventional spark ignition for different HCNG mixtures. However, this transition

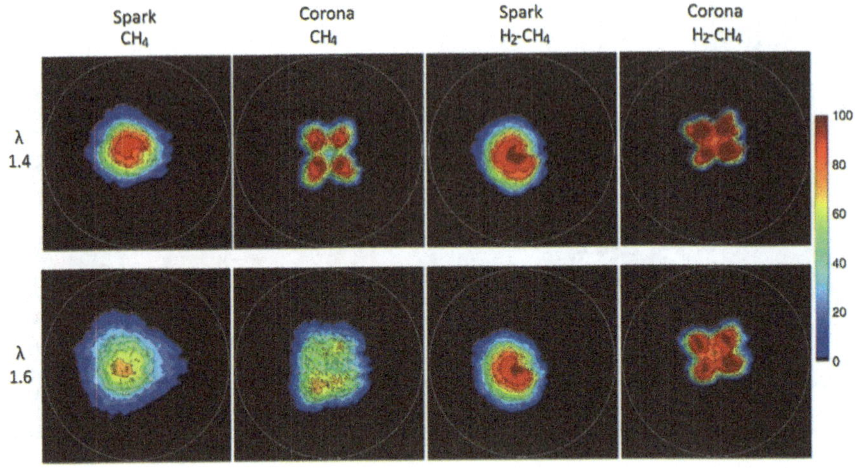

Fig. 9.15 Comparison of flame probability during 1% MFB for $\lambda = 1.4$ and $\lambda = 1.6$ for methane and hydrogen (Cruccolini et al. 2020)

was rather smoother in corona ignition. Hydrogen enrichment reduced cyclic variability with corona ignition; however, adding hydrogen increased the NOx emissions at stoichiometric conditions. Figure 9.15 shows the probability of flames at 9 mm of average equivalent radius (equivalent to 1% of MFB) for CNG and HCNG mixtures. Dark red contour represents 100% probability of the burned area while black represents the background. The figure shows that the corona ignition exhibited more dark areas than the spark ignition for both fuels and lambda conditions.

Ignition phenomena were captured using a high-speed camera. The shape of flame propagation was somewhat circular for spark. However, for corona ignition, a four-leaf flame structure was observed. Also, the flame presence probability increased upon adding hydrogen than the methane-fueled engine in corona ignition. The four-leaf flame structure was absent for lean mixtures of $\lambda = 1.4$–1.6 in methane fuelled engine. Marko et al. reported an increased EGR compatibility from 27% in spark ignition to 32% in corona ignition (Marko et al. 2017). Lower cyclic variations, stable combustion, and improved efficiency were observed with the corona ignition. Improved combustion characteristics of corona ignition are attributed to the faster flame propagation and leaner mixture operation. Mariani et al. experimented with the corona ignition in a spark-ignition engine and compared its efficiency and emissions (Mariani et al. 2013). They reported reduced combustion duration and increased engine efficiency in the range of 2–5%. There was a slight reduction in HC and CO emissions reported under some operating conditions. However, the NOx emission was considerably higher in corona ignition because of higher in-cylinder temperatures due to faster combustion.

9.4.2.3 Advantages, Limitations, and Future Prospects of Corona Ignition

Corona ignition covers a larger combustion chamber volume, resulting in faster flame propagation and smoother combustion. Corona can ignite leaner mixtures, which may not be ignitable by the conventional spark plugs otherwise. Also, the corona ignition system requires no separate ground electrode, which minimizes the heat loss associated with electrodes. However, there are associated challenges in the formation of the corona arc. When a metal surface is placed near the streamers (corona discharge), an arc is formed. However, it is essential to control the duration of the arc for the ignition of fuel–air mixtures, and further arcing may lead to excessive wear of sharp tips. The required input energy is higher in the corona ignition compared to the conventional spark ignition. Ignition events near the tip may be analyzed using optical diagnostic tools since corona ignition can ignite relatively leaner gaseous mixtures in the direct injection SI engines.

9.4.3 Turbulent Jet Ignition

Turbulent Jet Ignition (TJI) was first adopted in the two-stroke Ricardo Dolphin engines in the early twentieth century (Toulson et al. 2010). It uses a separate chamber where ignition starts with the help of a spark plug. Therefore it was also referred to as a pre-chamber spark ignition system. This ignition system facilitated to ignite leaner fuel–air mixtures, which are not easy to ignite using conventional spark plugs. However, this approach of igniting was not used widely in commercial engines because of associated complexity and the increased cost of the additional chamber in the cylinder head. In the recent past, the interest in lean combustion has increased. Hence TJI is being revisited by researchers from the automotive community.

9.4.3.1 Turbulent Jet Ignition Fundamental

TJI system consists of two chambers; the first one is known as prechamber, where the rich fuel–air mixture is made available for ignition with the help of a spark plug, and the second one is known as the main chamber where a lean fuel–air mixture is inducted. Here combustion is initiated in the prechamber, and due to increased chamber pressure, the flames and partially unburnt charge are forced to the main chamber via nozzle holes. The burning charge comes from the prechamber in the form of hot gas jets via a nozzle responsible for igniting the lean fuel–air mixture available in the main chamber. TJI is further referred to as stratified or homogeneous, depending on the mixture strength of the prechamber. In the stratified TJI system, a secondary injector injects the fuel into the prechamber, creating a rich mixture. The leaner mixture is inducted into the main chamber via the intake port. However, in the homogenous TJI, no separate injector is required for the prechamber, and fuel is injected only to the main chamber (either directly or via the port). Here the fuel–air mixture of the main chamber enters the prechamber via nozzle holes, and then ignition starts using a spark plug. Generally, in both modes of operation, the volume of the prechamber is kept very small (~3% or less than the clearance volume), and it does not affect the overall fuel consumption (Toulson et al. 2010).

9.4.3.2 Turbulent Jet Ignition Engine Characteristics

For the engine application of TJI, a compact prechamber is made in the vicinity of the spark plug in the cylinder head. Distaso et al. performed a numerical study to analyze a methane fuelled engine (Distaso et al. 2020). 3D CFD study is essential to understand the mixture characteristics and in-cylinder combustion behaviour. CFD study provides insights into physical and chemical phenomena in the prechamber, which is challenging to observe experimentally. Initially, they validated the CFD model against the experimental data. The model was developed using commercial software CONVERGE, and the prepared computational domain is shown in Fig. 9.16.

Figure 9.17a–f represents the simulated results of TKE and the streamline pattern. For differentiating the stream traces, combustion event was categorised as filling and scavenging (150–134 CAD bTDC), mixing (134–22 CAD bTDC), flame propagation (22–7 CAD bTDC), ejection (14.4–4 CAD bTDC), re-burning (4–16 CAD bTDC), and expulsion and Extraction (16–172 CAD bTDC) (Distaso et al. 2020). Figure 9.17a shows the fuel injection in the prechamber, where the scavenging happens to expel some trapped residual gas. In the mixing phase, the rich mixture in the prechamber gets diluted by a leaner charge. Flow entering the prechamber increases the turbulence slightly; however, the TKE value remains low (Fig. 9.17b). In a conventional SI engine, high turbulence levels negatively affect the flame kernel growth. However, the TJI system overcomes this issue because the kernel is developed in the upper part of the prechamber, where turbulence intensity is lower than the main chamber (cylinder). At -22 CAD, the spark is triggered, and the flame

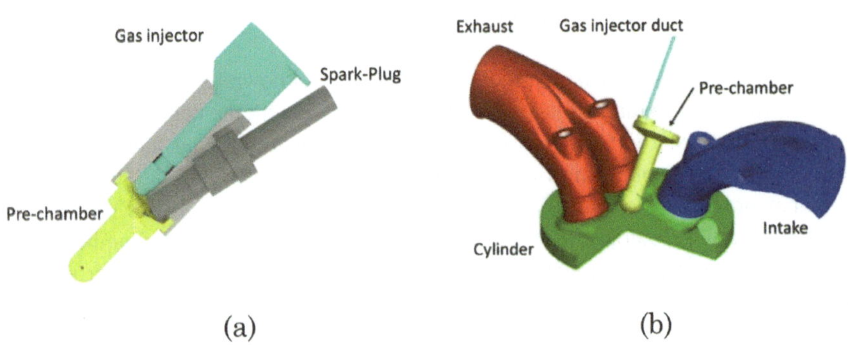

Fig. 9.16 **a** Proposed TJI configuration, **b** computational domain near TDC (Distaso et al. 2020)

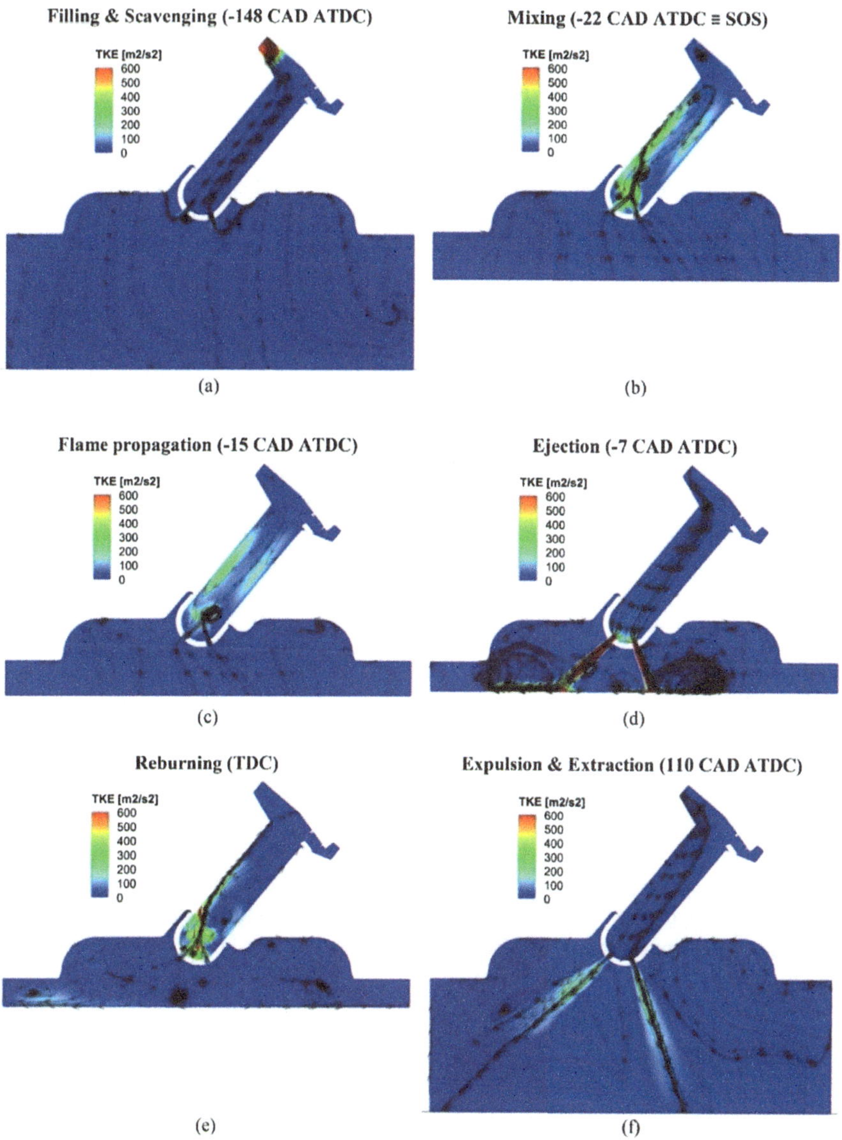

Fig. 9.17 Streamline traces and contour of TKE in a longitudinal plane for different events (methane, $\lambda = 1.3$) (Distaso et al. 2020)

propagation phase is initiated. In the ejection phase, multiple flame fronts are generated. Flow direction is inverted in the main chamber after the combustion events are initiated. This can be observed in Fig. 9.17e. The contour plot study showed that prechamber geometry and fuel injection strategy played the main role in mixing and combustion characterization. At the start of the spark, obtaining an optimized equivalence ratio distribution is important for improved performance and lower emissions. A higher in-cylinder pressure was reported with the TJI system compare to the conventional ignition system. Since a prechamber dominated the initiation of combustion, it promoted superior mixing and minimized the flame quenching in the main combustion chamber.

TJI system showed superior combustion stability, and its ignition energy requirement was lower for the lambda ranging from $\lambda = 1–1.8$. This may allow compact ignition coil and spark plug use, resulting in lower engine operating costs and compact packaging (Attard et al. 2010). The compression ratio of the engine can be increased without the possibility of knocking in the leaner mixture. It was predicted that an engine with a TJI system might achieve the indicated efficiency of up to 45%, which is more than the peak thermal efficiency achieved by the HCCI engines (Attard and Parsons 2010). Many researchers concluded that reduced NOx emissions would be possible for the TJI engine because of the lower temperature developed in the main chamber (Toulson et al. 2010; Alvarez et al. 2018 Jan). Furthermore, higher HC and CO emissions were reported with the TJI system. However, there is a scope to design an optimized prechamber that can promote even greater mixing and enhancing combustion parameters leading to lesser emissions and improved thermal efficiency.

9.4.3.3 Advantages, Limitations, and Future Prospects of Turbulent Jet Ignition

With increasing demands for an efficient engine with high power output and reduced emissions, the TJI engine has a significant role in a new generation engine with advanced technology. Engine with TJI system can offer several advantages over conventional ignition systems. The extension of lean limit operation is possible with the TJI system. Since it employs two combustion chambers (pre and main chamber), it offers greater control on the ignition events. TJI engine produces multiple and distributed sites in the main chamber, which provide a fast-burning rate in the main chamber with nominal cyclic variations. Also, the TJI system may work with a higher compression ratio since the tendency to knock is reduced by faster flame propagation. TJI provides flexibility to work in all ranges of engine loads without complex hardware systems. However, TJI involves challenges in handling fuel injection in the first chamber (prechamber). Therefore, research and development activities on injectors have to be taken up to inject the fuel effectively in the prechamber without any associated surface losses. An extensive computational study is also required to understand the underlying physics of mixing and flame growth from the prechamber to the main chamber. Higher improved combustion stability led to superior engine performance. However, HC emissions were also higher for the TJI system due to the additional surface area of the prechamber.

9.5 Concluding Remarks

Higher energy demand and stricter emission norms can be met by comprehensively analyzing the combustion events and parameters influencing them. The ignition system is responsible for igniting the fuel–air mixture in a SI engine. Ignition of the mixture has a significant effect on the flame evolution, growth, and propagation, consequently affecting the combustion. This chapter covers the importance of newer ignition systems and their evolution for IC engine applications. 1D and 3D modelling for SI engines have been included. The effect of the twin spark plug system promises a faster burn rate, leading to reduced combustion duration and improved combustion characteristics. Igniting leaner fuel–air mixture may not be possible with a conventional ignition system. Ignition of the lean mixture can be achieved by using advanced ignition systems such as laser-induced ignition, turbulent jet ignition (TJI), corona ignition. Laser-induced ignition is an electrodeless ignition system that gives the freedom to choose the ignition location anywhere in the combustion chamber, leading to improved combustion control. With laser-induced ignition and turbulent jet ignition, multi-point ignition is possible inside the engine combustion chamber. Therefore, considerable improvement in the flame propagation rate leading to lower combustion duration and improved combustion is possible. However, the practical application of advanced ignition systems faces several challenges that need to be resolved. In this chapter, the associated challenges and prospects are discussed.

References

Alvarez CE, Couto GE, Roso VR, Thiriet AB, Valle RM (2018) A review of prechamber ignition systems as lean combustion technology for SI engines. Appl Therm Eng 5(128):107–120. https://doi.org/10.1016/j.applthermaleng.2017.08.118

Attard W, Parsons P (2010) A normally aspirated spark initiated combustion system capable of high load, high efficiency and near zero NOx emissions in a modern vehicle powertrain. SAE Int J Engines 3(2):269–287. https://doi.org/10.4271/2010-01-2196

Attard W, Fraser N, Parsons P, Toulson E (2010) A turbulent jet ignition pre-chamber combustion system for large fuel economy improvements in a modern vehicle powertrain. SAE Int J Engines 3(2):20–37. https://doi.org/10.4271/2010-01-1457

Azarmanesh S, Targhi MZ (2021) Comparison of laser ignition and spark plug by thermodynamic simulation of multi-zone combustion for lean methane-air mixtures in the internal combustion engine. Energy 1(216):119309. https://doi.org/10.1016/j.energy.2020.119309

Birtas A, Boicea N, Croitoru G, Dinca M, Pavel N, Draghici F, Chiriac R (2019) On the possibility to improve petrol engine operation by laser ignition. Energy Proc 1(157):1022–1028. https://doi.org/10.1016/j.egypro.2018.11.269

Birtas A, Boicea N, Croitoru G, Dinca M, Dascalu T, Pavel N (2017) Combustion characteristics of a gasoline-air mixture laser ignition. In: Laser ignition conference 2017 Jun 20. Optical Society of America, pp LFA3-4. https://doi.org/10.1364/LIC.2017.LFA3.4

Bisetto F, Toniolo J, Menezes R (2006) Spark plugs for multi-fuel vehicles. SAE technical paper 2006-01-2630. https://doi.org/10.4271/2006-01-2630

Biswas S, Ekoto I (2019) Plasma ignition and turbulent jet ignition: advanced ignition system for automotive application. Sandia National Laboratory (SNL-CA), Livermore, CA (United States); 2019 Dec 1. https://www.osti.gov/servlets/purl/1643385

Chen Y, He LM, Fei L, Deng J, Lei JP, Yu H (2020) Experimental study of dielectric barrier discharge plasma-assisted combustion in an aero-engine combustor. Aerosp Sci Technol 1(99):105765. https://doi.org/10.1016/j.ast.2020.105765

Chou MS, Zukowski TJ (1991) Ignition of $H_2/O_2/NH_3$, H_2/air/NH_3, and $CH_4/O_2/NH_3$ mixtures by excimer-laser photolysis of NH_3. Combust Flame 87(2):191–202. https://doi.org/10.1016/0010-2180(91)90169-C

Cimarello A, Grimaldi C, Mariani F, Battistoni M et al (2017) Analysis of RF corona ignition in lean operating conditions using an optical access engine. SAE technical paper 2017-01-0673. https://doi.org/10.4271/2017-01-0673

Cruccolini V, Discepoli G, Cimarello A, Battistoni M, Mariani F, Grimaldi CN, Dal Re M (2020) Lean combustion analysis using a corona discharge igniter in an optical engine fueled with methane and a hydrogen-methane blend. Fuel 1(259):116290. https://doi.org/10.1016/j.fuel.2019.116290

Dale JD, Checkel MD, Smy PR (1997) Application of high-energy ignition systems to engines. Prog Energy Combust Sci 23(5–6):379–398. https://doi.org/10.1016/S0360-1285(97)00011-7

DeMichelis C (1969) Laser-induced gas breakdown: a bibliographical review. IEEE J Quantum Electron 5(4):188–202. https://doi.org/10.1109/JQE.1969.1075758

Discepoli G, Cruccolini V, Dal Re M, Zembi J, Battistoni M, Mariani F, Grimaldi CN (2018) Experimental assessment of spark and corona igniters energy release. Energy Proc 1(148):1262–1269. https://doi.org/10.1016/j.egypro.2018.08.001

Distaso E, Amirante R, Cassone E, De Palma P, Sementa P, Tamburrano P, Vaglieco BM (2020) Analysis of the combustion process in a lean-burning turbulent jet ignition engine fueled with methane. Energy Convers Manage 1(223):113257. https://doi.org/10.1016/j.enconman.2020.113257

Drake MC, Fansler TD, Lippert AM (2005) Stratified-charge combustion: modelling and imaging of a spray-guided direct-injection spark-ignition engine. Proc Combust Inst 30(2):2683–2691. https://doi.org/10.1016/j.proci.2004.07.028

Duan X, Zhang S, Liu Y, Li Y, Liu J, Lai MC, Deng B (2020) Numerical investigation of the effects of the twin-spark plugs coupled with EGR on the combustion process and emissions characteristics in a lean-burn natural gas SI engine. Energy 1(206):118181. https://doi.org/10.1016/j.energy.2020.118181

Kuang Z, Lyon E, Cheng H, Page V, Shenton T, Dearden G (2017) Multi-location laser ignition using a spatial light modulator towards improving automotive gasoline engine performance. Opt Lasers Eng 90:275–283. https://doi.org/10.1016/j.optlaseng.2016.10.024

Fernandes F, Lönarz M, Weyand P (2017) Development of an ignition coil integrated system to monitor the spark plugs wear. In: Günther M, Sens M (eds) Ignition systems for gasoline engines. CISGE 2016. Springer, Cham. https://doi.org/10.1007/978-3-319-45504-4_8

Forch BE, Miziolek AW (1991) Laser-based ignition of H_2O_2 and D_2O_2 premixed gases through resonant multiphoton excitation of H and D atoms near 243 nm. Combust Flame 85(1–2):254–262. https://doi.org/10.1016/0010-2180(91)90192-E

Gholamisheeri M, Wichman IS, Toulson E (2017) A study of the turbulent jet flow field in a methane fueled turbulent jet ignition (TJI) system. Combust Flame 1(183):194–206. https://doi.org/10.1016/j.combustflame.2017.05.008

Grzeszik R (2017) Impact of turbulent in-cylinder air motion and local mixture formation on inflammation in lean engine operation: is multiple point ignition a solution? In: Laser ignition conference 2017 Jun 20. Optical Society of America, pp LFA3-1. https://doi.org/10.1364/LIC.2017.LFA3.1

History of the Ignition System (2021). http://www.crankshift.com/ignition-system/#:~:text=The%20next%20major%20development%20in,%2C%20points%2C%20and%20a%20distributor. Access on 10 July 2021

Honold G, Bosch R (1902) Magneto-electric ignition apparatus for explosion-engines. United States patent US 802,291. 1905 Oct 17

Hwang J, Kim W, Bae C, Choe W, Cha J, Woo S (2017) Application of a novel microwave-assisted plasma ignition system in a direct injection gasoline engine. Appl Energy 1(205):562–576. https://doi.org/10.1016/j.apenergy.2017.07.129

Hwang J, Kim W, Bae C (2021) Influence of plasma-assisted ignition on flame propagation and performance in a spark-ignition engine. Appl Energy Combust Sci 1(6):100029. https://doi.org/10.1016/j.jaecs.2021.100029

Kumar D (2021) Simulation of high-pressure co-axial injection system for methanol adaptation in a locomotive engine, 2021. Thesis, Indian Institute of Technology Kanpur (India)

Kumar D, Agarwal AK (2020) Laser ignition technology for gaseous fuelled automotive engines. In: Singh A, Shukla P, Hwang J, Agarwal A (eds) Simulations and optical diagnostics for internal combustion engines. Energy, environment, and sustainability. Springer, Singapore. https://doi.org/10.1007/978-981-15-0335-1_10

Kumar D, Valera H, Agarwal A (2021) Numerical predictions of in-cylinder phenomenon in methanol fueled locomotive engine using high-pressure direct injection technique. SAE technical paper 2021-01-0492. https://doi.org/10.4271/2021-01-0492

Lavid M, Stevens JG (1985) Photochemical ignition of premixed hydrogen oxidizer mixtures with excimer lasers. Combust Flame 60(2):195–202. https://doi.org/10.1016/0010-2180(85)90007-0

Liedl G, Schuoecker D, Geringer B, Graf J, Klawatsch D, Lenz HP, Piock WF, Jetzinger M, Kapus P (2005) Laser-induced ignition of gasoline direct-injection engines. In: XV international symposium on gas flow, chemical lasers, and high-power lasers. Int Soc Opt Photon 5777:955–960). https://doi.org/10.1117/12.611324

Lorenz S, Bärwinkel M, Stäglich R, Mühlbauer W, Brüggemann D (2016) Pulse train ignition with passively Q-switched laser spark plugs. Int J Engine Res 17(1):139–50. https://doi.org/10.1177/1468087415597629

Mahabadipour H, Srinivasan KK, Krishnan SR (2019) An exergy analysis methodology for internal combustion engines using a multi-zone simulation of dual-fuel low-temperature combustion. Appl Energy 15(256):113952. https://doi.org/10.1016/j.apenergy.2019.113952

Mariani A, Foucher F, Moreau B (2013) The effects of a radio frequency ignition system on the efficiency and the exhaust emissions of a spark-ignition engine. SAE technical paper 2013-24-0053. https://doi.org/10.4271/2013-24-0053

Marko F, König G, Schöffler T, Bohne S, Dinkelacker F (2017) Comparative optical and thermodynamic investigations of high-frequency corona- and spark-ignition on a CV natural gas research engine operated with charge dilution by exhaust gas recirculation. In: Günther M, Sens M (eds) Ignition systems for gasoline engines. CISGE 2016. Springer, Cham. https://doi.org/10.1007/978-3-319-45504-4_18

Morsy MH (2012) Review and recent developments of laser ignition for internal combustion engines applications. Renew Sustain Energy Rev 16(7):4849–4875. https://doi.org/10.1016/j.rser.2012.04.038

Pal A, Agarwal AK (2015) Comparative study of laser ignition and conventional electrical spark ignition systems in a hydrogen-fuelled engine. Int J Hydrogen Energy 40(5):2386–2395. https://doi.org/10.1016/j.ijhydene.2014.12.030

Pavel N, Birtas A, Croitoru G, Dinca M, Boicea N, Dascalu T (2017) Laser ignition of a gasoline engine automobile. In: Laser ignition conference 2017 Jun 20. Optical Society of America, pp LWA4-3. https://doi.org/10.1364/LIC.2017.LWA4.3

Phuoc TX (2006) Laser-induced spark ignition fundamental and applications. Opt Lasers Eng 44(5):351–397. https://doi.org/10.1016/j.optlaseng.2005.03.008

Pischinger S, Heywood J (1990) How heat losses to the spark plug electrodes affect flame Kernel development in an SI-engine. SAE technical paper 900021. https://doi.org/10.4271/900021

Prasad RK, Agarwal AK (2021) Development and comparative experimental investigations of laser plasma and spark plasma ignited hydrogen enriched compressed natural gas fueled engine. Energy 1(216):119282. https://doi.org/10.1016/j.energy.2020.119282

Prasad RK, Agarwal AK (2021) Experimental evaluation of laser ignited hydrogen enriched compressed natural gas fueled supercharged engine. Fuel 1(289):119788. https://doi.org/10.1016/j.fuel.2020.119788

Saito T, Sugaya T, Takahashi E, Furutani H (2017) Influence of laser incident energy and focal length on multi-point laser ignition in an internal combustion engine at N_2 dilution. In: Laser ignition conference 2017 Jun 20. Optical Society of America, pp LWA4-2. https://doi.org/10.1364/LIC.2017.LWA4

Scarcelli R, Richards K, Pomraning E, Senecal P et al (2016) Cycle-to-cycle variations in multi-cycle engine RANS simulations. SAE technical paper 2016-01-0593. https://doi.org/10.4271/2016-01-0593

Schenk M, Schauer FX, Sauer C, Weber G, Hahn J, Schwarz C (2017) Challenges to the ignition system of future gasoline engines—an application oriented systems comparison. In: Günther M, Sens M (eds) Ignition systems for gasoline engines. CISGE 2016. Springer, Cham. https://doi.org/10.1007/978-3-319-45504-4_1

Singh AP, Kumar D, Agarwal AK (2020) Particulate characteristics of laser ignited hydrogen enriched compressed natural gas engine. Int J Hydrogen Energy 45(35):18021–18031. https://doi.org/10.1016/j.ijhydene.2020.05.005

Soldera FA, Mucklich FT, Hrastnik K, Kaiser T (2004) Description of the discharge process in spark plugs and its correlation with the electrode erosion patterns. IEEE Trans Veh Technol 53(4):1257–1265. https://doi.org/10.1109/TVT.2004.830977

Maly R (1984) Spark ignition: its physics and effect on the internal combustion engine. In: Hilliard JC, Springer GS (eds) Fuel economy. Springer, Boston, MA. https://doi.org/10.1007/978-1-4899-2277-9_3

Srivastava DK, Agarwal AK (2014) Comparative experimental evaluation of performance, combustion, and emissions of laser ignition with conventional spark plug in a compressed natural gas-fuelled single-cylinder engine. Fuel 1(123):113–122. https://doi.org/10.1016/j.fuel.2014.01.046

Tanoff MA, Smooke MD, Teets RE, Sell JA (1995) Computational and experimental studies of laser-induced thermal ignition in premixed ethylene-oxidizer mixtures. Combust Flame 103(4):253–280. https://doi.org/10.1016/0010-2180(95)00098-4

Taskiran OO (2020) Improving burning speed by using hydrogen enrichment and turbulent jet ignition system in a rotary engine. Int J Hydrogen Energy. https://doi.org/10.1016/j.ijhydene.2020.11.142

Tesla N (1898) Electrical igniter for gas-engines. United States patent US 609,250. 1898 Aug 16

Toulson E, Schock H, Attard W (2010) A Review of pre-chamber initiated jet ignition combustion systems. SAE technical paper 2010-01-2263. https://doi.org/10.4271/2010-01-2263

Valera H, Kumar D, Singh AP, Agarwal AK (2020) Modelling aspects for adaptation of alternative fuels in IC engines. In: Singh A, Shukla P, Hwang J, Agarwal A (eds) Simulations and optical diagnostics for internal combustion engines. Energy, environment, and sustainability. Springer, Singapore. https://doi.org/10.1007/978-981-15-0335-1_2

Vasile NT, Pavel N (2021) Multi-point laser-induced ignition of air-methane mixtures by a high peak-power passively Q-switched Nd: YAG/Cr4+: YAG laser. Opt Laser Technol 1(141):107169. https://doi.org/10.1016/j.optlastec.2021.107169

Vinogradov J, Rivin B, Sher E (2008) NOx reduction from compression ignition engines with pulsed corona discharge. Energy 33(3):480–491. https://doi.org/10.1016/j.energy.2007.06.004

Wermer L, Lefkowitz JK, Ombrello T, Im SK (2021) Spark and flame kernel interaction with dual-pulse laser-induced spark ignition in a lean premixed methane–air flow. Energy 15(215):119162. https://doi.org/10.1016/j.energy.2020.119162

Chapter 10
Methods in S.I. Engine Modelling: Auto-calibration of Combustion and Heat Transfer Models, and Exergy Analysis

Sami M. M. E. Ayad, Carlos R. P. Belchior, and José R. Sodré

Abstract This chapter reports on a workflow aimed at obtaining deeper insights into spark ignition engines using thermodynamic modelling. This workflow is divided into two steps: (i) Auto-calibration of combustion and heat transfer models using AVL Boost® and AVL Design Explorer; and (ii) in-cylinder exergy analysis using Wolfram Mathematica®. Model calibration is usually based on experimental pressure curve and combustion data. However, there is a gap in methods that generate accurate simulation results, while calibrating the combustion and heat transfer models without prior experimental results. Thus, the authors proposed an approach in their earlier work to address this gap. In this chapter, this proposed approach has been complemented with exergy analysis to form a complete workflow for engine research using simulation. This workflow was applied to a 4-cylinder gasoline engine template model as a case study. First, the combustion and heat transfer models are parametrized and used as design variables of an optimization problem. The objective functions in this problem are the combustion phasing, here defined as the crank angle in which the peak cylinder pressure occurs CA_{pp}, and the mechanical load, here defined as the indicated mean effective pressure IMEP. The appropriate temperature constraints were included to guarantee that the engine model was representative of the physical problem. The optimization problem is then solved for a target IMEP and CA_{pp} and the results are analyzed. Afterwards, we begin the exergy analysis of the engine to obtain deeper insights. The resulting curves for the thermodynamic state properties and species' molar fraction are exported from the simulation software to a program in Wolfram Mathematica that does the exergy analysis of the engine, providing deeper insights into the useful work available in the engine, losses due to heat transfer, losses in the exhaust gases, and combustion irreversibilities. This analysis can be very useful in determining the best fuel mixture, operating conditions, and areas for improvement.

S. M. M. E. Ayad (✉) · C. R. P. Belchior
Department of Mechanical Engineering, Federal University of Rio de Janeiro, Rio de Janeiro, Brazil
e-mail: samiayad@ufrj.br

J. R. Sodré
Department of Mechanical Engineering and Design, Aston University, Birmingham, UK

A. K. Agarwal et al. (eds.), *Engine Modeling and Simulation*, Energy, Environment, and Sustainability, https://doi.org/10.1007/978-981-16-8618-4_10

Keywords Engine simulation · Calibration model · Cylinder pressure · Engine modelling · Spark ignition engine · Exergy analysis

10.1 Introduction

Naturally, the development of engines that adopt new technologies, fuels, or configurations requires extensive investigations to evaluate the impact on their performance and emissions. Another purpose of these investigations is to calibrate the engine operation to these new conditions (Shivapuji and Dasappa 2017; Matuszewska et al. 2016). In that sense, the parametrization of data maps for optimal engine operation is defined as engine "calibration" (Grasreiner 2012). These, in turn, usually require significant labor, resources, and time (Faria et al. 2017). Therefore, the use of simulation is an economical way to address these issues as it allows extensive engine investigations with minimal physical infrastructure requirements (Caton 2016). In this chapter, a simple method to auto-calibrate the combustion and heat transfer model parameters is presented and applied. This method was first proposed by the authors and validated for engines using ethanol, hydrogen and natural gas as alternative fuels (Ayad et al. 2021). The validation consisted of a comparison between the pressure curve *versus* crank angle, which in turn, represent the engine operating cycle. However, further studies are still needed to verify if this approach adequately represents its burn and heat release rates.

Nevertheless, the evaluation of a power generation system should not be restricted to individual analyzes of energy. An approach that leverages information across a wide range of industries can provide a much better perspective (Jalil-Vega et al. 2020) and must include a technical overview of the different technologies and options. However, this technical overview cannot be limited to an energy analysis and must be able to apply technology efficiently and improve the conversion of energy resources in society (Jaber et al. 2008).

In this sense, exergy analysis is a tool for systems design and investigation. As such, exergy is a measure of the ability to perform useful work and is also known as availability (Heywood 2018). Since exergy is not a conserved property, it is destroyed by irreversibilities associated with entropy increase in actual processes. The key phenomena causing irreversibilities in spark ignition engines include heat transfer, combustion, friction, or mixing processes (Caton 2006). This exergy analysis approach differs from conventional first law analysis because it considers destruction of exergy in the evaluation of processes and systems (Mert et al. 2007). The use of exergy analysis helps to: (i) drive technological improvements (Baldi et al. 2014); (ii) present policy insights relevant to improving energy consumption and greenhouse gas emissions (Byers et al. 2015); (iii) identify the best paths for global decarbonization (Valero et al. 2018).

In this chapter, exergy analysis is used to identify possible improvements to a gasoline engine's operation using results from a thermodynamic simulation. These results were obtained by auto-calibrating the heat transfer and combustion model parameters to achieve the desired values for the mechanical load and combustion phasing. The following sections will present the theory, computational tools and methods behind the parameter auto-calibration method and exergy analysis. These methods will then be applied to an example case and the results will be discussed. This chapter aims to introduce the reader to a workflow that can accurately predict engine cycle operation and reveal deeper insights based on exergy analysis that can guide further engine research and development.

Therefore the two main novelties of this chapter include: (i) The validation of a novel simulation approach (Ayad et al. 2021) to a gasoline engine operating in fuel rich conditions. This approach had been validated with ethanol, hydrogen and natural gas, in stoichiometric and lean conditions (Ayad et al. 2021), but not with gasoline, nor in fuel-rich conditions. This chapter also shows mass fraction burned and heat release rate curves, further validating this simple and flexible method; and (ii) The extension of the simulation approach to include exergy analysis, thus constituting a framework for advanced studies in spark ignition engines.

10.2 Theoretical Basis

This section presents the theoretical basis of the methods described in this chapter. First, the engine cycle simulation is presented in Sect. 10.2.1, following a description of the optimization procedure in sub-Sect. 10.2.2. Finally, a theoretical description of exergy analysis is presented in sub-Sect. 10.2.3.

10.2.1 Thermodynamic Engine Model Description

The engine cycle was simulated using AVL Boost®. The equations and models are described in full in the software's theoretical manual, AVL Boost Version 2019—Theory (2019a). However, this section presents the central mathematical equations and simplifying hypotheses.

Figure 10.1 illustrates the 1st law energy balance within the cylinder modelled in the software where $p_c dV$ is the piston work, dQ_W is the rate of heat wall losses, $h_{BB} dm_{BB}$ is the blow-by enthalpy, m_c is the in-cylinder mass, p_c is the pressure inside the cylinder, T_c is the temperature inside the cylinder, V is the cylinder volume, $h_i dm_I$ is the enthalpy flow entering the control volume, and $h_e dm_e$ is the enthalpy flow exiting the control volume. Energy conservation yields Eq. (10.1) to calculate the rate of change of internal energy as a function of crankshaft angle (AVL 2019a):

$$\frac{d(m_c u)}{d\alpha} = \sum \frac{dm_i}{d\alpha} h_i - \sum \frac{dm_e}{d\alpha} h_e - p_c \frac{dV}{d\alpha} + \frac{dQ_F}{d\alpha} - \sum \frac{dQ_W}{d\alpha} - h_{BB} \frac{dm_{BB}}{d\alpha} \tag{10.1}$$

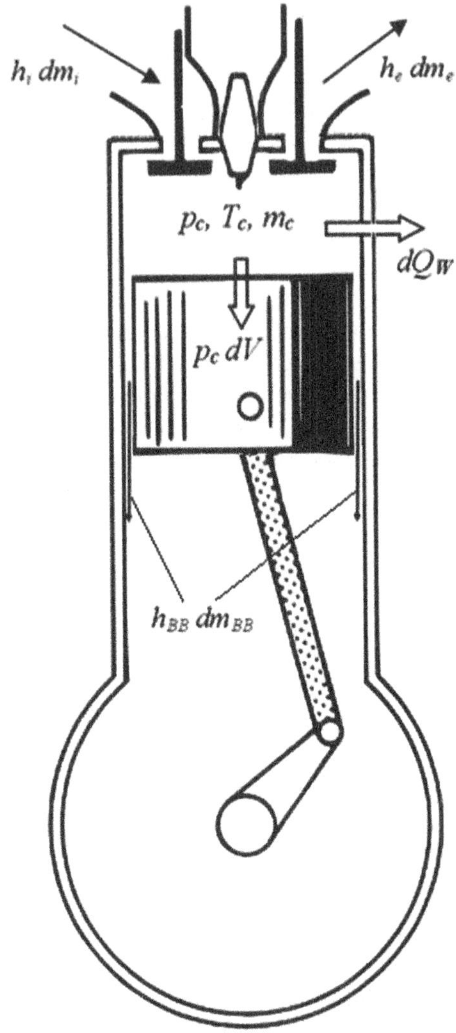

Fig. 10.1 Diagram illustrating the in-cylinder energy balance (AVL 2019a)

where:

$\frac{d(m_c u)}{d\alpha}$: rate of change of in-cylinder internal energy (kJ/°CA);
$-p_c \frac{dV}{d\alpha}$: piston work (kJ/°CA);
$\frac{dQ_F}{d\alpha}$: rate of fuel heat release (kJ/°CA);
$\sum \frac{dQ_W}{d\alpha}$: rate of wall heat losses (kJ/°CA);
$h_{BB} \frac{dm_{BB}}{d\alpha}$: blow-by enthalpy flow (kJ/°CA);
m_c: in-cylinder mass (kg);
u: specific internal energy (kJ/kg);
p_c: in-cylinder pressure (kPa);
T_c: in-cylinder temperature (K);

V: cylinder volume (m^3);
Q_F: fuel heat released (kJ);
Q_W: wall heat losses (kJ);
α: crank angle (°CA);
h_{BB}: blow-by gas specific enthalpy (kJ/kg).

AVL Boost adopts some simplifying hypotheses in Eq. (10.1) (AVL 2019a) listed below:

- The air fuel mixture is considered homogeneous;
- The mixture in the unburned zone is considered to have a constant air–fuel ratio during combustion;
- Based on the 2-zone modelling approach, gases in the unburnt and burnt zones have different compositions but the same temperature and pressure;
- There was no in-cylinder fuel evaporation.

The mass conservation is presented in Eq. (10.2), in which the in-cylinder mass variation is obtained from input and output mass flows (AVL 2019a):

$$\frac{dm_c}{d\alpha} = \sum \frac{dm_i}{d\alpha} - \sum \frac{dm_e}{d\alpha} - \frac{dm_{BB}}{d\alpha} \tag{10.2}$$

where:
$\frac{dm_c}{d\alpha}$: in-cylinder mass flow rate (kg/ °CA);
$\sum \frac{dm_i}{d\alpha}$: mass flow entering the control volume (kg/ °CA);
$\sum \frac{dm_e}{d\alpha}$: mass flow exiting the control volume (kg/ °CA);
$\frac{dm_{BB}}{d\alpha}$: mass flow rate due to blow-by (kg/°CA).

Equation (3) shows the ideal gas equation, which presents the relation between the universal gas constant (R_0), and the in-cylinder pressure, mass, volume, and temperature (Bueno 2016; Sun et al. 2019).

$$p_c = \frac{1}{V} m_c R_o T_c \tag{10.3}$$

One can then use the Runge–Kutta method to solve Eqs. (1) to (3) for the temperature. Inserting this temperature in Eq. (10.3) provides the in-cylinder pressure (AVL 2019a; Melo 2012; Bueno 2016; Cristo 2017).

If we only consider the high-pressure cycle period, the conservation of energy can be simplified as shown in Eq. (10.4). This period comprises the duration between the moment when the intake valve closes and the moment in which the exhaust valve opens.

$$\frac{d(m_c u)}{d\alpha} = -p_c \frac{dV}{d\alpha} + \frac{dQ_{Net}}{d\alpha} \tag{10.4}$$

The term $\frac{dQ_{Net}}{d\alpha}$ in Eq. (10.4) is the term for the rate of net heat release. This term is defined defined in Eq. (10.5).

$$\frac{dQ_{Net}}{d\alpha} = \frac{dQ_F}{d\alpha} - \sum \frac{dQ_W}{d\alpha} - h_{BB}\frac{dm_{BB}}{d\alpha} \tag{10.5}$$

The term $-p_c\frac{dV}{d\alpha}$ in Eq. (10.4) represents the piston work. The engine's indicated mean effective pressure, *IMEP* (kPa), is obtained by integrating the piston work as shown in Eq. (10.6).

$$IMEP = \frac{\oint p_c dV_\alpha}{V_d} \tag{10.6}$$

where:

V_d: swept volume (m^3).

The IMEP of the high-pressure cycle, *IMEPHP*, can be obtained using Eq. (10.7).

$$IMEPHP = \frac{\int_{\alpha_{IVC}}^{\alpha_{EVO}} p_c dV_\alpha}{V_d} \tag{10.7}$$

The brake mean effective pressure (BMEP), which is related to brake mechanical power output (P_e) and engine speed (N), is defined as shown in Eq. (10.8).

$$BMEP = \frac{P_e i}{V_d N} \tag{10.8}$$

The rate of net heat release rate is a function of the rate of fuel heat release, blow-by enthalpy flow, and rate of wall heat transfer. The blow-by enthalpy depends only on the effective blow-by gap, which was fixed at the default value of 0.0008 mm.

The term corresponding to the rate of fuel heat release $\frac{dQ_F}{d\alpha}$ is expressed in Eq. (10.9).

$$\frac{dQ_F(\alpha)}{d\alpha} = m_f LHV_F \frac{dx(\alpha)}{d\alpha} \tag{10.9}$$

where:

m_F : in-cylinder fuel mass (kg);

LHV_F : fuel's lower calorific value (kJ/kg);

$\frac{dx(\alpha)}{d\alpha}$: mass burning rate (kJ/°CA).

The mass burning rate is associated with both the mass fraction burned (MFB) and the combustion phasing. The most commonly used indicators for combustion phasing are the crank angle for the peak cylinder pressure (CA_{pp}) and a corresponding crank angle for the consumption of 50% of the fuel mass (CA_{50}). According to Ref. (Gong et al. 2020), CA_{pp} is a better option than CA_{50} for spark timing and combustion phasing control, and is thus used in this approach.

This chapter uses a two-zone combustion model, for which further details are available elsewhere (Caton 2016; Ayad et al. 2021, 2020; Heywood 2018; AVL 2019a). The combustion model used in this chapter is the Wiebe combustion model

(Vibe and Meißner 1970) which is a 0D model that has 4 tuning parameters. This combustion model was chosen due to its simplicity, reduced computational cost, and low number of tuning calibration constants.

10.2.1.1 Wiebe Combustion Model

The empirical Wiebe function is frequently used to model ICE's using both conventional fuels such as gasoline gasoline (Caton 2016; AVL 2019a; Melo 2012), diesel (Bueno 2016; Souza Junior 2009), as well as biofuels, such as ethanol, biodiesel and others (Shivapuji and Dasappa 2017; Faria et al. 2017; Melo 2012, 2007; Yeliana et al. 2011; Rocha 2016; Reyes et al. 2013, 2016). A more in-depth discussion of the Wiebe combustion model is not in the scope of this work, but can be found elsewhere (Ghojel 2010).

Equation (10) is used to calculate the fuel mass fraction burned ($x(\alpha)$).

$$x(\alpha) = 1 - e^{-a_{wiebe}\left(\frac{\alpha-\alpha_i}{\Delta\alpha}\right)^{m_{wiebe}+1}} \tag{10.10}$$

where:

a : adjustable parameter related to the combustion duration;
m_{wiebe} : form factor;
α_i: start of energy release or start of combustion;
$\Delta\alpha$: combustion duration.

The parameter a is directly related to the combustion duration and is not an independent parameter (Yeliana 2010). For a given duration, this parameter indicates how much fuel was burnt during the combustion process, and as such, works as an efficiency parameter. A variation of the adjustable parameters a and m_{wiebe} changes the shape of the curve significantly (Heywood 2018).

10.2.1.2 Heat Transfer Model

The amount of heat transferred through the walls affects the combustion phasing (Caton 2014). This shows how the term $-\sum \frac{dQ_W}{d\alpha}$ must be taken into consideration for accurate prediction of net heat release. Equation (11) is used to obtain the wall heat transfer for the surfaces in contact with the gases (AVL 2019a, b).

$$Q_w = \sum_{i=1}^{3} A_{w,i}(\alpha)h_w(\alpha)\big(T_c(\alpha) - T_{w,i}\big) \tag{10.11}$$

where:

$i = 1, 2,$ and 3: respectively wall surfaces of the piston, cylinder liner and head;
$h_w(\alpha)$: heat transfer coefficient (W/m^2 K);

$A_{w,i}(\alpha)$: surface area (m^2);
$T_c(\alpha)$: in-cylinder bulk gas temperature (K);
T_{wi}: cylinder wall temperature (K).

There are many models to calculate the heat transfer coefficient h_w. AVL Boost can use the following (AVL 2019a):

- Woschni 1978
- Woschni 1990
- Hohenberg
- Lorenz (for engines with divided combustion chamber nly)
- AVL 2000 Model
- Bargende

The Theory Manual (AVL 2019a) recommends Bargende's model (AVL 2019a, Bargende 1991) for spark ignition engines, so it was used in this research.

In any case, the calibration process happens independently from heat transfer model selected, and depends only on the surface wall temperatures of Eq. (10.11) and on some multipliers that are factors to scale the heat transfer. In summary, the 7 tuning parameters for the heat transfer model are:

- $htm_{Wpiston}$, htm_{Whead}, and htm_{Wliner}: heat transfer multipliers corresponding to the wall heat transfer through the piston, cylinder head, and liner, respectively.
- $T_{Wpiston}$, T_{Whead}, $T_{WlinerTDC}$, and $T_{WlinerBDC}$: surface temperatures corresponding respectively to the wall heat transfer through the piston, cylinder head, and liner at both top dead center (TDC), and at bottom dead center (BDC).

10.2.2 Optimization Method

The heat transfer and combustion model parameters were obtained using a multi-objective genetic algorithm (MOGA). The optimization problem aims to minimize the deviation between the desired and simulated values of the objective functions. For this purpose, the objective functions are *BMEP* and CA_{pp}. This algorithm uses a non-dominated ranking and selection scheme to compute an individual solution's fitness. For every generation, the fittest individuals are used to generate subsequent populations. A deeper discussion of the multi-objective genetic algorithm is beyond the scope of this paper, but the interested reader is referred to AVL, AVL Design Explorer Version 2019—DoE and Optimization, (2019) (AVL 2019c) for more information.

A multi-objective optimization problem is generally defined as:

Minimize $|\boldsymbol{F}(\boldsymbol{x})_{desired} - \boldsymbol{F}(\boldsymbol{x})_{simulated}|$

Given:

$$\boldsymbol{c}(\boldsymbol{x}) = 0$$

$$\boldsymbol{g}(\boldsymbol{x})_{lb} \leq \boldsymbol{g}(\boldsymbol{x}) \leq \boldsymbol{g}(\boldsymbol{x})_{ub}$$

$$\boldsymbol{x}_{lb} \leq \boldsymbol{x} \leq \boldsymbol{x}_{ub}$$

where:

$\boldsymbol{x} \in R^n$: decision variables vector.
$U(\boldsymbol{x})_{desired} \in R^l$: vector of the desired values of the objective functions.
$U(\boldsymbol{x})_{simulated} \in R^l$: vector of the simulated values of the objective functions.
$c(\boldsymbol{x}) \in R^m$: equality constraints vector.
$g(\boldsymbol{x}) \in R^n$: inequality constraints vector.
$g(\boldsymbol{x})_{lb} \in R^n$: vector of lower bound values of $g(\boldsymbol{x})$
$g(\boldsymbol{x})_{ub} \in R^n$: vector of upper bound values of $g(\boldsymbol{x})$
$\boldsymbol{x}_{lb} \in R^o$: vector of lower bound values of $\boldsymbol{x}$
$\boldsymbol{x}_{ub} \in R^o$: vector of upper bound values of $\boldsymbol{x}$
l: number of objective functions.
m: number of equality constraints.
n: number of unknown variables.
o: number of inequality constraints.

10.2.3 Exergy Analysis Theoretical Description

This sub-section will introduce the theoretical basis of exergy analysis as applied in this chapter.

10.2.3.1 Thermodynamic Property Calculation

The chemical species considered in these calculations are oxygen (O_2), nitrogen (N_2), carbon dioxide (CO_2), water (H_2O), carbon monoxide (CO), hydrogen (H_2), atomic oxygen (O), hydroxyl (OH), atomic hydrogen (H), atomic nitrogen (N), nitrogen oxide (NO) and a gasoline surrogate (C_8H_{16}). C_8H_{16} is chosen as a surrogate for gasoline, as it has similar molecular mass and air fuel ratio to the gasoline represented in AVL Boost (Azevedo Neto 2013). It is possible to calculate most of the thermodynamic properties of substances from their specific heat at constant pressure (c_p) (Moran et al. 2011). Thus, a single logarithmic 5th order polynomial equation proposed by Lanzafame and Messina was used (Lanzafame and Messina 2005). This equation was used to calculate the $\overline{c}_{\mathrm{p,I}}$ for each substance considered according to Eq. (10.12).

$$\overline{c}_{p,i}(T) = a_0 + a_1(\ln T) + a_2(\ln T)^2 + a_3(\ln T)^3 + a_4(\ln T)^4 + a_5(\ln T)^5 \tag{10.12}$$

where $\overline{c}_{\mathrm{p,I}}$ is the specific heat at constant pressure of substance I on a molar basis; and $a_0 - a_5$ are the coefficients particular to each of the substances.

The coefficients of all substances, except gasoline, were obtained from the work of Lanzafame and Messina (2002). Gasoline coefficients were obtained from Azevedo Neto (2013). The specific molar enthalpy of each substance ($\overline{h}_{\rm i}$) is calculated using Eq. (10.13).

$$\overline{h}_{\rm i}({\rm T}) = \overline{h}^{0}_{f,i} + \int_{T_0}^{T} \overline{c}_{\rm p,i}({\rm T})dT \tag{10.13}$$

where T_0 is the reference temperature equal to 25 °C (298 K) (Moran et al. 2011) and $\overline{h}^{0}_{f,i}$ is the enthalpy of formation at the reference temperature T_0 of substance i.

The specific molar entropy of each substance ($\overline{s}_{\rm i}$) can be calculated using Eq. (10.14).

$$\overline{s}_{\rm i}({\rm T}) = \overline{s}^{0}_{i} + \int_{T_0}^{T} \frac{\overline{c}_{\rm p,i}({\rm T})}{T}dT - \overline{R}\ln\left(\frac{P_i}{P_0}\right) \tag{10.14}$$

where P_0 is the reference temperature equal to 1 atm (Moran et al. 2011); $\underline{\overline{s}^{0}_{i}}$ is the absolute entropy of substance I at reference temperature and pressure; and $\overline{R}$ is the universal gas constant.

The reference enthalpy and entropy values of the substances listed above can be found in Azevedo Neto (2013).

The thermodynamic properties corresponding to the mixture on a molar basis, $\overline{c}_{\rm p}$,$\overline{h}$ e $\overline{s}$, can be calculated as follows:

$$\overline{c}_{\rm p}({\rm T}) = \sum_{i=1}^{n} y_{\rm i}\overline{c}_{\rm p,i}(T) \tag{10.15}$$

$$\overline{h}({\rm T}) = \sum_{i=1}^{n} y_{\rm i}h_{\rm i}(T) \tag{10.16}$$

$$\overline{s}({\rm T, P}) = \sum_{i=1}^{n} y_{\rm i}s_{\rm i}(T, P) \tag{10.17}$$

where $y_{\rm I}$ is the molar fraction of component i.

These properties can be converted to mass base. Thus, it is necessary to calculate the apparent molar mass of the mixture (MM), which is the sum of the molar masses of the substances ($MM_{\rm i}$) as shown in Eq. (10.18). Finally, the properties on a mass basis are shown in Eqs. (10.19)–(10.21).

$$MM = \sum_{i=1}^{n} y_{\rm i}MM_{\rm i} \tag{10.18}$$

$$c_p(T) = \frac{\overline{c}_p(T)}{M} \quad (10.19)$$

$$h(T) = \frac{\overline{h}(T)}{M} \quad (10.20)$$

$$s(T) = \frac{\overline{s}(T)}{M} \quad (10.21)$$

10.2.3.2 Exergy Analysis

Exergy is the maximum theoretical useful work possible for systems that interact until they reach equilibrium with their environment. Thus, exergy is a measure of the deviation of the state of a system in relation to the state of the environment (Bejan et al. 1996). In this way, a system that is in complete balance with its surroundings does not have exergy. Furthermore, a system carries more exergy the further its state is distanced from that of the environment (Sciubba and Moran 1995). In turn, the destruction of exergy occurs when there is a decrease in energy quality (Sciubba and Moran 1995). Therefore, exergy is an effective indicator of the potential of a substance to affect the environment (Rosen 2013).

So, exergy is the portion of energy that is useful to society. This property has several characteristics that suggest its use as a convenient meter for an energy accounting paradigm and, by reducing exergy losses, it is also possible to reduce the environmental impact caused by the system under analysis (Rosen and Dincer 2003). As a result, exergy analysis allows, among other applications: (i) to reveal whether there is a possibility of designing more efficient energy systems; (ii) identify and assess the environmental impact of the use of energy resources; and (iii) assist in sustainable development (Dincer 2002).

Total exergy is the maximum work obtained when a quantity of matter is brought into the unrestricted dead state (Azevedo Neto 2013). This state is characterized by mechanical, thermal, and chemical equilibrium with the environment. Thus, the total exergy of a system can be divided into two parts corresponding to thermomechanical and chemical equilibrium, respectively: (i) thermomechanical exergy; and (ii) chemical exergy (Azevedo Neto 2013). Therefore, the total exergy Ex can be obtained from the sum of the thermomechanical (Ex^{tm}) and chemical (Ex^{q}) exergies (Moran et al. 2011). This relationship is introduced in Eq. (10.22).

$$Ex = \text{Ex}^{tm} + \text{Ex}^{q} \quad (10.22)$$

Thermomechanical exergy corresponds to the maximum work obtained by a system when it interacts until it reaches thermal and mechanical equilibrium with the environment (Moran et al. 2011). Considering the negligible kinetic and potential energy, the specific thermomechanical exergy on a molar basis of a mixture of ideal gases ($\overline{ex}^{tm}$) is introduced in Eq. (10.23). This relationship is valid for a closed system.

$$\overline{ex}^{tm} = (\overline{u} - \overline{u}_0) + P_0(\overline{v} - \overline{v}_0) - T_0(\overline{s} - \overline{s}_0) \quad (10.23)$$

where:

$\overline{u}$: mixture's specific internal energy on a molar basis;

$\overline{u}_0$: mixture's specific internal energy on a molar basis at the reference temperature;

$\overline{v}$: mixture's specific volume on a molar basis;

$\overline{v}_0$: mixture's specific volume on a molar basis at the reference temperature;

$\overline{s}_0$: mixture's specific entropy on a molar basis at the reference temperature and pressure.

If a control volume is considered, the specific thermomechanical flow exergy on a molar basis of an ideal gas mixture, ($\overline{ex}_f^{tm}$) becomes (Azevedo Neto 2013):

$$\overline{ex}_f^{tm} = \left(\overline{h} - \overline{h}_0\right) - T_0(\overline{s} - \overline{s}_0) \quad (10.24)$$

where:

$\overline{h}$: mixture's specific enthalpy on a molar basis;

$\overline{h}_0$: mixture's specific enthalpy on a molar basis at the reference temperature.

In cases in which the chemical composition remains constant between two states, the variation of exergy, or flow exergy, corresponds only to thermomechanical exergy. However, it is necessary to consider the contribution of chemical exergy when there is a change in the chemical composition (Moran et al. 2011). In these cases, the chemical potential imbalance between the species of combustion products and the substances that make up the environment can be converted into more work (Azevedo Neto 2013; Kotas 2013).

In internal combustion engines, the exhaust gas composition differs from that of the environment and the chemical exergy contribution is significant. In fact, the more incomplete the combustion, the greater the chemical exergy due to the presence of unburned fuel (Azevedo Neto 2013; Gallo 1990). In this way, the specific chemical exergy on a molar basis of an ideal gas mixture ($\overline{ex}^q$) is obtained using Eq. (10.25) (Moran et al. 2011).

$$\overline{ex}^q = \sum_i y_i \overline{ex}_i^q + \overline{R}T_0 \sum_i y_i \ln y_i \quad (10.25)$$

where:

$\overline{ex}_i^q$ is the standard Chemical exergy for substance I in the mixture. The standard chemical exergy values of the substances considered in this chapter can be found in Ref. (Morris and Szargut 1986). The specific chemical exergy on a mass basis of an ideal gas mixture (ex^q) is determined in a manner analogous to the calculation of other thermodynamic properties. This relationship is represented in Eq. (10.26)

(Azevedo Neto 2013):

$$ex^q = \frac{\overline{ex^q}}{M} \tag{10.26}$$

The mixture's specific total flow exergy on a molar basis ($\overline{exf}$) and on a mass basis (exf) are obtained as shown in Eqs. (27) and (28), respectively (Azevedo Neto 2013). The total exergy and total flow exergy parameters are related as shown in Eq. (10.29) (Gallo 1990). The total exergy and total flow exergy, both on a mass basis, are given by Eqs. (30) and (31), respectively (Azevedo Neto 2013):

$$\overline{exf} = \overline{exf}^{tm} + \overline{ex^q} = \left(\overline{h} - \overline{h}_0\right) - T_0(\overline{s} - \overline{s}_0) + \overline{ex^q} \tag{10.27}$$

$$exf = (h - h_0) - T_0(s - s_0) + ex^q \tag{10.28}$$

$$\overline{ex} = \overline{exf} + \mathrm{T}\overline{R}\left(\frac{P}{P_0} - 1\right) \tag{10.29}$$

$$Ex = m\frac{\overline{exf}}{M} \tag{10.30}$$

$$exf = \frac{\overline{exf}}{M} \tag{10.31}$$

where:

m: mixture's mass inside the control volume.

Exergy variation between two states 1 and 2, ΔEx, is given by Eq. (10.32).

$$\Delta Ex = Ex_1 - Ex_2 = Ex_i - Ex_e + Ex_Q - Ex_W - Ex_{dest} \tag{10.32}$$

where:

Ex_1: exergy at state 1;

Ex_1: exergy at state 2;

Ex_i: exergy due to the flow that goes in the control volume;

Ex_e: exergy due to the flow that exits the control;

Ex_Q: exergy transfer via heat between the system and the reference environment;

Ex_W: exergy that is converted into net work;

Ex_{dest}: part of the exergy that is destroyed due to irreversibilities.

The exergy transfer via heat (Ex_Q) is presented in Eq. (10.33).

$$Ex_Q = Q_P\left(1 - \frac{T_0}{T_P}\right) \tag{10.33}$$

where:

Q_P: heat exchanged through the cylinder walls;

T_P: cylinder wall temperature.

The transfer of exergy via work corresponds to the net work (Ex_W). Thus, it can be defined as the difference between the shaft work and the surrounding work (Azevedo Neto 2013). This relationship is given in Eq. (10.34).

$$Ex_W = W - W_{atm} = \int_1^2 (P - P_0)dV \tag{10.34}$$

where:

W: shaft work;

W_{atm}: surrounding work.

The exergy destroyed due to irreversibilities is given by Eq. (10.35).

$$Ex_{dest} = Ex_e - Ex_s + Ex_Q - Ex_W - \Delta Ex \tag{10.35}$$

If the transient regime is considered, the control volume exergy balance is rewritten as a function of the crankshaft angle (α). This is introduced in Eq. (10.36).

$$\dot{E}x_{dest} = exf\frac{dm}{d\alpha} + \dot{E}x_Q - \dot{E}x_W - \frac{dEx}{d\alpha} \tag{10.36}$$

Finally, the terms in the exergy balance in the closed phase of the engine's operating cycle can be presented as ratios of the ingoing exergy (Azevedo Neto 2013). These ratios are given in the form of percentage in Eqs. (37)–(40).

$$Ex_W = \frac{Ex_{W,\mathrm{HP}}}{Ex_i}100\% \tag{10.37}$$

$$Ex_Q = \frac{Ex_{Q,HP}}{Ex_i}100\% \tag{10.38}$$

$$Ex_e = \frac{Ex_{e,HP}}{Ex_i}100\% \tag{10.39}$$

$$Ex_{dest} = \frac{Ex_{dest,HP}}{Ex_i}100\% \tag{10.40}$$

where the terms *HP* correspond to the parameters calculated during the closed phase of the motor operating cycle.

10.3 Computational Tools and Methods

Here, the computational tools and methods used in this research will be presented, initially introducing the tools employed in Sect. 3.1, then continuing with a description of the approach in Sect. 3.2. This will be followed with a discussion of the output responses of interest in sub-Sect. 3.3, and end with the case study description in Sect. 3.4.

10.3.1 Computational Tools

The proffered simulation framework is composed of two parts: the predictive simulation and exergy analysis. Within the simulation approach, there are two major steps: the internal combustion engine simulation using AVL Boost (2019b), and the tuning of the parameters of the combustion and heat transfer models using the optimization features of AVL Design Explorer software (2019c).

AVL Design Explorer alters the decision variables of the optimization problem, which in turn are input in the simulation software. The results of the simulation are then compared to the desired values to evaluate the objective functions. This process is iterated until the termination criteria are met. It is worth noting that this simulation approach is independent of the software used, and only requires that the simulation model and its parameters can be integrated to and modified by a software that can perform multi-objective optimization.

In the second part of this framework, the method for exergy analysis consists of applying the thermodynamic properties and exergy calculations outlined above in the commercial software Wolfram Mathematica®. To that end, the thermodynamic output from AVL Boost is exported to Wolfram Mathematica and the calculation procedure is then implemented.

10.3.2 Methods

Here, the methods used in this paper are described, presenting the method for auto-calibration of heat transfer and combustion models. Then, the exergy analysis method is shown, followed by a description of the simulation and exergy analysis workflow.

10.3.2.1 Auto-calibration of Combustion and Heat Transfer Models

The simulation approach utilized to calibrate the combustion and heat transfer models is detailed in this sub-section. It begins with an introduction of the objective functions. Afterwards, it presents the decision variables and their lower and upper bound values and then the constraints of the optimization problem. Finally, the parameters used in the MOGA algorithm are given. This method has been successfully applied and validated in spark ignition engines operating on hydrogen, natural gas and ethanol (see Ayad et al. 2021 for further details) and will be validated with a gasoline engine in the case study presented in Sect. 3.4 of this chapter. To perform this validation, the output responses described in Sect. 3.3 will be compared and discussed.

10.3.2.2 Objective functions

Based on the previous discussion, $BMEP$ and CA_{pp}, were chosen to represent the mechanical load and combustion phasing parameters, respectively. Thus, it is possible to define the objective functions in the following way:

$$|BMEP_{desired} - BMEP_{simulated}|$$

$$|CA_{PP\,desired} - CA_{PP\,simulated}|$$

10.3.2.3 Design variables

The aforementioned heat transfer and combustion model parameters are used to tune the net heat release rate according to Eq. (10.5). Accordingly, the optimization problem's design (or decision) variables are defined as these tuning parameters (Ayad et al. 2021). Table 10.1 describes the Wiebe model parameters, while Table 10.2 describes the Bargende heat transfer model parameters (AVL 2019a, b; Bargende 1991; Perini et al. 2010). As design variables, these parameters will be altered and adjusted in order to minimize the objective functions, while respecting the constraints. To the end user, this tuning can be likened to black-box optimization. Lastly, Table 10.3 presents the lower and upper bound values of the decision variables (AVL 2019a, b).

Table 10.1 Parameters of the Wiebe model

Parameter	Description
α_s	Crank angle corresponding to start of combustion (°CA)
$\Delta\alpha$	combustion duration (°CA)
m_{wiebe}	Form factor (−)
a_{wiebe}	adjustable parameter related to the combustion duration (−)

Table 10.2 Heat transfer model parameters

Parameter	Description
Tw_{piston}	Piston temperature (°C)
Tw_{head}	Cylinder head temperature (°C)
$Tw_{linerTDC}$	Cylinder liner temperature at TDC position (°C)
$Tw_{linerBDC}$	Cylinder liner temperature at BDC position (°C)
$Tw_{exhaust}$	Exhaust port temperature (°C)
$htm\,w_{piston}$	Wall heat transfer multiplier corresponding to piston (−)
$htm\,w_{head}$	Wall heat transfer multiplier corresponding to cylinder head (−)
$htm\,w_{liner}$	Wall heat transfer multiplier corresponding to cylinder liner (−)

Table 10.3 Design variables' values range

Model	Parameter	x_{lb}	x_{ub}
Wiebe combustion model	α_s (°CA)	−40	−10
	$\Delta\alpha$ (°CA)	20	180
	$m_{wiebe}(-)$	0.1	2.6
	$a_{wiebe}(-)$	2.303	6.908
Bargende heat transfer model	Tw_{piston}(°C)	140	290
	Tw_{head}(°C)	120	270
	$Tw_{linerTDC}$(°C)	120	240
	$Tw_{linerBDC}$(°C)	90	220
	$Tw_{exhaust}$(°C)	140	270
	$htm\,w_{piston}(-)$	0.5	5
	$htm\,w_{head}(-)$	0.5	5
	$htm\,w_{liner}(-)$	0.5	5

10.3.2.4 Constraints

The inclusion of heat transfer parameters as decision variables together with those of the combustion model is an important feature of this simulation method (Ayad et al. 2021). For the simulation model to be realistic, the in-cylinder temperature was constrained according to empirical observations across a variety of engines (Prah et al. 2016). An additional restriction was added to ensure that CA_{pp} happens after CA_{50}. The constraints used and their boundaries are shown in Table 10.4.

Table 10.4 Equations corresponding to the inequality constraints and their respective limits

Constraint	$g(x)_{lb}$	$g(x)_{ub}$
$(Tw_{piston} - Tw_{head})$ (°C)	15	–
$(Tw_{piston} - Tw_{exhaust})$ (°C)	20	40
$(Tw_{piston} - Tw_{linerTDC})$ (°C)	20	50
$(Tw_{head} - Tw_{exhaust})$ (°C)	0	–
$(Tw_{exhaust} - Tw_{linerTDC})$ (°C)	0	–
$(Tw_{linerTDC} - Tw_{linerBDC})$ (°C)	0	25
$(CA_{pp} - CA_{50})$ (°CA)	0	–

Table 10.5 Values for the optimization parameters

Parameter	Value
Crossover probability distribution (−)	10
Mutation probability distribution (−)	30
Number of generations (−)	75
Population size (individuals)	40
Probability for crossover (−)	0.9
Probability for mutation (−)	$\frac{1}{n}$

10.3.2.5 Optimization parameters

Table 10.5 shows the parameters applied in the MOGA algorithm as set in AVL Design Explorer (AVL 2019c). The mutation probability is defined as $\frac{1}{n}$, in which n is the decision variables' number (Deb et al. 2002). The remaining values are either empirical values or chosen based on recommendations found in the literature(AVL 2019c). According to the AVL Design Explorer Theory Manual (AVL 2019c), crossover probability distribution defines a probability function for crossover operation. The same applies to the mutation probability distribution and the mutation operation. These parameter values determine how global (when small) or local (when large) the design space search will be. The crossover and mutation probabilities are numbers between 0 and 1 that define the probability of these operators being applied. The population size defines the number of design points, which are generated by the algorithm in every generation, and each generation consists of a set of design points with the size of the set defined by population size (AVL 2019c).

10.3.2.6 Exergy Analysis

Initially, some parameters from the simulation model are exported from AVL Boost as.dat data files and imported to Wolfram Mathematica for use in calculating the thermodynamic properties (Azevedo Neto 2013). These parameters are functions of crank angle and are recorded for the entire engine operation cycle and are listed as follows: Pressure, temperature, volume, mass, work per volume, net work, wall heat flow corresponding to the cylinder piston, head, and wall, and mole fraction of the species considered (O_2, O, N_2, N, H_2, H, H_2O, OH, CO_2, CO, NO, C_8H_{16}).

Once the thermodynamic properties are calculated, the exergy analysis is applied using the calculations described above and the output responses are obtained (Azevedo Neto 2013).

10.3.2.7 Simulation and Exergy Analysis Workflow Overview

Figure 10.2 includes a flowchart of the proposed workflow. It starts with the definition of the case to be analyzed, which includes selection of the operating conditions, fuel and engine. Following this, the simulation software AVL BOOST is used to build the corresponding model. The heat transfer and combustion are then defined and parametrized in order for them to be input as design variables in AVL Design

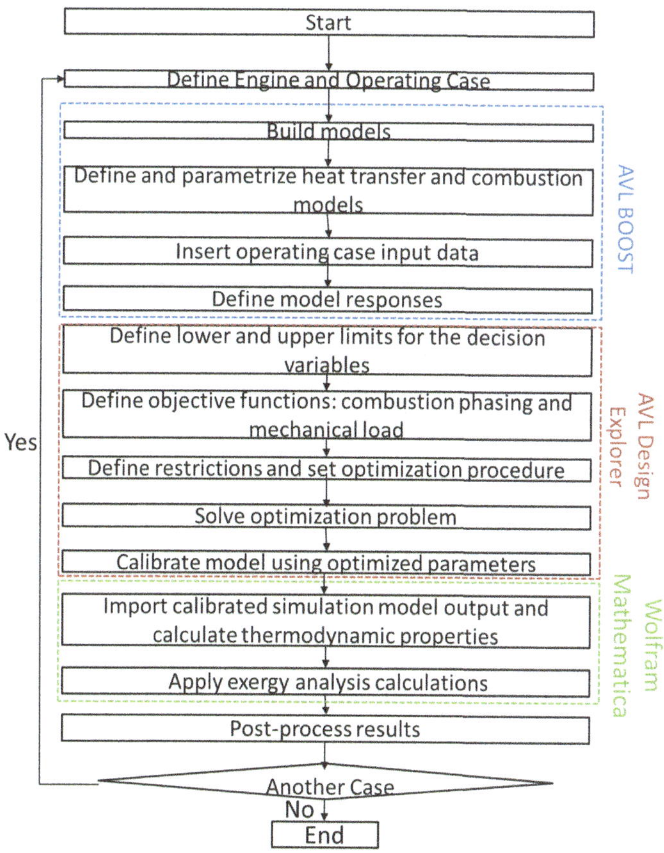

Fig. 10.2 Flowchart of the proposed workflow

Explorer. Subsequently, the relevant input data from the operating case is entered in the model, output responses are defined in AVL Boost. These responses will be described later in this chapter.

The following step is to define the objective functions, design variables, and constraints on AVL Design Explorer, effectively determining the optimization problem. This problem is then solved using the MOGA algorithm, so that the optimization of the design variables corresponds to the calibration of the tuning parameters in order for the simulated objective functions to approach the desired values (Ayad et al. 2021).

The second part of the workflow is done in Wolfram Mathematica. First, output from the calibrated simulation model in AVL Boost is imported. This output is used to calculate the thermodynamic properties as outlined above. Then, exergy analysis calculation is applied, and the results are then postprocessed. Where other cases require simulation, the workflow returns to the case definition step, else, the process is finalized.

10.3.3 Output Responses

The output responses are separated into two main categories: output responses from the simulation method and output responses from the exergy analysis.

The first output responses are obtained by using the simulation method to calibrate the model and are compared with the template case results. They are *BMEP* and CA_{pp}. Also, CA_{10}, CA_{50} and CA_{90} represent 10, 50 and 90% of the mass fraction burned, the peak pressure inside the cylinder (P_{max}), and brake power. The primary purpose in comparing the MFB responses is to determine whether it is possible to predict CA_{50} c when CA_{pp} is used the combustion phasing parameter. Moreover, a comparison of CA_{10} and CA_{90} is added to quantify deviations in the combustion process between the model and the template case. The brake power and *IMEPHP* are included to show how different definitions of the mechanical load parameter relate.

Additionally, the predicted pressure curves inside the cylinder, mass fraction burned, and heat release rates, are compared to the template case. Each curve is recorded between intake valve closing and exhaust valve opening, corresponding to the interval between −110° CA ATDC and 130° CA ATDC at 1° CA intervals. Then, the coefficient of determination R^2 between the observed template case and simulated datasets was calculated using Eq. (10.11). This coefficient is interpreted as a measure of fit adequacy. The calculations in Eq. (10.11) will be performed to the points corresponding to the in-cylinder pressure, mass fraction burned, and heat release rate.

$$R^2 = \left[\frac{\sum\left(y_{temp} - \overline{y_{temp}}\right)\left(y_{sim} - \overline{y_{sim}}\right)}{\sum\left(y_{temp} - \overline{y_{temp}}\right)^2 \sum\left(y_{sim} - \overline{y_{sim}}\right)^2}\right]^2 \tag{10.41}$$

where:

y_{sim}: point in the predicted curve;

y_{exp} : point observed in the template case curve;

$\overline{y_{sim}}$: mean of all points in the predicted curve;

$\overline{y_{exp}}$: mean of all points observed in the template case curve.

In addition to the metrics, there are a set of output parameters from the simulation model that are not intended for comparison between observed and predicted but are used as input for the exergy analysis (Azevedo Neto 2013). These parameters are functions of crank angle and are recorded for the entire engine operation cycle and are listed as follows: Pressure, temperature, volume, mass, work per volume, net work, wall heat flow corresponding to the cylinder piston, head, and wall, and mole fraction of the species considered (O_2, O, N_2, N, H_2, H, H_2O, OH, CO_2, CO, NO, C_8H_{16}).

Finally, the output responses of interest in the exergy analysis are Ex_W, Ex_Q, Ex_e, and Ex_{dest}.

10.3.4 Case Study

To demonstrate the proposed workflow, it is applied to an example case available in the AVL Boost template library. This refers to a 4-cylinder gasoline engine operating in fuel rich conditions at 5000 RPM. The stoichiometric air fuel ratio is 14.6, meaning this case had an air excess ratio (λ) of 0.836. Table 10.6 shows the data of the engine used in this example case.

The original template with all the default parameters will be referred to as the base case. BMEP and combustion phasing are the only parameters used as input for the auto-calibration method. For the base set, curves for in-cylinder pressure, mass fraction burned, and heat release rate were obtained using the original base case with data measured at 1° CA interval.

The proposed simulation approach was then applied to the base case. Thus, the heat transfer and combustion model parameters were auto-calibrated using multi-objective optimization for a desired CA_{pp} and *BMEP.* These objective functions represent the phase of combustion and mechanical load, respectively. Consequently, this model with optimized parameters will be referred to as the auto-calibrated model, and its output responses were compared to those of the base case for validation. Once validated, exergy analysis was applied to the auto-calibrated model to obtain deeper insights into the engine operation and possible improvements. This increases the potential capabilities of simulation in engine research.

Table 10.6 Data of the engine used in the case study

Parameter	Description
Model name	4-cylinder gasoline engine example
Engine	Naturally aspirated, 4-cylinder gasoline engine
Fuel	Gasoline
Engine speed (RPM)	5000
BMEP (bar)	11.15
CA_{pp} (°ATDC)	18.38
Air–fuel ratio	12.21
Bore (m)	0.086
Stroke (m)	0.086
Connecting rod length (m)	0.1435
Compression ratio (−)	10.5:1
IVO (°CA)	340
IVC (°CA)	610
EVO (°CA)	130
EVC (°CA)	390

10.4 Results and Discussion

This section will introduce and discuss the results corresponding to the validation of the auto-calibration method, and to the exergy analysis applied to the validated case. In order to validate the simulation approach, the aforementioned output responses from the auto-calibrated simulation were compared with the base case. Subsequently, the following output responses were analyzed for the exergy analysis: Ex_W, Ex_Q, Ex_e, and Ex_{dest}.

10.4.1 Validation of the Auto-calibrated Model

Figure 10.3 shows a comparison between the in-cylinder pressure curves of the base case and the auto-calibrated model corresponding to the example case. The auto-calibrated pressure curve shows good correlation with that of the base case, with R^2 higher than 0.99. Despite this, one can observe that the peak cylinder pressure of the auto-calibrated model is slightly lower than the base case. This is seen clearly in Table 10.7, that presents a comparison between the results for the base case and those predicted with the model calibrated with optimization. A deviation below 2.6% can be observed between the calibrated and the base case of *BMEP*, CA_{pp}, and P_{max} using the Wiebe combustion model. It is likely that the lower peak pressure is related

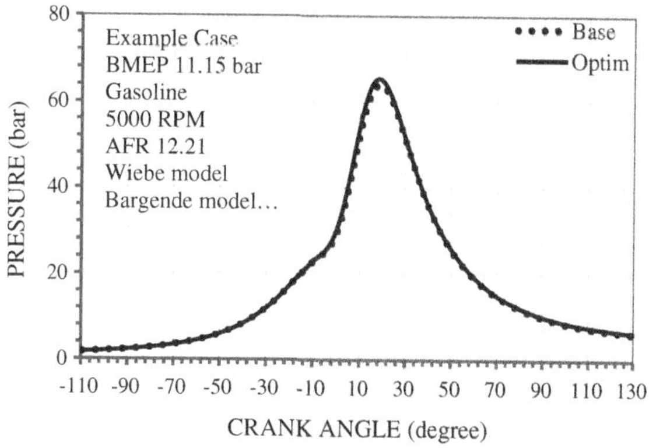

Fig. 10.3 Simulated cylinder pressure curves corresponding to the auto-calibrated model (Optim) and the base case (Base) for the example case using Wiebe combustion model and Bargende wall heat transfer model

Table 10.7 Comparison between simulated results for the base case and the auto-calibrated model

Case	Parameter	Base case	Auto-calibrated model	Relative percentual difference (%)	Relative actual difference
Gasoline Engine Ethanol 5000 RPM 11.15 bar Wiebe Bargende	BMEP (bar)	11.15	10.87	2.52	0.28
	IMEPHP (bar)	13.91	13.63	1.99	0.28
	Brake power (kW)	90.52	92.86	2.59	2.34
	CA_{pp} (°)	18.38	18.75	2.01	0.37
	Peak pressure (Bar)	64.69	63.33	2.10	1.36
	CA_{10} (°)	2.05	2.45	–	0.40
	CA_{50} (°)	11.61	12.01	–	0.39
	CA_{90} (°)	22.04	22.53	–	0.49

to the BMEP, as the deviation in CA_{pp} is too small in absolute terms to influence peak pressure. The deviation in the objective functions BMEP and CA_{pp} can be attributed to the relatively small number of generations and population size used in this problem. Increasing these parameters might improve convergence and result in better optimization performance but should be balanced out with computational cost and time required for simulation.

Nevertheless, these errors denote minor differences in absolute values. Furthermore, it can be seen that CA_{50} falls within the 5 to 11° ATDC bounds expected for conventional engines, but CA_{pp} is slightly above the expected values of the 9–16° ATDC range (Caton 2014).

In addition, one can see that with an accurate friction model and representation of the gas exchange processes (ie. intake and exhaust), it is possible to use $BMEP$ or brake mechanical power as input parameters, instead of just $IMEPHP$, as done in Ayad et al. (2021). Using $IMEPHHP$ would require a priori knowledge of the engine operation cycle, and additional instrumentation in experiments. $BMEP$, on the other hand, can be more readily used in operating case definition both experimentally and computationally.

However, it is important to note that the proper representation of the friction model and gas exchange processes in this example case happened because they were unchanged from those in the base case. This might not be possible in every case as there are many factors that affect model accuracy and fidelity. This discussion is not in the scope of this chapter, but more details can be found elsewhere (Ayad et al. 2021).

Finally, the results for combustion phasing matched well to the base case, showing better correlation than those found in Ayad et al. (2021) using the same method. This is even more evident when examining the mass fraction burned curve shown in Fig. 10.4, which shows R^2 higher than 0.99. These improved results can be explained by the accurate representation of the studied engine's intake and exhaust processes. This is not always possible, as was the case in Ayad et al. (2021), because the data required for this representation was not available. This, in turn, can result in an incorrect representation of the air and fuel quantities present in the cylinder or fuel properties in the simulation. Thus, even if the pressure curves match well, the mass fraction burned curves may not.

To further validate the auto-calibrated model, the net heat release rate curves are compared and presented in Fig. 10.5. The curve shows good correlation between the predicted model and base case with R^2 higher than 0.99, but peak net heat release rate was slightly underpredicted. This can be explained by the same reason as the one given in relation to peak in-cylinder pressure. In any form, these results are competitive with other approaches (Gutiérrez et al. 2020), if not better (Rocha 2016).

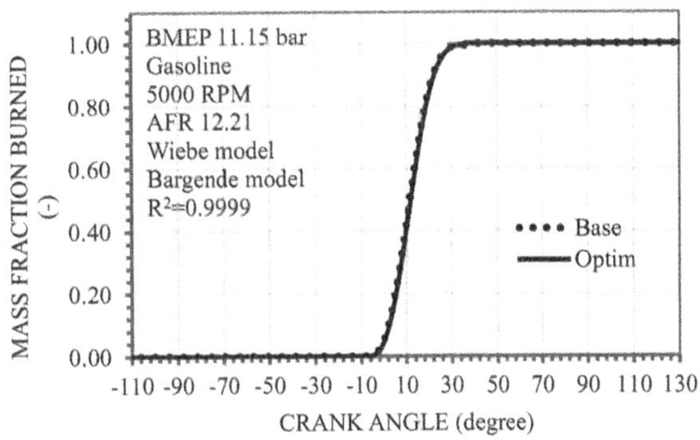

Fig. 10.4 Mass fraction burned curves corresponding to the auto-calibrated model (Optim) and the base case (Base) for the example case using Wiebe combustion model and Bargende wall heat transfer model

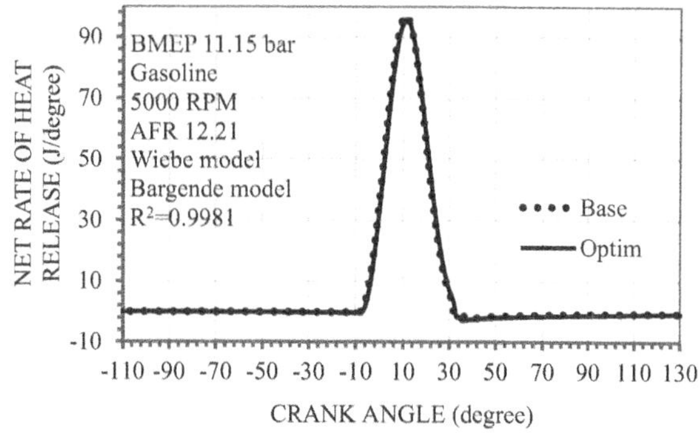

Fig. 10.5 Net rate of heat release curves corresponding to the auto-calibrated model (Optim) and the base case (Base) for the example case using Wiebe combustion model and Bargende wall heat transfer model

10.4.2 Exergy Analysis of the Auto-calibrated Model

Here, the results corresponding to the application of exergy analysis to the auto-calibrated model are presented and discussed.

Figure 10.6 presents both the mass fraction burned and the terms of the exergy balance *versus* crankshaft angle. This graph shows that prior to the combustion start, the cylinder charge's exergy increases due to the compression work being converted into thermomechanical exergy. However, during the combustion and expansion

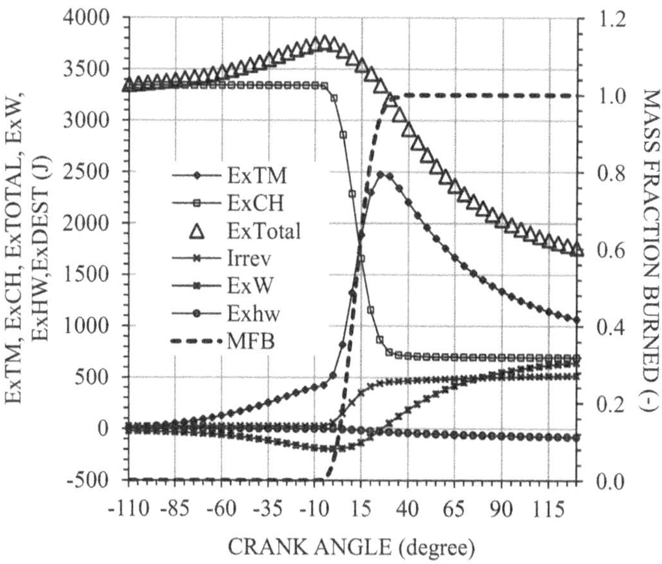

Fig. 10.6 Terms of the exergy balance (J) *versus* crankshaft angle for the auto-calibrated model. Mass fraction burned is plotted on the secondary axis

phases, the chemical exergy of the air fuel mixture is converted into thermomechanical and chemical exergy of the burned gas mixture, and transferred via work, via heat, and destroyed due to irreversibilities.

Since the exhaust gases exergy is lost to the atmosphere when the exhaust valve opens, they represent available work that is wasted (Azevedo Neto 2013). In the same graph, one can verify that the charge's total exergy remains relatively high at exhaust valve opening, due in part to its chemical and thermomechanical exergies. The chemical exergy is still high because of the incomplete combustion associated with fuel rich operation, meaning that there is unburned fuel, carbon monoxide and other unburned hydrocarbons present in the exhaust gas mixture. The thermomechanical exergies, on the other hand, are associated with the high temperatures and pressures at the opening of the exhaust valve.

The exergy destroyed due to irreversibilities, on the other hand, shows the contribution of the mixing and combustion irreversibilities and cannot be identified by using traditional first law analysis (Rakopoulos 1993). It is clear from the significant increase in the exergy destruction term during the fuel burning period, that the combustion process is the primary contributor to the destruction of exergy. It illustrates the importance of improving the combustion processes in order to potentially recover some of the exergy that would have been otherwise destroyed, and convert it to useful work. Such improvements must also consider other factors that cause exergy destruction, such as the mixing process (Heywood 2018), or even the fuel composition (Rakopoulos et al. 2008).

Figure 10.7 shows a Sankey diagram that illustrates how much of the ingoing exergy is transferred due to work (Ex_W), via heat transfer (Ex_Q), due to the exhaust gases (Ex_e) and destroyed exergy due to irreversibilities (Ex_{dest}) in the high-pressure phase of the engine operating cycle.

The main purpose of an internal combustion engine is to convert the chemical exergy of the fuel into shaft work, which is favored by higher compression ratios (Lior and Rudy 1988). This parameter, however, is limited in spark ignition engines due to detonation and knock, and altering it requires physical modifications to the engine (Ravi et al. 2017). Similarly, turbocharging and supercharging are widely recognized alternatives to recover some of the exergy contained in the exhaust gases and convert it into Ex_W (Heywood 2018). Both are also knock limited and their use must be carefully assessed for the purpose, operating conditions and fuel used. Bottoming the internal combustion engine with other cycles to form combined cycles presents multiple possibilities for further recovery of exhaust gas exergy, which would be wasted otherwise.

One can see from Ex_Q that only a small part of the ingoing exergy in this case is transferred via heat transfer. This can, again, be due to the fuel rich operation that decreases in-cylinder temperatures. This is also reflected in the high value of Ex_{dest} that indicates how much of the ingoing exergy is destroyed by irreversibilities, including those caused by the incomplete combustion and low in-cylinder temperatures. Operating slightly-lean and turbocharging might result in higher in-cylinder temperatures and lower destruction of exergy due to irreversibilities. On the other hand, operating with high excess air may also result in exergy destruction due to

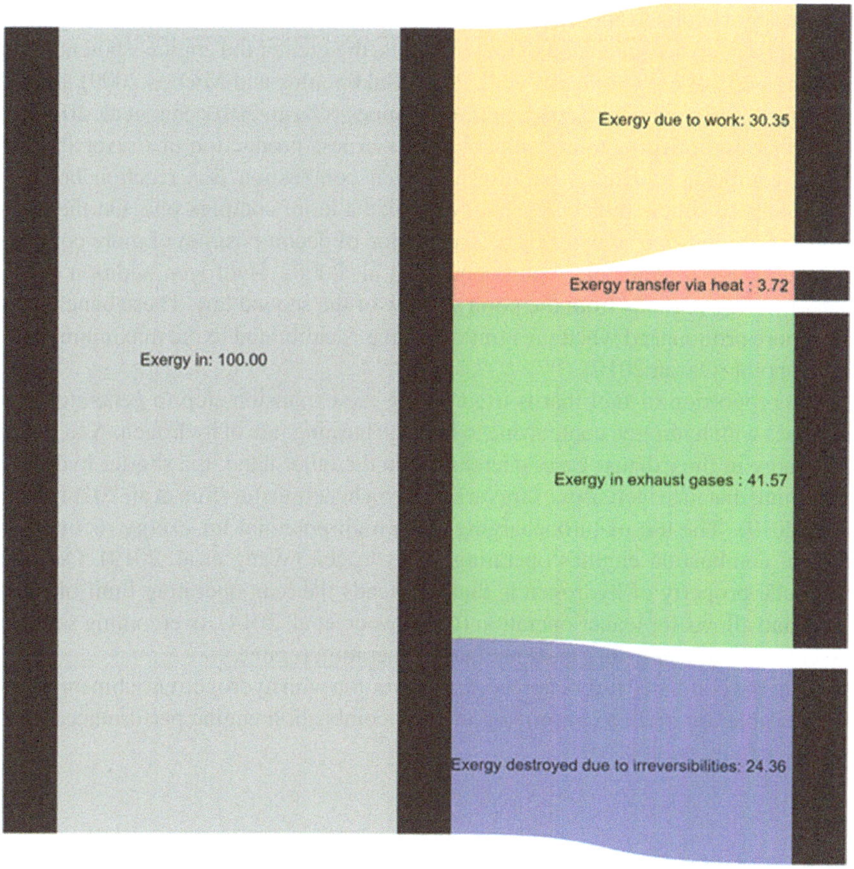

Fig. 10.7 Sankey diagram illustrating how the ingoing exergy is transferred due to work, via heat transfer, via the exhaust gases, and destroyed due to irreversibilities in the high-pressure phase of the engine operating cycle

mixing of the fuel–air combustion products with the excess air and low in-cylinder temperatures (Heywood 2018). Furthermore, excess air operation is conducive to an increase in Ex_W, while also decreasing Ex_e.

Therefore, operating with leaner mixtures might be quite beneficial from a second law perspective. However, the narrow range for the lean operating limit of gasoline and other conventional fuels, and risks of misfire and knock, are barriers to lean operation. Finally, the supplementation of the main fuel with one that has beneficial properties might also contribute to improving the exergy transfer in the engine. This is the case with hydrogen addition.

Hydrogen as a fuel has higher second-law efficiency due to its greater compressibility, more efficient combustion, and lower specific fuel consumption. However, an engine powered only by hydrogen may have less brake work output than engines powered by gasoline (Nieminen and Dincer 2010). On the other hand, the greater the amount of hydrogen added to fuel mixtures, the greater the engine's potential for carrying out work (Rakopoulos et al. 2008; Rakopoulos and Michos 2009) and the efficiency of second law in spark ignition engines (Ozcan 2010; Sun et al. 2019).

Furthermore, hydrogen enrichment results in less production of irreversibilities during combustion. This is because hydrogen combustion is a reaction between two relatively simple molecules that synthesize a more complex one. On the other hand, the combustion of other fuels is a reaction of decomposition of more complex molecules into simpler ones (Rakopoulos et al. 2008). Hydrogen addition favors lean operating regimes from the point of view of the second law. These benefits are even more pronounced when the ignition timing is calibrated to the maximum brake torque point (Ozcan 2010).

The proportion of fuel that is used during the expansion step to generate work increases with hydrogen content due to the high burning rate of hydrogen. As a result, the exergy in the exhaust gases decreases. On the other hand, the shorter hydrogen quenching distance increases exergy transfer via heat transfer (Sun et al. 2019; Wang et al. 2019). The use of turbocharging has a high potential for exergy recovery in internal combustion engines operating on hydrogen (Wang et al. 2019). One last beneficial property of hydrogen is that it extends the lean operating limit of other fuels, and allows for leaner operation (Greenwood et al. 2014), overcoming some of the concerns of using pure gasoline in that operation regime.

Thus, the combined turbocharged, lean operation with hydrogen enrichment might be a feasible approach to improving internal combustion engine performance from the second law point of view.

10.5 Conclusion

A simulation and exergy analysis workflow is presented that includes: (i) a simulation method to calibrate the combustion and heat transfer model parameters, and (ii) exergy analysis to obtain deeper insights into the working processes happening in-cylinder in the engine. This workflow was demonstrated by applying it to an example case available in AVL Boost templates. The main conclusions drawn from the work presented in this chapter are:

- results obtained using the simulation method proposed compared well with those in the base case;
- a good representation of the friction model and intake and exhaust processes allows for freedom in determining the mechanical load parameter. Furthermore, it might lead to better representation of cylinder charge and, consequently, better predicted models. This accurate representation might not always be possible, however;
- inclusion of exergy analysis allows for additional insights and to further help guide both engine development, as well as engine calibration; and

- exergy analysis suggests that hydrogen enrichment, combined with turbocharging and lean operation can improve operation of this gasoline engine.

This workflow can be a powerful aid in developing and optimally calibrating spark ignition engines. The predictive simulation requires knowledge of parameters that can be mapped in the electronic control unit, thus simplifying the modelling process. This approach is flexible and has been shown here and elsewhere to work with a range of engines, fuels and operating conditions, while still achieving good accuracy (Ayad et al. 2021). The addition of exergy analysis to the workflow facilitates greater insights into sources of losses and possible enhancements and modifications. Consequently, this framework both reduces the extent of experimental testing required and steers further research.

Acknowledgements This study was financed in part by the Conselho Nacional de Desenvolvimento Científico e Tecnológico (CNPq), Brazil.

References

AVL (2019) AVL BOOST Version 2019 – Theory

AVL (2019) AVL BOOST Version 2019—user guide

AVL (2019) AVL Design Explorer Version 2019—DoE and Optimization

Ayad SMME, Vago CL, Belchior CRP, Sodré JR (2021) Cylinder pressure based calibration model for engines using ethanol, hydrogen and natural gas as alternative fuels. Energy Rep. https://doi.org/10.1016/j.egyr.2021.06.015

Ayad SMME, Belchior CRP, da Silva GLR, et al (2020) Analysis of performance parameters of an ethanol fueled spark ignition engine operating with hydrogen enrichment. Int J Hydrogen Energy 45. https://doi.org/10.1016/j.ijhydene.2019.05.151

Azevedo Neto RM (2013) Simulação Computacional e Análise Exergética de um Motor de Motocicleta de Baixa Cilindrada com Misturas de Gasolina e Etanol. UFRJ/COPPE

Baldi F, Johnson H, Gabrielii C, Andersson K (2014) Energy and exergy analysis of ship energy systems—The case study of a chemical tanker. In: Proceedings of 27th International conference on efficiency, cost, optimization, simulation and environmental impact of energy systems, ECOS 2014, vol 18, pp 82–93. https://doi.org/10.5541/ijot.70299

Bargende M (1991) Equations for calculating the non-steady state wall heat losses in the high pressure part of petrol engines; Ein Gleichungsansatz zur Berechnung der instationaeren Wandwaermeverluste im Hochdruckteil von Ottomotoren. Technische Hochschule Darmstadt (Germany). Fachbereich 16 - Maschinenbau

Bejan A, Tsatsaronis G, Moran MJ (1996) Thermal design and optimization. Wiley

BUENO J (2016) Estudo numérico da influência das caracter{\'\i}sticas de injeção de misturas óleo diesel-biodiesel-etanol nas emissões de NOx. Tese de D. Sc., COPPE/UFRJ, Rio de Janeiro, RJ, Brasil

Byers EA, Gasparatos A, Serrenho AC (2015) A framework for the exergy analysis of future transport pathways: application for the United Kingdom transport system 2010–2050. Energy 88:849–862. https://doi.org/10.1016/j.energy.2015.07.021

Caton JA (2006) Utilizing a cycle simulation to examine the use of EGR for a spark-ignition engine including the second law of thermodynamics. Am Soc Mech Eng Intern Combust Engine Div ICE 2006:1–16. https://doi.org/10.1115/ICEF2006-1508

Caton JA (2014) Combustion phasing for maximum efficiency for conventional and high efficiency engines. Energy Convers Manag. https://doi.org/10.1016/j.enconman.2013.09.060
Caton JA (2016) An introduction to thermodynamic cycle simulations for internal combustion engines. Wiley, USA
de Cristo BEB (2017) Análise Dos Parâmetros De Desempenho De Um Motor De Ignição Por Centelha Operando Com Gasolina Ou Etanol Com Adição De Hidrogênio
de Faria MMNMMN, Vargas Machuca Bueno JPJP, Ayad SMMESMME, Belchior CRPCRP (2017) Thermodynamic simulation model for predicting the performance of spark ignition engines using biogas as fuel. Energy Convers Manag 149:1096–1108. https://doi.org/10.1016/j.enconman.2017.06.045
Deb K, Pratap A, Agarwal S, Meyarivan T (2002) A fast and elitist multiobjective genetic algorithm: NSGA-II. IEEE Trans Evol Comput. https://doi.org/10.1109/4235.996017
Dincer I (2002) The role of exergy in energy policy making. Energy Policy 30:137–149
Gallo WLR (1990) Análise exergética de motores a gasolina e a álcool
Ghojel JI (2010) Review of the development and applications of the Wiebe function: A tribute to the contribution of Ivan Wiebe to engine research. Int J Engine Res 11:297–312. https://doi.org/10.1243/14680874JER06510
Gong C, Li Z, Yi L et al (2020) Comparative analysis of various combustion phase control methods in a lean-burn H2/methanol fuel dual-injection engine. Fuel 262:116592. https://doi.org/10.1016/j.fuel.2019.116592
Grasreiner S (2012) Combustion modeling for virtual si engine calibration with the help of 0D 3D methods. Tech Univ Bergakademie Freib 179
Greenwood JB, Erickson PA, Hwang J, Jordan EA (2014) Experimental results of hydrogen enrichment of ethanol in an ultra-lean internal combustion engine. Int J Hydrogen Energy 39:12980–12990. https://doi.org/10.1016/j.ijhydene.2014.06.030
Gutiérrez RHR, Monteiro UA, Vaz LA (2020) Predictive thermodynamic model of the performance of a stationary spark-ignition engine running on natural gas. J Brazilian Soc Mech Sci Eng 42:1–16
Heywood JB (2018) Internal combustion engine fundamentals, 2nd ed. McGraw-Hill Education
Jaber JO, Al-Ghandoor A, Sawalha SA (2008) Energy analysis and exergy utilization in the transportation sector of Jordan. Energy Policy 36:2995–3000. https://doi.org/10.1016/j.enpol.2008.04.004
Jalil-Vega F, García Kerdan I, Hawkes AD (2020) Spatially-resolved urban energy systems model to study decarbonisation pathways for energy services in cities. Appl Energy 262:114445. https://doi.org/10.1016/j.apenergy.2019.114445
Kotas TJ (2013) The exergy method of thermal plant analysis. Elsevier
Lanzafame R, Messina M (2002) Experimental data extrapolation by using V order logarithmic polynomials. In: Design, operation, and application of modern internal combustion engines and associated systems. ASME International
Lanzafame R, Messina M (2005) New gases thermodynamic properties models to predict combustion phenomena. In: SAE Technical Paper Series. SAE International
Lior N, Rudy GJ (1988) Second-law analysis of an ideal Otto cycle. Energy Convers Manag 28:327–334. https://doi.org/10.1016/0196-8904(88)90054-4
Matuszewska A, Owczuk M, Zamojska-Jaroszewicz A et al (2016) Evaluation of the biological methane potential of various feedstock for the production of biogas to supply agricultural tractors. Energy Convers Manag 125:309–319
Melo TCC (2007) Thermodynamic modeling of an Otto cycle flexible fuel type engine. Working with Gasoline, Ethanol and Natural Gas, p 169
Melo TC (2012) Análise experimental e simulação computacional de um motor flex operando com diferentes misturas de etanol hidratado na gasolina. UFRJ
Mert SO, Dincer I, Ozcelik Z (2007) Exergoeconomic analysis of a vehicular PEM fuel cell system. J Power Sour 165:244–252. https://doi.org/10.1016/j.jpowsour.2006.12.002
Moran MJ, Shapiro HN, Boettner DD, Bailey MB (2011) Fundamentals of engineering thermodynamics, 7th edn

Morris DR, Szargut J (1986) Standard chemical exergy of some elements and compounds on the planet earth. Energy. https://doi.org/10.1016/0360-5442(86)90013-7
Nieminen J, Dincer I (2010) Comparative exergy analyses of gasoline and hydrogen fuelled ICEs. Int J Hydrogen Energy 35:5124–5132. https://doi.org/10.1016/j.ijhydene.2009.09.003
Ozcan H (2010) Hydrogen enrichment effects on the second law analysis of a lean burn natural gas engine. Int J Hydrogen Energy 35:1443–1452. https://doi.org/10.1016/j.ijhydene.2009.11.039
Perini F, Paltrinieri F, Mattarelli E (2010) A quasi-dimensional combustion model for performance and emissions of SI engines running on hydrogen-methane blends. Int J Hydrogen Energy 35:4687–4701
Prah I, Trenc F, Katrašnik T (2016) Innovative calibration method for system level simulation models of internal combustion engines. Energies 9:708. https://doi.org/10.3390/en9090708
Rakopoulos CD (1993) Evaluation of a spark ignition engine cycle using first and second law analysis techniques. Energy Convers Manag 34:1299–1314
Rakopoulos CD, Michos CN (2009) Generation of combustion irreversibilities in a spark ignition engine under biogas-hydrogen mixtures fueling. Int J Hydrogen Energy 34:4422–4437. https://doi.org/10.1016/j.ijhydene.2009.02.087
Rakopoulos CD, Scott MA, Kyritsis DC, Giakoumis EG (2008) Availability analysis of hydrogen/natural gas blends combustion in internal combustion engines. Energy 33:248–255. https://doi.org/10.1016/j.energy.2007.05.009
Ravi K, Pradeep Bhasker J, Porpatham E (2017) Effect of compression ratio and hydrogen addition on part throttle performance of a LPG fuelled lean burn spark ignition engine. Fuel 205:71–79. https://doi.org/10.1016/j.fuel.2017.05.062
Reyes M, Melgar A, Pérez A, Giménez B (2013) Study of the cycle-to-cycle variations of an internal combustion engine fuelled with natural gas/hydrogen blends from the diagnosis of combustion pressure. Int J Hydrogen Energy 38:15477–15487. https://doi.org/10.1016/j.ijhydene.2013.09.071
Reyes M, Tinaut FV, Melgar A, Pérez A (2016) Characterization of the combustion process and cycle-to-cycle variations in a spark ignition engine fuelled with natural gas/hydrogen mixtures. Int J Hydrogen Energy 41:2064–2074. https://doi.org/10.1016/j.ijhydene.2015.10.082
Rocha HMZ (2016) Determinação dos efeitos da utilização de hidrogênio em grupos geradores a diesel operando com diferentes misturas diesel-óleo vegetal. UFRJ/COPPE
Rosen MA (2013) Using exergy to correlate energy research investments and efficiencies: Concept and case studies. Entropy 15:262–286
Rosen MA, Dincer I (2003) Exergy–cost–energy–mass analysis of thermal systems and processes. Energy Convers Manag 44:1633–1651
Sciubba E, Moran MJ (1995) Second law analysis of energy systems: towards the 21-st century: workshop. School of Engineering, University of Roma 1 "La Sapienza", 5–7 July 1995. CIRCUS, Roma
Shivapuji AM, Dasappa S (2017) Quasi dimensional numerical investigation of syngas fuelled engine operation: MBT operation and parametric sensitivity analysis. Appl Therm Eng. https://doi.org/10.1016/j.applthermaleng.2017.06.086
Souza Junior GC (2009) Simulação termodinâmica de motores diesel utilizando óleo diesel e biodiesel para verificação dos parâmetros de desempenho e emissões. Dissertação de Mestrado em Engenharia Mecânica. Universidade Federal do Rio de Janeiro, RJ, 2009, 139 p
Sun P, Liu Z, Yu X et al (2019) Experimental study on heat and exergy balance of a dual-fuel combined injection engine with hydrogen and gasoline. Int J Hydrogen Energy 44:22301–22315. https://doi.org/10.1016/j.ijhydene.2019.06.149
Valero A, Valero A, Calvo G et al (2018) Global material requirements for the energy transition. An exergy flow analysis of decarbonisation pathways. Energy 159:1175–1184. https://doi.org/10.1016/j.energy.2018.06.149
Vibe II, Meißner F (1970) Brennverlauf und kreisprozess von verbrennungsmotoren. Verlag Technik

Wang X, Sun B gang, Luo Q he (2019) Energy and exergy analysis of a turbocharged hydrogen internal combustion engine. Int J Hydrogen Energy 44:5551–5563. https://doi.org/10.1016/j.ijhydene.2018.10.047

Yeliana (2010) Parametric combustion modeling for ethanol-gasoline fuelled spark ignition engines

Yeliana Y, Cooney C, Worm J et al (2011) Estimation of double-Wiebe function parameters using least square method for burn durations of ethanol-gasoline blends in spark ignition engine over variable compression ratios and EGR levels. Appl Therm Eng. https://doi.org/10.1016/j.applthermaleng.2011.01.040

Chapter 11
Investigation on the Effect of Injection Timings on Combustion, Performance and Emissions of a Pure Methanol Fuelled DISI Engine Through 1-D Simulations

Vikas Kumar Sahu, Inderpal Singh, Atul Dhar, Parmod Kumar, and Pravesh Chandra Shukla

Abstract The engine researchers and auto makers are putting out significant effort to develop an alternative to petroleum fueled internal combustion (IC) engines for the production of energy in the automotive. Various emerging technologies like electric vehicles (EVs), fuel cells, hydrogen fueled engines etc. are being used as alternative option for IC engines. Biofuels utilization have shown advantages in terms of minimum modification in the existing engine technologies. On the other hand, spark ignition (SI) engine is being used for two-wheelers, lawn movers, aircraft engines, pumping and electricity generating engines and it may be challenging to replace these working engines in short time and it is expected that SI engine would continue to serve as power generating unit for the coming couple of decades. Alcohols have been treated as alternative fuel for internal combustion engines for a long time especially in SI engines by blending. However, alcohols have a lot of potential to be utilized independently in SI engines. In the present study, a detailed modeling work would be performed to investigate the effect of injection timings on the methanol fueled direct injection (DI) SI engine through one dimensional (1-D) simulations. This study suggested as the start of injection (SOI) situation retarded, the heat release rate (HRR) curve shifted to the left. There is hardly any difference in NOx emission depending on injection timing. With a -31° CA SOI at 3.5 kJ fuel energy content, there is a significant quantity of HC observed due to lower fuel efficiency at much advanced SOI conditions.

Keywords Methanol · Spark ignition · Direct injection · 1-D simulation · GT-SUITE

V. Kumar Sahu · P. Chandra Shukla (✉)
Department of Mechanical Engineering, Indian Institute of Technology Bhilai, Raipur, India
e-mail: pravesh@iitbhilai.ac.in

I. Singh · A. Dhar · P. Kumar
School of Engineering, Indian Institute of Technology Mandi, Mandi, India

A. K. Agarwal et al. (eds.), *Engine Modeling and Simulation*, Energy, Environment, and Sustainability, https://doi.org/10.1007/978-981-16-8618-4_11

Abbreviations

IC	Internal combustion
SI	Spark ignition
CFD	Computational fluid dynamics
MVEM	Mean value engine model
DI	Direct injection
DoE	Design of experiment
TDC	Top dead center
SOI	Start of injections
EVs	Electric vehicles
IVC	Inlet valve close
EVO	Exhaust valve open
HRR	Heat release rate
CPOA	Cylinder pressure analysis mode
1-D	One dimensional
EGR	Exhaust gas recirculation
SPI	Stochastic pre-ignition
HCCI	Homogeneous charge compression ignition
URANS	Unsteady reynolds averaged Navier–Stokes
0-D	Zero dimensional
3-D	Three dimensional
BSFC	Break specific fuel consumption
SITurb	Spark-ignition turbulent flame model
M + P	Measured and predicted
SOC	Start of combustion
CA2	The crank angle at which 2% of the total heat has released
CA90	The crank angle at which 90% of the total heat has released

11.1 Introduction

Gasoline-fueled spark ignition (SI) engines have been the finest option for mopeds and scooters since their introduction. However, the scarcity of petroleum fuel in future and concerns for dependency to petroleum rich countries compelled netizens to consider alternatives of gasoline. Methanol is the most promising alternatives to gasoline fuel in terms of efficiency, availability and renewability. Numerous engine designs are being investigated, including traditional as well as new engine concepts. Methanol has been used as fuel in engines for a long period of time. Their functioning in a traditional engine is also reliant on a number of characteristics such as injection timing, combustion duration and ignition delay period. Experimentation of these parameters on conventional gasoline engines required lot of resources. Hence simulation using software tools to predict the nature of combustion for alternative fuels before experimentation is the trend of the town now a days. Computational fluid dynamics (CFD) has led to the reduction in the lead time as compared to experimental work. CFD has become a vital tool in the field of engine research and development. Accurate development of the physical models and combustion kinetics has led to correct prediction of the parameters via simulations of internal combustion (IC) engines using CFD. In order to obtain precise/accurate results, understanding of the physical phenomenon, beforehand, as well as correct selection of the combustion models is essential. Complex combustion phenomenon in IC engines also requires fine tuning of certain numerical model parameters due to inaccessibility of certain parameters while experimentation that are necessary in setting up the one-dimensional (1-D) model.

Some of the previous works on modeling of engines and simulation have been discussed to understand the prerequisite of these operations. Piano et al. (2016) compared the results of engine experiments with non-predictive and predictive combustion model of same engine in 1-D GT-POWER software. They used a four-cylinder, common rail diesel engine, equipped with a variable geometry turbocharger and short route exhaust gas recirculation (EGR) system to validate in-cylinder pressure, heat release rate, NO_x and soot emissions of multitoned combustion model (Piano et al. 2016). Blumreiter et al. (2019) performed the simulations and suggested that in-cylinder design modifications, injection technique, and thermal barrier coating would boost the efficiency of an alcohol-fueled engine. Phenomenological combustion and emission models along with air path dynamics are developed by Uppalapati et al. (2019) in GT-SUITE using several engine test points. Further, they analyzed the design of experiments data generated using 1-D validated models and final calibration were again validated in engine test cell. They observed that, this process helped in reducing error induced by engine aging, test cell variation and engine to engine variations leading to better statistical models developed for calibration optimization and development (Uppalapati et al. 2019). By Bellis et al. (2021) a numerical analysis was conducted to examine the possible increase of thermal efficiency for ultra-lean operations in SI engine at a constant speed and load with varying air/fuel ratio. The commercial simulation software (GT-Power™) includes in-built templates that simulate the engine's internal-combustion, heat transmission, and emission levels. Their assessments showed that ultra-lean conditions result in both improved thermal efficiency and decreased NO_x emissions, while increasing HC formation (Bellis et al. 2021). Splitter et al. (2017) analyses the uncertainty in the time of igniting of the gasoline fuel in the presence of stochastic pre-ignition (SPI). Their results demonstrated that increasing inlet temperature and backpressure both triggered rises in pre-spark heat release and ignition sensitivity. Back pressure does, however, make more of a difference. They discovered that there are basic difficulties related to fuel ignition in highly-boosted downsized engines, which may lead to SPI (Splitter et al. 2017). By premixing, diluting, and adjusting the combustion timing, it is possible to improve operating efficiency of SI engine. Svensson and Verhelst (2019) investigated a multi-cylinder direct injection compression ignition (CI) engine simulation in which methanol and gasoline were used. Methanol was shown to be on average 5.5% more efficient than gasoline, due to the reduced exhaust loss. They reported that as a result of methanol's greater octane rating, the fuel injection can be completed before the beginning of ignition resulting in partly premixed combustion (Svensson and Verhelst 2019). Methanol has a greater oxygen content and combustion efficiency than gasoline. Kang et al. (2018) developed a model of a methanol-gasoline engine with a variable fuel capability, based on GT-Power software. The study found that compared to the original engine, the flexible hybrid fuel engine decreased fuel consumption from 3 to 10% (Kang et al. 2018). Methanol's high evaporation energy leads to low combustion temperature which results in low exhaust energy and the wall heat losses. Negus et al. (2021) designed a four-cylinder, spark ignited, direct injection passenger car engine model using 0/1-D methods in GT-POWER software combined with FKFS user cylinder object. Their results showed that the replacement of gasoline with methanol increased the efficiency by 8% at lower load and by 25% at full load. Methanol allowed increase in compression ratio by 4 units and still be operated with the optimal 50% mass fraction burned (mfb50%) of 8° CA at all times due to high knock resistance (Negüs et al. 2021). Jandl et al. (2016) set up a zero-dimensional (0-D) vaporization model to study the mixture formation and cold start capability of ethanol, methanol, and 2-butanol blends with gasoline. Their model was successfully able to calculate the vaporization behavior of different mixtures based on the chemical composition and the single component properties of fuels (Jandl et al. 2016). Methanol has a higher-octane number (102–107), which helps to decrease knocking despite of increasing the compression ratio in engines. Pucilowski et al. (2017) investigated the combustion behavior of methanol with Unsteady Reynolds averaged Navier–Stokes (URANS) simulations. Their simulations used a precise chemical kinetic mechanism and a well-stirred reactor methodology (Pucilowski et al. 2017). Zhang et al. (2016) constructed a semi-detailed chemical reaction mechanism for simulating the combustion of acetone and alcohols with diesel mixtures in a constant volume combustion chamber. They obtained reasonable agreement between experimental and simulated data of ignition delay, cylinder pressures, and heat release rates using shock tube and constant volume chamber simulations (Zhang et al. 2016). Pelucchi et al. (2017) explored the essential impact of chemical kinetics in the operability and efficiency maps of homogeneous charge compression ignition (HCCI) engines fueled with n-butanol and n-pentanol. They created a multi-zone kinetic model of the HCCI engine and simulated between inlet valve close (IVC) and exhaust valve open (EVO). The "onion-skin" method was used to design their multizone combustion chamber configuration (Pelucchi et al. 2017). Based on data from engine tests, mean value engine model (MVEM) is the basis of control design, which can predict engine dynamics with reasonable approximation by ignoring the heat loss and sub-cycle events for advanced internal combustion engines. Wu et al. (2011) worked on air path MVEM modeling method based on the 1-D detail model for a turbocharged diesel engine. In the heat release process is modelled by Tartakovsky et al. (2013), the velocity of fuel burning as a feature of fuel composition and air excessiveness is taken into consideration for methanol fuel. Based on the outcome of in-cylinder pressure data analysis Michos and Bikas (2020) developed a multi-zone combustion assessment tool for homogeneous charge SI engines using an analytically derived quasi-dimensional model. Bozza et al. (2019)

updated 0-D Turbulence Model for SI Engines' Tumble and Turbulence. Their results of 0-D derived mean flow velocity, turbulence intensity, and tumble velocity have excellent internal consistency with three dimensional (3-D) results. Hu et al. (2014) ran a 1-D model validation simulation of a substantially reduced SI engine, and scavenge valve timing including the lift was adjusted to achieve lowest break specific fuel consumption (BSFC). In a study conducted by Vieira et al. (2014), the experimental results of Baeta (2006) were compared to CFD simulations based on the parameters of combustion (i.e. BSFC etc.) of Baeta (2006). They reported that the suggested model performs well at matching the engine's testing outcomes (Baeta 2006).

The aim of this chapter was optimization, validation and modeling of a one-cylinder, four-stroke, direct injection spark ignition (DISI) engine to investigate the effect of injection timings on the methanol fueled DISI engine through 1-D simulations with the help of GT-SUITE software. Out of the available options of the combustion models available, predictive combustion model namely, Spark-Ignition Turbulent Flame model (SITurb). The predictive model is selected keeping in mind the interest of the study. Since, present work revolves around the effect of various SOI conditions which has direct or indirect impact on the burn rate, thus predictive model is most appropriate choice. The utmost need of the predictive combustion models is the requirement of the calibration to measured data to provide accurate results. The model was firstly calibrated using the experimental data from the published literature and four attributes (dilution effect multiplier, flame kernel growth multiplier, turbulent flame speed multiplier and Taylor length scale multiplier) optimized using integrated design optimizer tool of GT-POWER to obtain single set of constant values for all operating parameters. Further, to check the viability of the model the simulation was run at varying fuel energy content for the three different SOI conditions taken for the validation case. The results were compared and examined further in relation to methanol combustion with similar conditions.

11.2 Model Development

Modeling of engine setup require complete understanding of predeveloped templates in the software. In this chapter GT-SUITE software developed by Gamma Technologies have been utilized. GT-SUITE; a commercial simulation software is used for the modelling. It comprises of the GT-POWER engine library which is used as the engine and vehicle simulation tool (GammaTechnologiesLLC 2020). The solution is based on the 1-D fluid dynamics, solving the fluid flow and heat transfer for the engine system. It comprises of certain specialized models which helps to model the entire engine system accurately for analysis. Its predictive and non-predictive combustion models can be used for the combustion modelling for the prediction of the burn rate which is almost impossible to calculate while experimentation. Appropriate selection of the combustion model is needed as per simulation requirement. But for these

Table 11.1 Engine parameters (Li et al. 2019)

Scania D13 heavy-duty engine	
Description	Parameter values
Compression ratio	17.3:1
Displacement volume	2124 cm^3
Bore	130 mm
Stroke	160 mm
Connecting rod	255 mm
Speed	1200 rpm
Intake valve closing (IVC)	−141° CA TDC
Exhaust valve opening (EVO)	137° CA TDC
Number of nozzles	10
Nozzle hole diameter	0.175 mm
Distance between spark and injector	37 mm

models to do precise prediction of the parameters, beforehand, calibration and validation of the models is required. The numerical model developed is validated using literature published data.

11.2.1 Engine Parameters, Assumptions and Boundary Conditions

The study models a Scania D13 heavy-duty engine with only one cylinder fired. This study is based, keeping in mind the high knock resistance tendency of methanol. Thus, the compression ratio of 17.3 is used with direct injection and spark ignition fuel strategy. Single injection with constant injection pressure throughout the injection duration is used. The injector with 10 number of nozzle holes is used for the injection of the methanol fuel. Other specifications of the engine are as specified in Table 11.1.

The initial and boundary conditions for the simulation are taken as cited in the literature Li et al. (2019). The simulated pressure data is matched with the experimental data reported in the literature (Li et al. 2019). Correct matching of the pressure trace is required to consider the validation of the model for precise and accurate measurement of the heat release rate (HRR), heat transfer and emissions.

11.2.2 Modeling Matrix

In this chapter, the measured and predicted combustion profiles were compared thoroughly for consistency checks. Then multipliers for the predicted combustion profile was optimized. With these optimized multipliers the forward run simulations were

Table 11.2 Modeling matrix

Parameter	Value		
(A) *Validation modeling matrix*			
Fuel energy content	3.5 kJ		
Start of injection	−27° CA TDC	−29° CA TDC	−31° CA TDC
(B) *Test modeling matrix*			
Fuel energy content	3.5 kJ	3.8 kJ	4.2 kJ
Start of injection	−27, −29, and −31° CA TDC	−27, −29, and −31° CA TDC	−27, −29, and −31° CA TDC

validated with respect to the measured data. Further, the same set of inputs were applied to evaluate the effect of varying injection timing on combustion, performance and emission parameters. For the same process modeling matrix of considered variables are specified in Table 11.2.

11.2.3 Engine Model Preparation

Engine model consists of major components, their connections and references. This model needs to filled by accurately measured parameters for more accurate prediction. Here, engine model preparation was performed in two steps. First step constitutes preparation of optimization model for closed volume analysis. Second model was utilized for the validation and test simulations.

As shown in Fig. 11.1 optimization model comprises of fuel injector, engine cylinder and engine crank train. This model applied cylinder pressure analysis mode (CPOA) for the closed volume combustion chamber (IVC to EVO). This was used to obtained the combustion multipliers for the methanol fuel in specified DISI engine. The engine used in the study is a modified Scania D13 one-cylinder, four-stroke direct injection spark ignition diesel engine taken from literature. Table 11.1 summarizes the major parameters of the engine. In injector amount of mass injected, injection timing, nozzle specifications and injection profile has been defined. Engine crank train was prepared with the cylinder geometry and IVC. Further initial condition for air inlet was defined, so that based on defined volumetric efficiency density of air can be corelated. EVO was defined in the engine cylinder template.

In the optimization, engine cylinder initials (volumetric efficiency, residual and exhaust gas recirculation, air trapping ratio and fuel trapping ratio) have been entered (Venkateshmohan and Kumar 2015). SITurb, (a predictive combustion and emission model for DISI engines) was used to simulate combustion, hence multipliers were calculated for the SITurb approach. In the mathematical formulas below (Eqs. 11.1–11.3), four multipliers employed for calibration in the 'SITurb' model was given (Heywood 2018);

Fig. 11.1 Optimization model from GT SUITE software

$$(f_{\text{Dillution}}) = 1 - 0.75 * C_{\text{DE}}(1 - (1 - 0.75 * C_{\text{DE}} * \text{Dillusion})^7) \quad (11.1)$$

$$S_T = C_{TFS}u'\left(1 - \frac{1}{1 + C_{FKG}\left(\frac{R_f}{L_i}\right)^2}\right) \quad (11.2)$$

$$\lambda = \frac{C_{TLS}L_i}{\sqrt{Re_t}} \quad (11.3)$$

Here;
Dilution = residual mass fraction in unburned zone
C_{DE} = Dilution Effect Multiplier
C_{TFS} = Turbulent Flame Speed Multiplier
C_{FKG} = Flame Kernel Growth Multiplier
C_{TLS} = Taylor Length Scale Multiplier
λ = Taylor microscale length

S_T = turbulent flame speed
u′ = turbulent intensity
L_i = integral length scale
Re_t = turbulent Reynolds number

The obtained value of $f_{Dillution}$, ST, and λ from above mentioned equation was further used to calculate the entrained mass and burned mass inside the combustion chamber. The Woshini model (which uses a formula that roughly emulates the traditional Woschni correlation without the swirl) was used to simulate in-cylinder heat transfer. Formula for average gas velocity without swirl is mentioned in Eq. (11.4) below (Woschni 1967);

$$w = \left[C_1 S_p + C_2 \frac{V_S T_1}{P_1 V_1} (P - P_0) \right] \tag{11.4}$$

Here; S_P = Mean piston speed
P = Instantaneous cylinder pressure
C_1 and C_2 = Constant dependent at process
$P_1 T_1 V_1$ = working fluid pressure, temperature and volume
P_0 = Motored cylinder pressure at same crank angle as P
w = average gas velocity
V_S = Swept volume

Emission components (CO, CO_2 and HC) was calculated based on the equilibrium of mole fractions as described by Olikara and Borman (1975). NO_x component was calculated from the extended zeldovich mechanism (Mitchell 1976; Nelson 1976).

Calculated burn rate for a thermodynamic cycle have been entered as an input parameter for approximation of in-cylinder pressure and heat release rate. This engine cylinder was set for reverse run at first step. In this, case setup was created for obtaining the combustion profile for the reverse run, which was compared with the measured combustion profile input given. Further passing the consistency checks of reverse run the same model was run again with some modifications in the engine cylinder (CPOA was changed to measured and predicted (M + P)). In this, case setup was changed from combustion profile to the engine combustion multiplies for obtaining the optimized value of those. A separate optimization module has been created to run the iteration. The optimization will be run for 1021 cycles iteration reaching a set of multipliers fit for using the parameters that have been specified in advance. The obtained optimized multipliers for three SOIs have been validated in the forward run model.

11.2.4 Test Model Preparation

The model validation tests were carried out for above mentioned three SOIs at a speed of 1200 rpm with fuel energy content of 3.8 kJ. In this engine model as shown in Fig. 11.2, the intake air pressure and temperature was considered as the atmo-

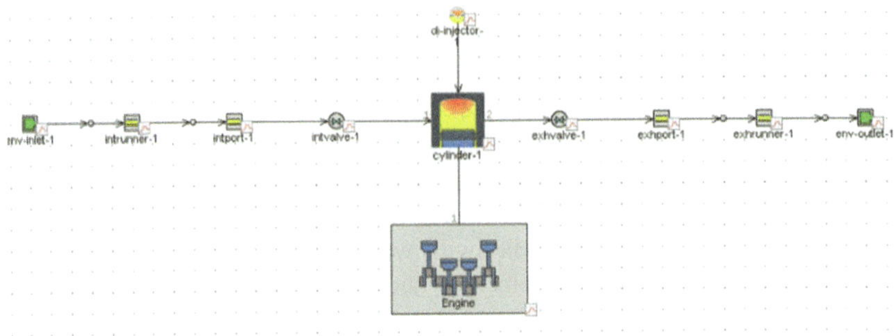

Fig. 11.2 Test model from GT SUITE software

spheric environmental conditions in end environment template. This end environment template was connected to the inlet runner and then inlet port. Dimensions of the inlet port and inlet runner was defined with the initial air conditions. Lift array data with respect to cam angle has been provided in the inlet valve. The engine cylinders was connected to the injector and engine crank train as in the previous model. In the cylinder heat transfer object, wall temperature was defined as previously. However, the engine cylinder initials was replaced by air inlet conditions. Flow object was defined as per the piston dimensions. and combustion object SITurb was defined with optimized multiplier values. Using this technique, the cylinder's shape, spark and injection positions and timing, air turbulence and fuel characteristics, as well as other combustion variables was considered to predict the combustion profile. This model was used for validation, in which the burn rate parameters were not required and forward simulation was applied by setting off the reverse run parameters. The fuel is tracked as it is injected, evaporates, mixes with the surrounding gas, and burns in this model. This model may be used to monitor single or many injection events, and each injection event is described as an injection pulse that is recorded independently of all other pulses. SITurb is more efficient in terms of both predicted accuracy and runtime compared to nonpredictive and semi predictive SI module. Similar to the intake valve, exhaust valve for cylinder was defined with its lift array. Further, the engine combustion simulations were carried out on GT-SUITE software by means of 1-D gas dynamics simulation code.

11.3 Methanol Fueled Direct Injection Spark Ignition (SI) Engine 1-D Modelling

The initial and boundary conditions for the simulation are taken as cited in the literature Pucilowski et al. (2017). The simulated pressure data is matched with the experimental data reported in the literature (Pucilowski et al. 2017). Correct matching of the pressure trace is required to consider the validation of the model for precise and

Table 11.3 Comparison of experimental and modelling values for base case simulation

Comparison of experimental and modelling values for base case simulation		
Parameter	Experimental	Simulation
Speed	1200 rpm	1200 rpm
Cylinder head temperature	390 K	390 K
Cylinder liner temperature	390 K	390 K
Piston temperature	420 K	420 K
Lambda	1.5	1.55
Intake air temperature	358 K	358 K
Intake air pressure	1 bar	1 bar

accurate measurement of HRR, heat transfer and emissions. To check the accuracy of the model the pressure trace for three different SOI conditions is matched with the experimental data reported at one fuel energy content i.e., constant mass of fuel injected for all three SOI conditions. Small tuning of the parameters such as lambda was done well within 3% range to exactly tune the volumetric efficiency and to pass the consistency check for the reverse run simulation (CPOA). The spark timing and injection pressure are used as cited in the literature with value of −2° CA TDC and 2000 bar, respectively. The boundary temperature conditions, and other parameters tuned are shown in Table 11.3 with comparison to that reported in the literature.

11.3.1 Comparison of the In-Cylinder Pressure and HRR

Figure 11.3 shows the pressure trace along with the heat release rate obtained from the forward run simulation after calibrating the combustion SITurb model. For the calibration of the combustion model, the single optimized set of the multipliers (dilution effect multiplier, flame kernel growth multiplier, turbulent flame speed

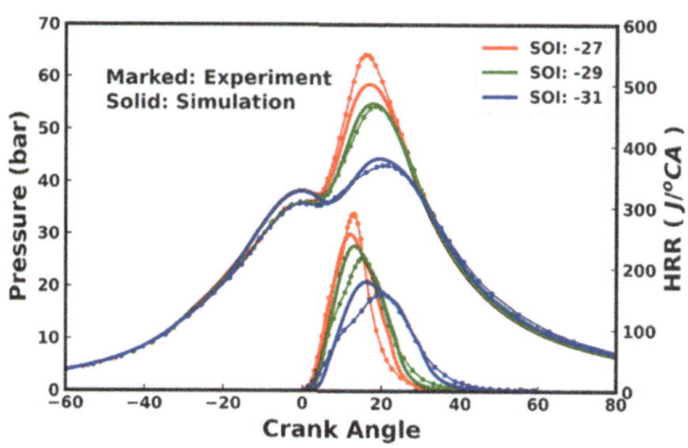

Fig. 11.3 Validation case pressure and HRR vs crank angle trace for 3.5 kJ fuel energy content at different SOI conditions

multiplier and Taylor length scale multiplier as explained in the earlier sections is obtained by running the integrated design optimizer tool of GT-POWER. The optimized set of values for the attributes; dilution effect multiplier, flame kernel growth multiplier, turbulent flame speed multiplier and Taylor length scale multiplier are 0.7401, 2.8505, 0.7976 and 1.1782 respectively. The plot is shown in comparison to the experimental results obtained in the Li et al. (2019). On comparison of the model results under the fuelled conditions they show a good match with that of the experimental results. Little deviation in the pressure value at the end of the compression stroke observed for the simulated curve is due to little tuning in the lambda value done as specified in the Table 11.4 to pass the consistency check of reverse run simulation, to make zero the value of cumulative burn-rate during the compression stroke to avoid inaccuracies. This was done due to some assumptions taken in the model for certain unknown parameters during experiment run like air trapping ratio and residual mass fraction. For the available fuel mass in the cylinder to match the predicted fuel burned this cumulative error needs to be zero, therefore little tuning to the value of lambda was done, but well within the range as specified in the literature corresponding to the constant value of indicated mean effective pressure of 8 bar used during experimentation. This pressure trace comparison is done for the fuel energy content of value 3.5 kJ, kept constant, for all three SOI conditions. The deviation in the peak pressure value for the all the SOI conditions is below 8%. With best match got for the SOI condition of −29 and −31° CA TDC and all follow the similar trend as that of the experimental data. These findings helped us to validate the model and to further run the simulations at other different engine conditions.

11.3.2 Effect of Injection Timings

Similarly, keeping other parameters same the other set of simulations were run on changing the fuel energy input content in the cylinder with one set being at 3.8 kJ and other at 4.2 kJ and comparing the results with that of 3.5 kJ fuel energy content for all three SOI conditions.

11.3.2.1 Effect of Injection Timing on In-Cylinder Pressure, Temperature, and Heat Release Rate

Figure 11.4 shows the in-cylinder temperature along with the heat release rate for all three cases of fuel energy content. The injection timing in swept from −31° to −27° TDC. There is an increase in the pressure value with the increase in the fuel content input in the cylinder. On comparing for one case, the pressure increases with the retarding of the injection timing. The SOI condition of −31 shows the lower pressure and correspondingly the heat release rate for all the cases with least value of 44.391 bar at −31° CA TDC injection timing. The maximum value of peak pressure is found for the injection timing of −27° TDC with the value of 81.433 bar for 4.2 kJ

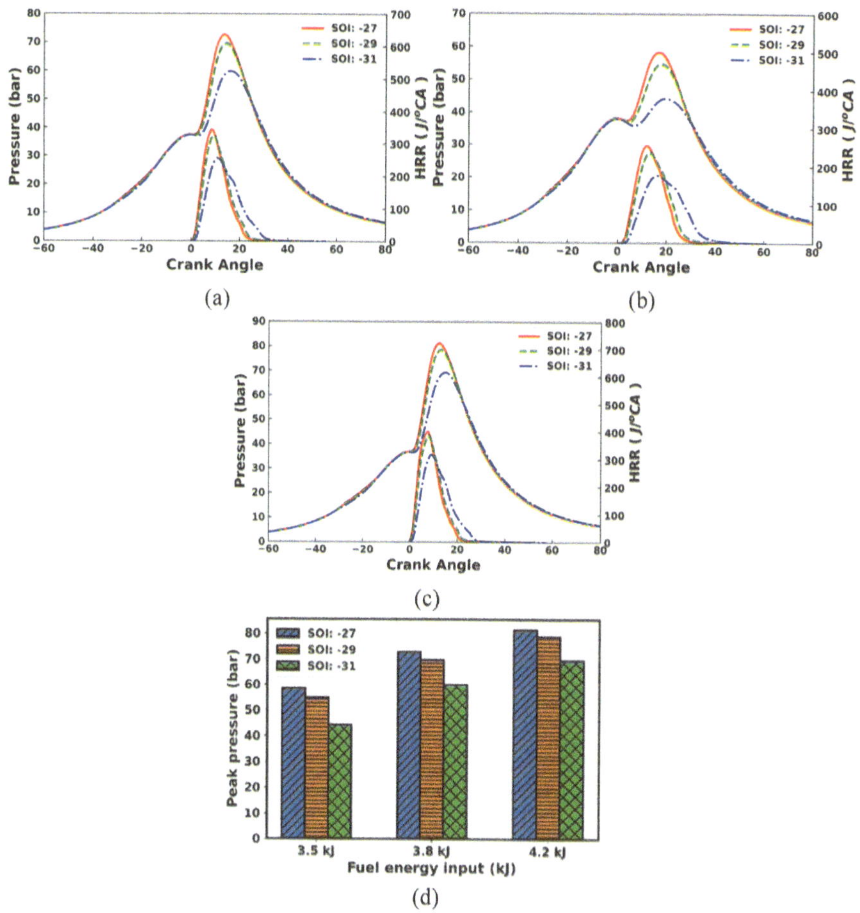

Fig. 11.4 Variation of pressure and heat release rate for different SOI conditions at **a** 3.5 kJ, **b** 3.8 kJ, **c** 4.2 kJ fuel energy input and **d** peak pressure value for the respective conditions

as the fuel energy input. The trend of the HRR curve can be seen to shift towards left with retardation of the SOI condition. The reason behind the increase in the peak in-cylinder pressure due to retarded SOI is due to late injection, methanol does not get enough time for mixing and forming high stratified mixture near to the spark at the time of ignition (Li et al. 2010), thus on combustion leading to higher pressure values.

With advance injection, fuel gets higher time to mix with air and thus forming a quite lean mixture at the time of ignition. Furthermore, since for the first case mass of fuel injected was already on the lesser side and on advance injection or higher mixing time, it got leaner at the time of ignition as well as uniformly distributed leading to lower combustion efficiency and poor flame propagation. Thus, lower peak in-cylinder pressure for SOI condition of −31° TDC (Li et al. 2010). Studies reveal that turbulent kinetic energy influences the flame propagation speed (Kim et al. 2015). It was found that −31° TDC, showed lower TKE value as compared to other SOI conditions, thus further giving inferences of lower fuel efficiency of this SOI condition as compared others. There is improvement seen in the combustion characteristics for higher fuel injected cases, for advanced −31° TDC SOI condition as the fuel/air mixture does not remain too lean now during the time of ignition i.e., there is a little stratified mixture near spark plug at time of ignition. This can also be related to the fact that high fuel energy input keeping other parameters same the fuel/air mixture now tends to move towards the stoichiometric mixture giving better combustion. Thus, leading to higher peak in-cylinder pressure for advanced SOI conditions as well as seen in Fig. 11.4b, c.

Peak pressure value for the SOI condition of −27° TDC is highest for all three cases with the value of 58.516 bar, 72.837 bar and 81.433 bar, which is 24.13, 17.7 and 14.72% higher than its counterpart i.e., −31 SOI in the respective fuel content condition. Peak pressure rise rate also shows similar trend as that of the peak pressure and HRR as should be case. Figure 11.5a shows the variation of the peak pressure rise rates for all the injection timing conditions. The highest value of 6.29 bar/degree is found for the SOI condition of −27° TDC. The peak pressure rise rate data is required for the analysis of the knock tendency at a particular SOI condition. Peak pressure rise rate location and peak pressure location CAD moves towards the TDC with retarding of the injection timing. Table 11.4 shows the location of the peak pressure along with its value.

Figure 11.5b shows the variation of the peak in-cylinder temperature values at different injection timings. The highest value of cylinder temperature is obtained for SOI condition of −27° TDC with the value of 2083.42, 2200.61 and 2297.4 K for respective fuel energy content. The reason behind it is the presence of high stratified charge during the time of spark due to retardation of the injection timing thus sudden combustion leading to high in-cylinder temperature.

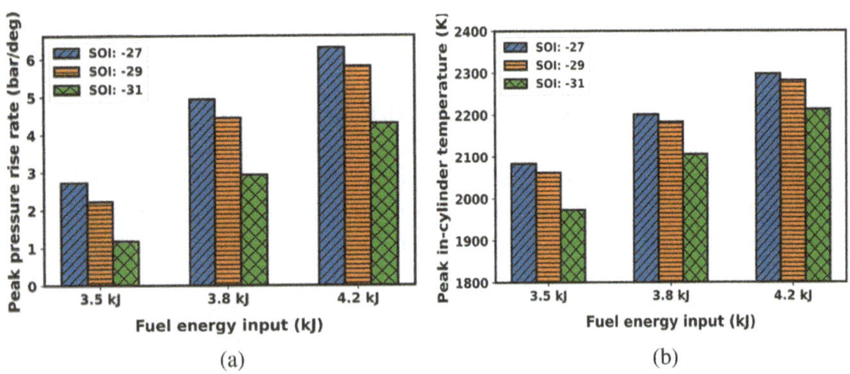

Fig. 11.5 Effect of SOI condition at different fuel energy content on **a** peak pressure rise rate (bar/deg) and **b** peak in-cylinder temperature (K)

Table 11.4 Location of peak pressure CAD and along with its value

Fuel energy input	Peak pressure (bar)			Peak pressure location (CAD)		
	SOI: −27	SOI: −29	SOI: −31	SOI: −27	SOI: −29	SOI: −31
3.5 kJ	58.51665	54.828053	44.39126	16.699066	17.83517	19.552729
3.8 kJ	72.83795	69.735115	59.92228	13.844577	14.607818	16.040384
4.2 kJ	81.43321	78.71064	69.441956	12.131549	12.903793	14.768547

11.3.2.2 Effect of Injection Timing on Ignition Delay and Combustion Duration

The start of combustion and end of combustion parameters are decided depending upon the criteria setup for the mass fraction of fuel burned and might vary differently with the study. Crank angle at which 2% of the total heat has released (CA2) is taken as the start of combustion (SOC). With fixed spark timing of −2° CA, ignition delay is calculated as the difference between the CA2 value and the spark timing. End of combustion is taken at crank angle at which 90% of the heat has released (CA90). Combustion duration, in this study is taken as the burn duration from 10 to 90% of the mass fraction burned. Figure 11.6a, b shows the ignition delay period and the combustion duration at three different fuel energy input to cylinder when the injection timing is swept from −31° CA to −27° CA.

SOC seems to occur earlier as the injection timing is swept from −31° CA to −27° CA i.e., the SOC location tends to move towards to the TDC leading to decrease in ignition delay period for injection timing of −27° CA. This difference in the SOC location at lower fuel energy content is more prominent leading to higher difference in the value of ignition delay period at lower fuel content than higher fuel content. This extend of SOC towards TDC at higher fuel energy content is found to be lower when injection timing is swept from −31° CA to −27° CA, leading to marginal difference in ignition delay period for 4.2 kJ fuel energy content. This further explains the increasing combustion duration as we advance the injection

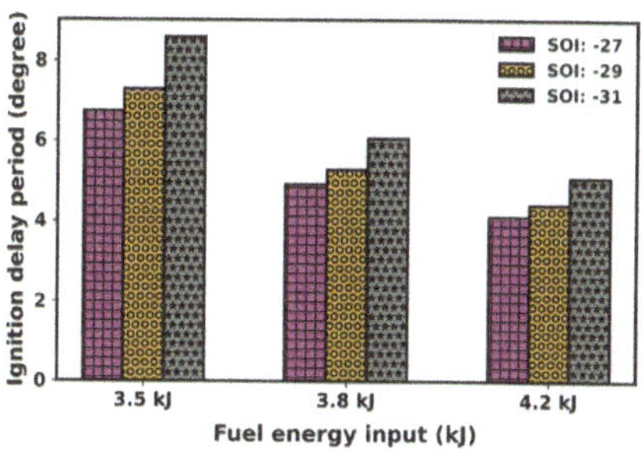

(a)

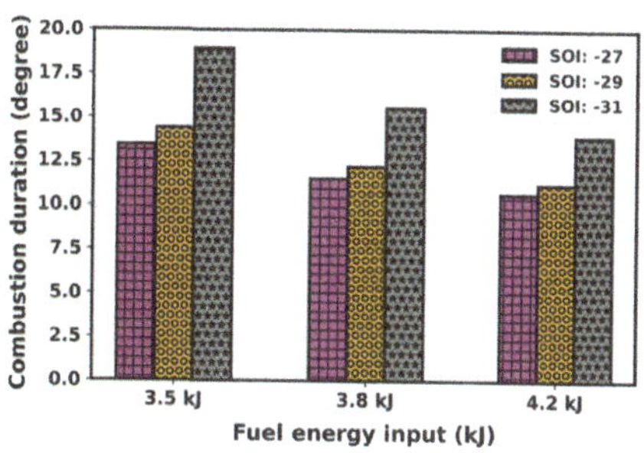

(b)

Fig. 11.6 Effect of SOI condition at different fuel energy content on **a** ignition delay period and **b** combustion duration timing. Lower ignition period at retarded SOI conditions leads to higher in-cylinder temperature, further increasing the burning velocity (Li et al. 2016). Combustion duration is in inverse relation with the in-cylinder temperature; therefore, we found a decrease in combustion duration as we retard the injection timing. The difference in combustion duration is lower at higher fuel content case since the fuel/air mixture tends towards the stoichiometric mixture thus fuel distribution becomes better and more homogeneous for early injection as well.

11.3.2.3 Effect of Injection Timing on Performance

Figure 11.7a shows the variation of the indicated thermal efficiency with changing injection timings. There is no as such proper trend that the parameter follows with the change of injection timing. The highest thermal efficiency is found for the −27° CA for the fuel energy input of 3.5 kJ with value of 42.75%. The entire range of thermal efficiency for all the cases is found in the range of 41–43% which is good as compared to its counterpart diesel engine at this compression ratio. These higher power output and efficiencies of methanol due to its high knock resistance tendency helps for its usage in the high compression ratio DISI engines. Least thermal

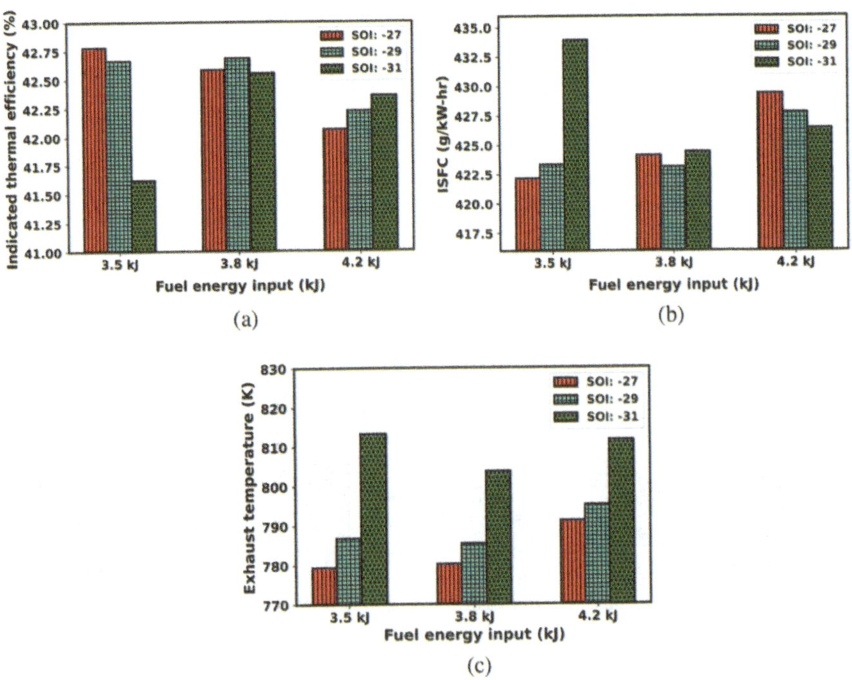

Fig. 11.7 Effect of SOI condition at different fuel energy content on **a** indicated thermal efficiency, **b** indicated specific fuel consumption, **c** exhaust temperature

efficiency of 41.62% is found for the injection timing of −31° CA due to lesser area under the P-θ curve for this case. Figure 11.7b, c shows the indicated specific fuel consumption and exhaust/outlet temperature for all the cases. And it is well known fact that higher the exhaust temperature lower is the thermal efficiency, thus it can be seen that higher temperatures are found for −31° CA injection timing for all the fuel input content. Hence, lower is its thermal efficiency. ISFC is found to be least with the value of 422.16 g/kW-h for −27° CA for 3.5 kJ fuel energy content input. The justification for the indicated thermal efficiency can be better understood based on the energy distribution discussed in the later section. The heat transfer losses and exhaust losses monitor the indicated thermal efficiency (Caton 2018).

11.3.2.4 Effect of Injection Timing on Emission

Figure 11.8a, b shows the HC and NO_x emissions with the changing SOI conditions for three different fuel energy input content. There is a marginal difference in the NO_x concentration with the change of injection timing for a particular case. But there is a decrease in the NO_x (ppm) concentration with the increase of the fuel energy input. The reason behind this is as all other parameters are kept constant while changing only the fuel energy input to cylinder making a leaner mixture at 3.5 kJ fuel content, while 4.2 kJ makes a lesser lean air/fuel mixture. This tends to make the mixture more towards the stoichiometric air/fuel ratio, leading to better combustion and lesser emissions. The NO_x concentration is highest for the 3.5 kJ fuel energy content lying within the range of 1030–1035 ppm. Similarly, HC concentration too show marginal difference as we swept the injection timing from −27° CA to −31° CA TDC for 3.8 and 4.2 kJ energy content. There is a drastic higher amount of HC concentration with value of 6.55 ppm for the SOI condition of −31° CA as could be seen in Fig. 11.8a. This is due to fact that at this condition the flame propagation is low and fuel mixture is lean as well leading to low in-cylinder and thus making lower fuel efficiency. This tends to this higher concentration of HC (ppm) at this condition.

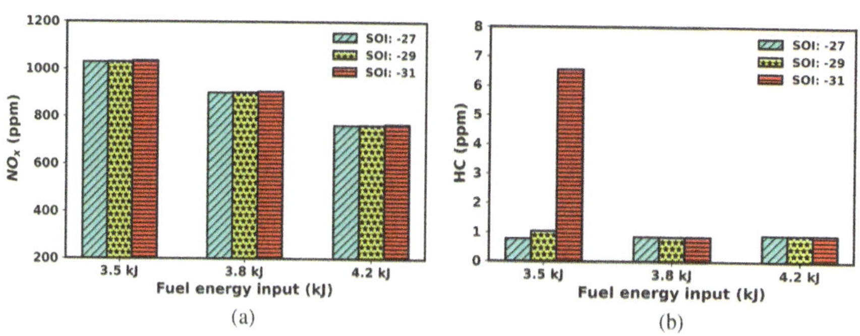

Fig. 11.8 Effect of SOI condition at different fuel energy content on **a** NO_x emissions, **b** HC emissions (ppm)

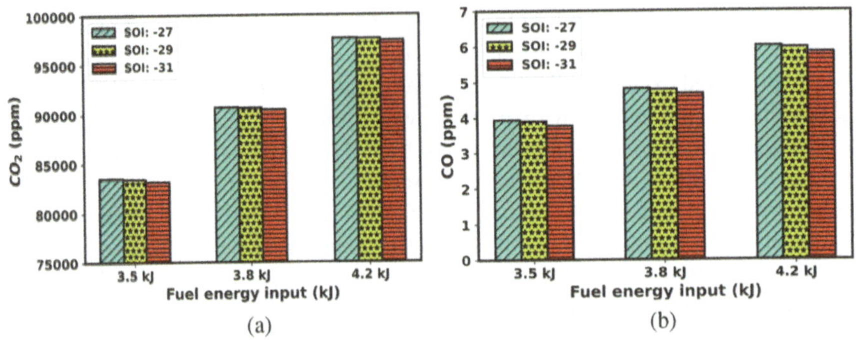

Fig. 11.9 Effect of SOI condition at different fuel energy content on **a** CO_2 emissions, **b** CO emissions (ppm)

There is not much difference in the value of these emissions as we retard the injection timing as shown in Fig. 11.9. These emissions are the result of complete combustion and will show a trend opposite to that of HC. The maximum value of CO_2 and CO emission is found for the SOI condition of -27° CA TDC with the value of 97,638 ppm and 5.83 ppm respectively.

11.3.2.5 Energy Distribution with Varying SOI Conditions for Different Fuel Energy Input

Energy distribution comparison for all the SOI conditions under investigation for three different fuel energy input is shown in Fig. 11.10. At lower fuel energy input of 3.5 kJ, the exhaust losses overpower the heat transfer losses. Though the heat transfer losses for the advanced SOI condition decreases gradually as compared to early injection, but higher exhaust losses become dominant causing decreased indicated thermal efficiency for −31° CA TDC. The least being for this case with the value of 41.62% At higher fuel energy content of 4.2 kJ (Fig. 11.10c), there is negligible difference in the losses either exhaust or heat transfer, hence corresponding little variation in the indicated thermal efficiency. From all the plots in Fig. 11.10 whenever there is a decrease in the heat transfer loss there is corresponding increase in the exhaust losses i.e., there is rather no conversion of it to work output. Thus, building up scope for the waste heat recovery and after-treatment of exhaust waste heat source.

11.4 Summary

1D numerical simulations supported by the experimental data have been carried out on the high compression ratio DISI methanol fueled engine. The utmost requirement

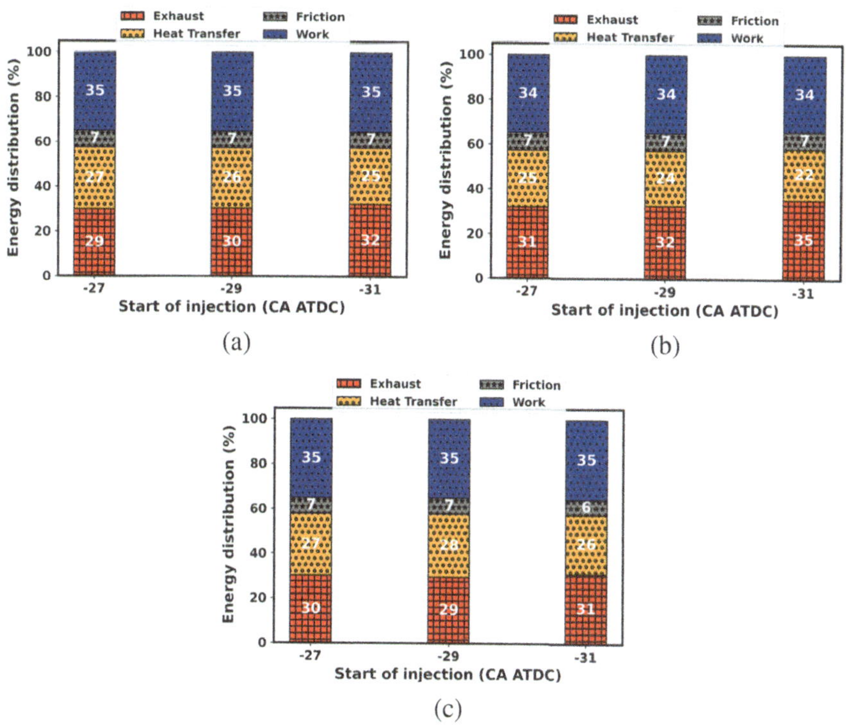

Fig. 11.10 Energy distribution with varying SOI condition **a** 3.5 kJ, **b** 3.8 kJ and **c** 4.2 kJ fuel energy input

of the 1D model is to select the appropriate combustion model out of the available options of non-predictive, semi-predictive and predictive based on the interest of the study. Predictive combustion model was taken up for this study as model was developed to study the effect of injection timing, since it has a significant or direct influence on the burn rate. These models required calibration to the measured data for which the model was validated on taking up the experiment data from the published literature to provide accurate results. Further, the investigation was taken up with varying fuel energy content to find the viability of the model at more operating parameters. The combustion characteristics of methanol fuel is found to be sensitive to the injection timings. The study reveals that too much advanced injection duration i.e., at −31° TDC leads to very lower flame propagation speed and thus lower in-cylinder turbulence kinetic energy, which leads to quite lower in-cylinder pressure. It was observed that HRR curve shifts to the left as the SOI condition retarded. The reason behind which was, less time available for mixing of the fuel, resulting in greater in-cylinder pressure and temperature. The maximum cylinder temperature and pressure is achieved at −27° TDC owing to the existence of a highly stratified charge at the moment of ignition, resulting in rapid combustion. The SOI condition of −31° TDC showed improvement in fuel efficiency at higher fuel energy input as the equivalence ratio tended towards the stoichiometric mixture leading to higher stability. Not much variation in NO_x emission is observed with the change in injection time. Yet when the fuel energy input increases, the NO_x (ppm) concentration decreases because of the leaner mixture at 3.5 and 3.8 kJ fuel energy input, while mixture closer to the stoichiometric at 4.2 kJ fuel energy content resulted in improved combustion and reduced emissions. There is a much large amount of HC with $-31°$ CA SOI at 3.5 kJ fuel energy content, lesser homogeneous mixing due to much leaner fuel/air mixture, resulting in a low in-cylinder pressure and low fuel efficiency. With much advancement in the SOI condition, there exists prolonged ignition delay, thus causing leaner fuel/air mixture to abnormal combustion, higher unburned emissions. This phenomenon though show improvement when the fuel/air mixture is taken closer to the stoichiometric mixture. It was our best attempt to model the higher compression ratio engine with DISI mode of combustion for methanol as the fuel. Since, the utmost requirement of the 1D combustion model is to calibrate it with the existing measured data, due to limited experimental data existing till date, there was some shortcoming left in its fully accurate calibration. Thus, exists a few quantitative discrepancies but the consistent trend of in-cylinder pressure and HRR in the validation results show reliability of numerical model for qualitative analysis. Greater range of measured data available will further help in increasing the accuracy as well as add better viability to the existing model for more operating range.

It is a worldwide urge for creation of simulation tools to accurately predict the combustion, performance and emission of IC engines. However, with more than hundred operating variables, inclusion of new techniques, addition of variety of fuels and increasing complexity of ICE, prediction within acceptable accuracy limits seems to be far from reach.

References

Baeta JGC (2006) Metodologia experimental para a maximização do desempenho de um motor multicombustível turboalimentado sem prejuízo à eficiência energética global

Blumreiter J, Johnson B, Zhou A, Magnotti G, Longman D and Som S (2019) Mixing-limited combustion of alcohol fuels in a diesel engine. SAE Technical Paper 2019-01-0552, 2019. https://doi.org/10.4271/2019-01-0552

Bozza F, Teodosio L, De Bellis V, Fontanesi S, Iorio A (2019) A refined OD turbulence model to predict tumble and turbulence in SI engines. SAE Int J Engines 12:15–30

Caton JA (2018) Maximum efficiencies for internal combustion engines: Thermodynamic limitations. Int J Engine Res 19:1005–1023

De Bellis V, Malfi E, Bozza F, Kumar D, Serrano D, Dulbecco A and Zaccardi J-M (2021) Experimental and 0D numerical investigation of ultra-lean combustion concept to improve the efficiency of SI engine. SAE Int J Adv Curr Prac Mobil 3(4):1993–2008. https://doi.org/10.4271/2021-01-0384

GammaTechnologiesLLC (2020) GT SUITE Engine Performance Application Manual

Heywood JB (2018) Internal combustion engine fundamentals. McGraw-Hill Education

Hu B, Akehurst S, Brace C, Copeland C and Turner J (2014) 1-D simulation study of divided exhaust period for a highly downsized turbocharged SI engine-scavenge valve optimization. SAE Int J Engines 7:1443–52

Jandl S, Schacht H-J, Schmidt S, Dawin U, Kölmel A, Leiber S (2016) Evaporation and cold start behavior of bio-fuels in non-automotive applications. SAE Int J Engines 9:2381–2395

Kang Y, Shi X, Ni J and Qi H (2018) Experiments of methanol-gasoline SI Engine performance and simulation of flexible fuel characteristic field. SAE Technical Paper 2018-01-0927. https://doi.org/10.4271/2018-01-0927

Kim T, Song J and Park S (2015) Effects of turbulence enhancement on combustion process using a double injection strategy in direct-injection spark-ignition (DISI) gasoline engines Int J Heat Fluid Flow 56:124–36

Li J, Gong C-M, Su Y, Dou H-L, Liu X-J (2010) Effect of injection and ignition timings on performance and emissions from a spark-ignition engine fueled with methanol. Fuel 89:3919–3925

Li Y, Jia M, Chang Y, Kokjohn SL, Reitz RD (2016) Thermodynamic energy and exergy analysis of three different engine combustion regimes. Appl Energy 180:849–858

Li Y, Bai X-S, Tunér M, Im H G and Johansson B (2019) Investigation on a high-stratified direct injection spark ignition (DISI) engine fueled with methanol under a high compression ratio. Appl Thermal Eng 148:352–62

Michos KN, Bikas G (2020) Quasi-dimensional multi-zone combustion diagnostic tool for si engines with novel NOx and CO emissions models. SAE Int J Adv Curr Pract Mobil 2:1818–1848

Mitchell RE (1976) Nitrogen oxide formation from chemically-bound nitrogen during the combustion of fossil fuels [Extended Zeldovich reactions]. Sandia Labs, Livermore

Negüs F, Grill M, Bargende M (2021) Efficiency potential of SI engines with gasoline and methanol: a 0D/1D investigation. SAE Technical Paper 2021-01-0385. https://doi.org/10.4271/2021-01-0385

Nelson HF (1976) Nitric oxide formation in combustion. AIAA J 14:1177–1182

Olikara C, Borman GL (1975) A computer program for calculating properties of equilibrium combustion products with some applications to IC engines. SAE Technical Paper 750468:1975. https://doi.org/10.4271/750468

Pelucchi M, Bissoli M, Rizzo C, Zhang Y, Somers K, Frassoldati A, Curran H, Faravelli T (2017) A kinetic modelling study of alcohols operating regimes in a HCCI engine. SAE Int J Engines 10:2354–2370

Piano A, Millo F, Boccardo G, Rafigh M, Gallone A and Rimondi M (2016) Assessment of the predictive capabilities of a combustion model for a modern common rail automotive diesel engine. SAE Technical Paper 2016-01-0547. https://doi.org/10.4271/2016-01-0547

Pucilowski M, Jangi M, Shamun S, Li C, Tuner M and Bai X-S (2017) Effect of start of injection on the combustion characteristics in a heavy-duty DICI engine running on methanol. SAE Technical Paper 2017-01-0560. https://doi.org/10.4271/2017-01-0560

Splitter D, Kaul B, Szybist J, Jatana G (2017) Engine operating conditions and fuel properties on pre-spark heat release and SPI promotion in SI engines. SAE Int J Engines 10:1036–1050

Svensson E, Verhelst S (2019) Simulation based investigation of achieving low temperature combustion with methanol in a direct injected compression ignition engine. SAE Technical Paper 2019-01-1152. https://doi.org/10.4271/2019-01-1152

Tartakovsky L, Baibikov V, Veinblat M (2013) Comparative performance analysis of SI engine fed by ethanol and methanol reforming products. SAE Technical Paper 2013-01-2617. https://doi.org/10.4271/2013-01-2617

Uppalapati LR, Vernham B, Wei Y (2019) Development and validation of engine calibration using 1D predictive models. SAE Technical Paper 2019-01-1135. https://doi.org/10.4271/2019-01-1135

Venkateshmohan V, Kumar M (2015) Predictive diesel combustion using DI-pulse in GT-power (Master's thesis)

Vieira TR, Baeta JGC, Sodré JR (2014) Computer simulation of turbocharged si engine running on ethanol. SAE Technical Papers 2014-36-0366. https://doi.org/10.4271/2014-36-0366

Woschni G (1967) A universally applicable equation for the instantaneous heat transfer coefficient in the internal combustion engine. SAE Technical Paper 670931:1967. https://doi.org/10.4271/670931

Wu H, Wang X, Winsor R, Baumgard K (2011) Mean value engine modeling for a diesel engine with GT-power 1D detail model. SAE Technical Paper 2011-01-1294. https://doi.org/10.4271/2011-01-1294

Zhang S, Xu Z, Lee T, Lin Y, Wu W and Lee C-F (2016) A semi-detailed chemical kinetic mechanism of acetone-butanol-ethanol (ABE) and diesel blends for combustion simulations. SAE Int J Engines 9:631–40

Part V
Exhaust Heat Recovery

Chapter 12
Automotive Exhaust Thermoelectric Generator Unit Integrated to Exhaust Noise Muffler: Heat Recovery and Noise Attenuation Simulations

Sarthak Nag, Atul Dhar, and Arpan Gupta

Abstract This study aims at energy harvesting mufflers that utilize thermoelectric generators (TEG) to convert waste heat from the engine exhaust into electricity and simultaneously reduce engine noise. The recovered electricity can be used to power the auxiliary units in the automobile and can effectively improve the overall efficiency of the system. In this study, an automotive exhaust thermoelectric generator (AETEG) unit is fabricated to extract the waste exhaust heat from the engine. To further enhance the performance of AETEG unit, changes in the internal geometry of the unit are proposed, and a systematic computational study is carried. Investigation results indicate that transformations in the internal geometry enhance the heat transfer rate due to lower flow stratification and induced turbulence. This shows increasing potential for higher electricity generation. However, an increase in backpressure due to flow obstruction remains a problem with increasing baffle count. The internal geometry with six baffles gave the best thermal performance with backpressure within acceptable limits for the studied cases. Additionally, that AETEG unit serves as a noise muffler and attenuates the exhaust noise by 53 dBA.

Keywords Waste heat recovery · Thermoelectric generator · Automotive muffler · Sound attenuation

12.1 Introduction

The exponential growth in the human population in the last few hundred years, coupled with a multi-fold increase in the global economy, has led to accelerated environmental degradation. The increasing human population is stressing the earth's environment due to excessive global pollution, majorly due to transportation, power plants, and industries (Mulla and Rabinal 2019; Birol 2016; Huang et al. 2017). Moreover, due to the increasing energy demands to sustain the economic rise, the demand for conventional fuels has gone up in recent times, which has pushed us

S. Nag · A. Dhar (✉) · A. Gupta
School of Engineering, Indian Institute of Technology Mandi, Mandi 175005, India
e-mail: add@iitmandi.ac.in

A. K. Agarwal et al. (eds.), *Engine Modeling and Simulation*, Energy, Environment, and Sustainability, https://doi.org/10.1007/978-981-16-8618-4_12

towards the cliff of the global energy and environmental crisis (Smil 2017; Tripathi et al. 2018). Renewable and cleaner energy has been in the limelight due to the advances made in the past decade in terms of the economics involved (Karthick et al. 2019; Gu et al. 2019). Even though there is a strong need to shift to cleaner alternative fuel to fulfil future energy demands, there is also a necessity to minimize heat losses in the present and prospective future systems to increase the overall efficiency and extractability in these systems. Waste heat is a by-product in any thermal process which is dissipated unused in the environment. Despite having a lower exergy, waste heat recovery, in recent years, has been in spot-light for the improvement it brings to the system efficiency and the gains in terms of pollution reduction (Nag et al. 2018; Shu et al. 2018; Muralidhar et al. 2018; Nour Eddine et al. 2018; Thacher et al. 2007; Deng et al. 2013; Liu et al. 2016; Orr et al. 2016; Omer et al. 2020; Angeline et al. 2019).

Although internal combustion (IC) engines are much celebrated due to their high-power output and spread throughout the transportation industry, they are not one of the most efficient systems, and their overall efficiency is typically around 30%. A typical IC engine only converts 25 to 30 percent of the available fuel energy into useful work (Orr et al. 2016; Tripathi et al. 2017; Demir and Dincer 2017). Around 40% of fuel energy goes out unutilized through the tailpipe of the engine in the form of heat in exhaust gases. Another 30% of the fuel energy passes to the coolant through engine walls in the form of heat. Figure 12.1 shows the energy split diagram in a typical engine (Nag et al. 2018). This points towards high potential of the waste heat recovery from an automotive engine for improving the sustainability and overall efficiency of the internal combustion engines.

Waste heat is low-grade energy, having low exergy, and thus requires an efficient device for its conversion into useful energy. Broadly, exhaust heat recovery (EHR) can be classified into two segments: (i) solid state-based EHR and (ii) fluid-based EHR (Tripathi et al. 2017). The Carnot efficiency limits all the thermodynamic processes, and despite the massive development in fluid-based devices, their upper limit will always be less than the Carnot efficiency as it is based on the sensible heat input and the space for the advancement is limited (Henry and Prasher 2014). On the other hand, the solid-state energy conversion devices are independent systems having so moving parts and thus no thermal limitations. Importantly, they can utilize a constant temperature input, which is very similar to the Carnot cycle (Henry and Prasher 2014). For the solid-state, the employment of thermoelectric generator (TEG) modules to recover

Fig. 12.1 Energy split diagram in a typical engine

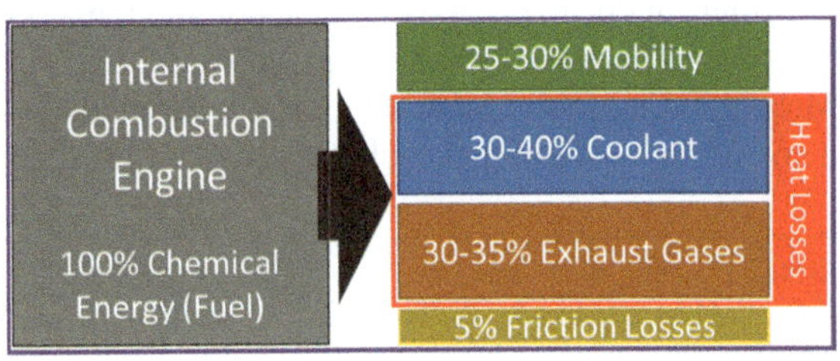

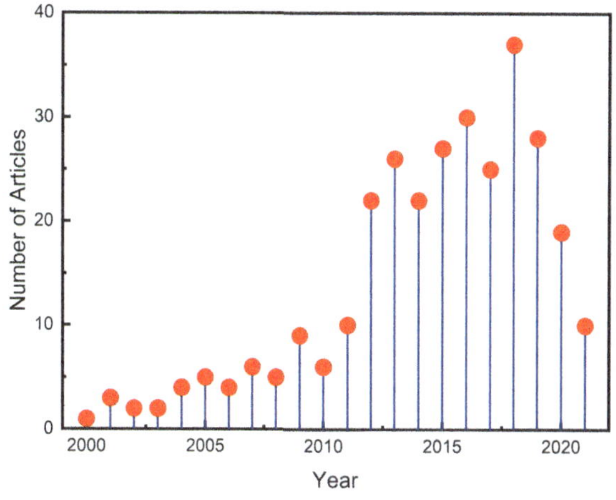

Fig. 12.2 Data of the number of articles published in the domain of automotive thermoelectric generation. The data is obtained by searching for keywords "automotive" and "thermoelectric" on the Web of Science portal on 19-June-2021

exhaust waste heat and convert it into useful electricity has gained an increased pace in the past decade (Pang et al. 2018). Figure 12.2 shows the substantial increase in the number of articles in the automotive thermoelectric domain published since year 2000, which justifies the interest in the TEG for recovering automotive exhaust heat. Thermoelectric generation works on the principle of Seebeck effect which states that when there is a temperature difference between two different conductors or semiconductors, a potential difference is produced between them, which is proportional to the difference in temperature. A TEG consists of several thermocouples depending on the size and capacity of the TEG module. A thermocouple is a small unit that is made of *P*-type and *N*-type semiconductors. These thermocouples are then connected; electrically in series and thermally in parallel, and placed between two highly thermally conductive ceramic wafers to form a TEG (DiSalvo 1999; Soleimani et al. 2020).

The application of TEG to automobiles dates back to early 1900s when Leigh Hale invented a thermo-electric based battery for motor vehicles (Hale 1913). Since then, there has been continuous effort to improve both the efficiency of thermoelectric materials as well as that of the heat exchanger. In 1976, Serksnis reported a TEG unit utilizing exhaust gas as a heat source to replace the existing charging system in vehicles and proposed a design for its development. The author also highlighted high reliability and lower maintenance in their utilisation due to the absence of any moving parts in the presence of high thermal loads (Serksnis 1976). Furthermore, to replace the shaft driven alternators, Hi-Z Technology developed a 1 kW output thermoelectric generator for class eight Diesel truck engines under U.S. Department of Energy and California energy commission funding since 1992 (Bass et al. 1994). It employs 72 TEG modules made of bismuth telluride (Bi_2Te_3). The power output increased from 400 to 1068 W with an improvement in the internal design of the heat exchanger. In specific, their study highlighted the importance of internal geometry optimization for extracting high power output. Many automotive organizations also entered the domain of EHR with Nissan using a Si-Ge TEG modules (Crane et al. 2013). They were able to produce 36 W on a 3-L road vehicle at 60 kmph uphill drive. The design of the AETEG unit is most critical in improving their efficiencies in actual deployment, and the general design issues in their development for automotive applications, such as highly fluctuating flow and temperatures, a necessity to increase the radiator capacity or selection of TEG material were highlighted by Stabler (Stabler 2009). In 2009, BMW group expanded their efficient dynamics by developing a TEG unit for their vehicles (Liebl et al. 2009). They reported a fuel reduction of around 5–7% and power generation of 200 W at the speed of 135 kmph in their prototype testing, which seems to highlight the potential of TEG deployment. They also restated the importance of advancements in thermoelectric materials in terms of economics and power generation for successful power generation. Espinosa and co-workers took the modelling approach for thermoelectric generators applied to the diesel automotive heat recovery (Espinosa et al. 2010). They found a strong correlation between the generator architecture and the operating conditions of the engine. They also highlighted the importance of a bypass line in the exhaust to save the thermoelectric materials from high temperatures at heavy loads. Saqr and co-workers reviewed the potential replacement of alternator or assistance in charging the battery unit by thermoelectric generation and presented various objectives and challenges to define the future path (Saqr et al. 2008). The study concluded the importance of heat exchanger geometry, material, insulation site, and cooling system for success of TEG in automotive. Karri, Thacher and Helenbrook tested the deployment of TEG on two cases: a sports utility vehicle and a stationary compressed natural gas-fuelled engine generator set (Karri et al. 2011). They also used two different materials for their study. The quantum well thermoelectric material proved to be highly effective against the Bi_2Te_3 material. The relative fuel savings using quantum well material were 1.25 and 3% for sports utility vehicles and stationary generator engines. They also discussed about the parasitic losses in their study, with power required to transport the TEG unit being dominant in sports utility vehicle case. Su et al. made significant progress in optimizing the AETEG unit by testing the modifications both experimentally and using simulations. They also highlighted the importance of striking a balance between heat recovery and acoustic attenuation for future studies. In 2014, researchers from Wuhan University made a big leap and developed AETEG unit, which meets the automotive electrical requirement with a power output of 944 W (Liu et al. 2015). Their AETEG is called "four-TEG system" and is made from 240 thermocouples. It showed the feasibility of the replacement of alternator with AETEG unit. To enhance the heat transfer through modification in internal geometry, Lu et al. studied three geometries: empty cavity, 1-inlet 2-outlet, and 2-inlet 2 outlet (Lu et al. 2013). They measured it by supplying the hot cases through blower to mimic engine test conditions. The heat transfer enhancement in 1-inlet 2-outlet was best for the studied cases. Liu et al. also carried a similar study with fishbone-shaped heat exchangers and chaos-shaped heat exchangers and found chaos-shaped to be better in terms of heat transfer (Liu et al. 2014). Lee et al. designed a heat exchanger which incorporated fins for enhanced heat transfer between TEG modules and the exhaust gases (Lee and Bae 2008). Hatami et al. used the multi-objective optimization-based approach and proposed a cylindrical structure with radial fins for optimized thermal performance (2015). However, the addition of components in the exhaust line increases exhaust backpressure and system complexity. The low figure of merit of thermoelectric materials translates into the lower efficiency of the AETEG unit (Vining 2009; Sharma et al. 2014). Therefore, the consumption of additional power for cooling the AETEG unit by a standalone system has been a problem and more often inconvenient. A study by Kim et al. (2011) have suggested the integration of AETEG cooling with the engine cooling loop. To integrate muffler with AETEG unit, Deng et al. studied the acoustic attenuation of AETEG unit (Deng et al. 2016). The study identifies a need for attention to the muffler-AETEG integrated systems for reducing the backpressure. Wang and co-workers used the dimpled surface instead of fins to reduce the backpressure and reported a reduction of 20.57% in the AETEG unit (Wang et al. 2018). They also found a net power improvement by massive 173% by using dimples in their study. Pacheco et al. focused on the thermal control of the automotive TEG units (2020). They embedded heat pipes in their TEG unit to spread the heat in the longitudinal direction to avoid overheating in specific regions. They reported 5.4% reduction in fuel consumption and extracted an average power of 532 W during a driving cycle. Zhao et al. proposed an intermediate fluid thermoelectric generator system to separate the exhaust line with the thermoelectric modules (Zhao et al. 2019). In their system, the temperature distribution across the modules was more uniform due to the heat transfer being achieved using phase change of the intermediate fluid. He et al. modelled a multilayer thermoelectric generator and matched their study with the heavy-duty diesel engine (2020). They studied the effects of key parameters on the hot and cold sides and measured its performance under transient cycle. Luo et al. also did a numerical and experimental study on the performance of converging TEG unit where they modified the geometry of the system (2020). They reported an even temperature distribution in their proposed design and verified the same using experimental study.

Therefore, to study the effect of the addition of internal modifications on the thermal performance, pressure drop, and exhaust noise attenuation, this work simultaneously analyses the thermal and acoustic noise attenuation performance of the AETEG unit. This work also includes experimental investigation for maximum power output at three tested loads. The addition of internal baffle structures enhances the thermal performance of the AETEG unit. These inclusions of internal structures also showed exhaust noise attenuation by multiple reflections of the noise wave inside the AETEG cavity. The results show the promising potential of AETEG integrated with mufflers for energy harvesting. Its implementation to the IC engine-powered vehicles may lead to the component reduction in the exhaust line components and potential alternator unit replacement.

12.2 Methods and Procedures

12.2.1 Power Output and Voltage Measurements

The experimental setup was designed to study the net potential of waste heat recovery from engine exhaust. Moreover, the other important feature was its implementation in cooling the exhaust gases for exhaust gas recirculation (EGR) for limiting emissions of oxides of nitrogen (NOx), which has been published elsewhere (Nag et al. 2019). Figure 12.3 shows the schematic diagram of the experimental setup for exhaust heat recovery. The AETEG unit was fabricated using an aluminium sheet of thickness 3 mm, which served as a medium of heat exchange between the exhaust and the TEG modules. Iron and Brass are other potential contenders, but the primary reason for choosing aluminium was its high thermal conductivity, thermal uniformity and the economics involved for manufacturing. Iron has low thermal uniformity (Deng et al. 2013), and brass is quite expensive when compared with aluminium. The complete setup was installed on the exhaust line of the test engine (make: Kirloskar; model: TV1). The basic details of the test engine have been explained elsewhere (Nag et al. 2019). The load on the engine was varied using a transient dynamometer controlled using a variable frequency drive (VFD).

To measure the exhaust inlet and outlet temperature, two p-type thermocouples were installed at their respective positions (see Fig. 12.4a). For the energy recovery, a total of 8 modules (make: Hi-Z, model: 14 V) were used. These TEG modules were chosen as they are bismuth telluride (Bi_2Te_3) based and can operate at high temperatures of 250 °C for continuous use and 400 °C for intermittent use. 2 parallel arrays containing 4 modules in series in each array were used for experimental studies to maximize both current and voltage in the circuit. These TEG modules were placed on the aluminium box using thermal grease to minimize the thermal contact resistance (Fig. 4b).

Cooling of the cold side of TEG module is important to maintain high-temperature gradient for better output. Water, at room temperature (32 °C), was used as the coolant fluid. A cooling plate with internal channels for the flow of water was fabricated using

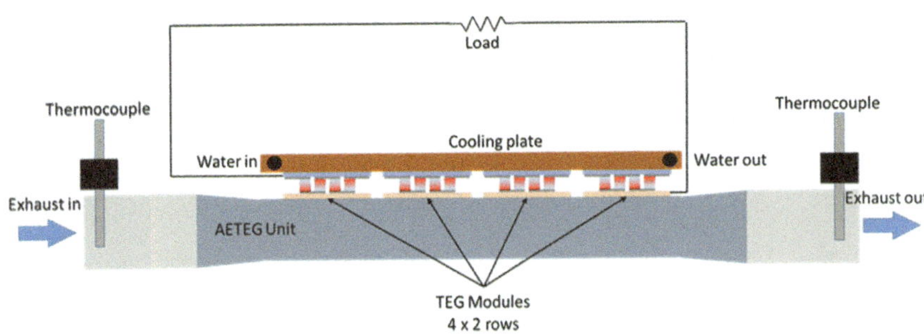

Fig. 12.3 Schematic diagram of the experimental setup for exhaust heat recovery

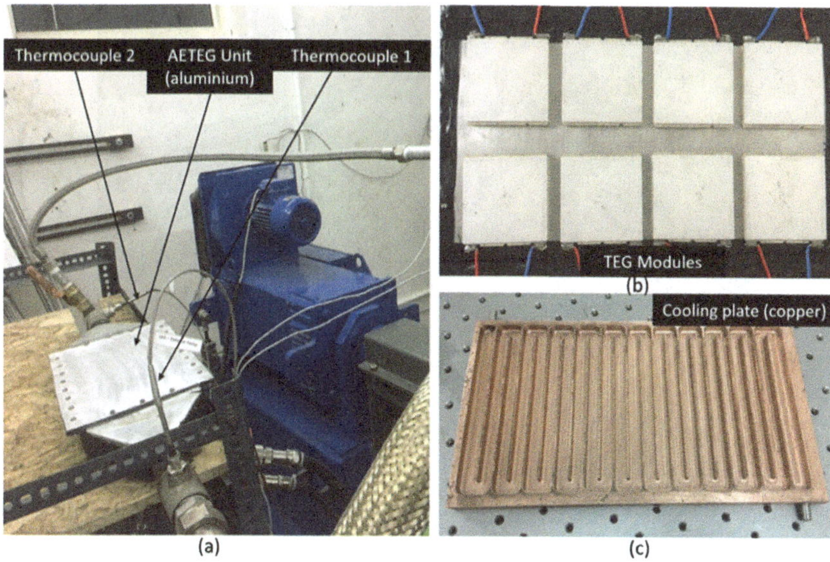

Fig. 12.4 **a** AETEG placement on engine exhaust line; **b** TEG modules placed on AETEG unit; **c** copper cooling plate

Table 12.1 Design parameters for AETEG unit

Parameter	Value
No. of modules used	8
Thermoelectric material	Bismuth Telluride (Bi_2Te_3)
Area of each module	4.34×10^{-3} m^2
Available area	0.55 m^2
Material of AETEG unit	Aluminium
Material of cooling plate	Copper
Dimensions of AETEG unit	0.325 × 0.175 × 0.05 m
Dimensions of cooling plate	0.32 × 0.18 × 0.008 m
Ambient temperature	32 °C
Cooling water flow rate	1 lpm
Cooling water temperature	R.T. (32 °C)

copper block (Fig. 12.4c). Copper was used to manufacturing the cooling plate due to the requirement of very high thermal conductivity on the cooling side to maintain a steady temperature gradient across the TEG module. The heat transport can further be improved by inducing surface roughness to the channels or by depositing porous material which may lead to the two-phase heat transport, but remain beyond the scope of present study. Table 12.1 shows the design parameters of the AETEG unit.

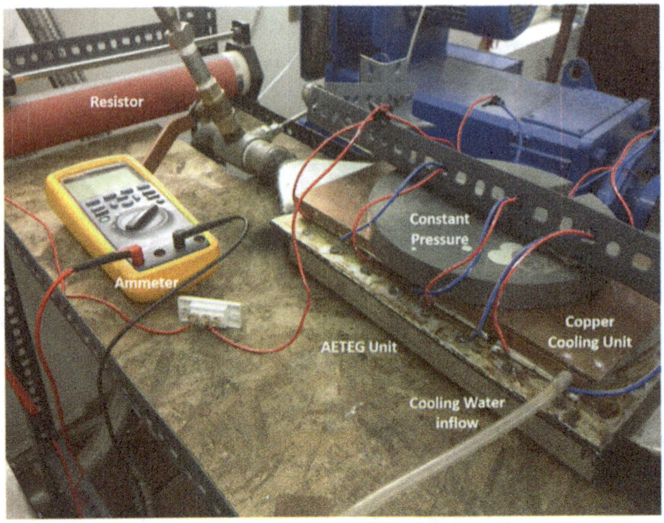

Fig. 12.5 Experimental setup of AETEG unit used for power production

A uniform load was placed on the AETEG unit to minimize the vibrations and further reduce the thermal contact resistance. This additional load also helped to avoid any TEG module displacement in the AETEG unit due to the excessive vibrations at higher loads and lubrication caused by the thermal paste. Figure 12.5 shows the complete experimental setup for the study. To examine and estimate the net power output produced from the AETEG unit by recovering waste heat of the exhaust, a basic electrical circuit was developed by deploying a 4-Ω resistor to dissipate the power. The current was measured using an ammeter, from which the net voltage and power were calculated.

With the aim to enhance the power output of the AETEG unit, the temperature on the interface of the TEG modules and aluminum surface of AETEG unit was measured using the thermal imaging camera (make: TESTO, model: 875-2i) to check its response to further internal modifications. After the successful power and voltage measurements, the cooling plate and TEG modules were removed from the aluminum AETEG unit, and the thermal imaging camera was used to measure the interface temperatures. Black conducting tape was applied on the top surface of the AETEG heat exchanger during the measurement to reduce the scope of error in the measurement. Further, these measured temperature values also served the purpose of tuning the heat transfer coefficient for the computational model. Figure 12.6 shows the setup view while measuring interface temperatures.

Fig. 12.6 Experimental setup for interface temperature measurements

12.2.2 Simulation Case Setups and Validations

12.2.2.1 Simulation Model for Thermal Studies

The details of the studied geometry are given in Table 12.1. The simulations were performed using a CFD simulation package Star CCM. The *k-epsilon* turbulence model was used to compute the fluid flow and heat exchange rate. The exhaust gas was modelled as ideal gas with a density of 1.249 kg/m^3 (The Automotive Research Association of India 2010). No-slip boundary conditions were given to the walls of the aluminium AETEG unit. The flow velocity at the inlet was calculated and kept at 4.85 m/s. The model setup and the boundary conditions are given in Table 12.2. The inlet temperature boundary conditions were given as per recorded from thermocouple 1 installed at the entry of exhaust gases. To model the convection from the interface to the surrounding environment, the heat transfer coefficient (*htc*) was tuned iteratively for load 100% with the help of the experimentally measured temperatures from the infrared thermometry. On tuning the model, the value of *htc* was 7.9 W/m^2 K for 100% load, and for convenience, was assumed to be the same for all the other tested loads. The closeness of the mean temperature and the temperature variation along the central line justify the consideration of 7.9 W/m^2 K for all tested loads.

Table 12.3 shows the variation for average temperatures between our experimental results and the simulation results. The computational model tends to over-predict the mean temperature; however, the maximum deviation is only 4.14%. The deviation was higher at lower loads, but the model performed well at high loads. Figure 12.7 shows the variation of the temperature at the central line of the AETEG unit.

Table 12.2 Simulation conditions and models

Conditions	Model
Model state	3-dimensional, steady state
Flow type	Segregated flow (subsonic)
Equation of state	Ideal gas ($\rho = 1.249$ kg/m^3)
Energy	Segregated fluid temperature
Equation of motion	Reynolds averaged Navier stokes
Turbulence model	k-epsilon
Boundary conditions	Velocity inlet (4.85 m/s) Pressure outlet (1 atm) No-slip wall Convection ($h = 7.9$ W/mK)

Table 12.3 Comparison of experimental findings and the simulation results

Case	Average temperature (°C)		% deviation
	Experimental	Simulations	
Load 25%	98.2	102.269	−4.14
Load 50%	121.6	124.721	−2.56
Load 75%	145.9	147.862	−1.34
Load 100%	179.1	180.434	−0.74

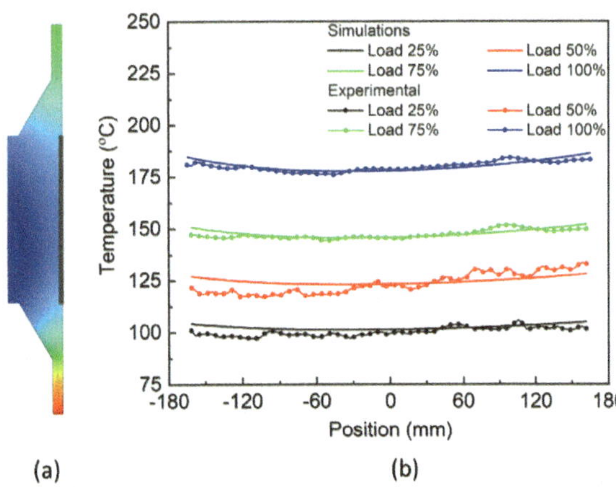

Fig. 12.7 **a** Central line where temperatures are measured for validation; and **b** variation of temperature at central line for both experimental findings and simulation results

After tuning the computational model for the baffle-less geometry, the variations in the internal geometry were made to quantify the enhanced heat transfer. Figure 12.8 shows the different internal geometries for the simulation. The number of baffles was varied in three steps: 3, 4, and 6. The limiting case was chosen as 6 baffles because increasing the number of baffles beyond 6 increased the back pressure beyond 1 kPa, which was considered as the upper limit for the present study.

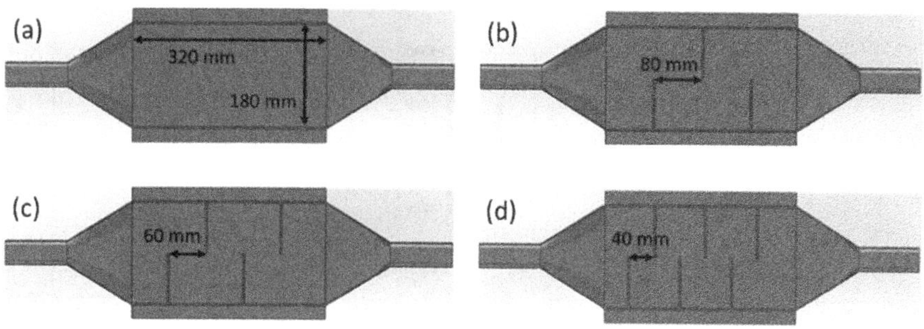

Fig. 12.8 Internal geometries for computational study **a** no baffles; **b** 3 baffles; **c** 4 baffles and **d** 6 baffles

12.2.2.2 Simulation Model to Study Acoustic Attenuation

Present day automobiles have dedicated muffler to reduce exhaust noise (Nag et al. 2017). The governments have set different norms and standards for the maximum allowable noise from an automobile. Moreover, excess noise from the automobile has severe health problems and can cause nausea, tiredness, and headache after long travels.

To study the acoustic advantage of the proposed geometries, Pressure acoustic analysis was carried out using the finite element approach. The dynamics of sound propagation is governed by (12.1).

$$\nabla \cdot \left(-\frac{\nabla p}{\rho}\right) - \left(\frac{\omega}{c}\right)^2 \frac{p}{\rho} = 0 \tag{12.1}$$

where p is the acoustic wave pressure, ρ is the density of medium, c is the speed of sound, and ω is the angular frequency.

Transmission loss (TL), an important parameter to determine the performance of a muffler, is the ratio of incoming acoustic energy to outgoing acoustic energy. The TL is given by (12.2).

$$\mathrm{TL} = 10 \log_{10} \frac{w_{\mathrm{in}}}{w_{\mathrm{out}}} \tag{12.2}$$

where, $w_{\mathrm{in}} = \int \frac{{p_o}^2}{2\rho c} \mathrm{d}A$ and $w_{\mathrm{out}} = \int \frac{|p|^2}{2\rho c} \mathrm{d}A$ are the incoming acoustic energy at inlet side and outgoing acoustic energy at the outlet side, respectively.

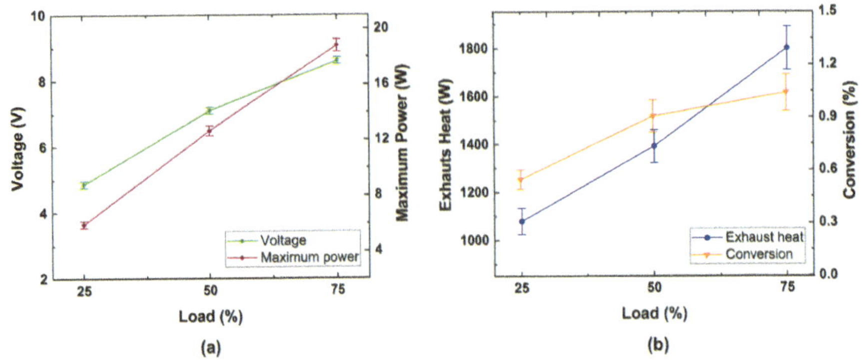

Fig. 12.9 **a** Peak voltage and maximum power at different loads; and **b** total exhaust heat and % conversion to electricity at different loads

12.3 Results and Discussions

12.3.1 Power and Voltage Output

Figure 12.9a shows the peak voltage and maximum power output for the setup. A maximum power output of 18.83 W was obtained at a load of 75%. For the same load, the peak voltage was 8.63 V. As the temperature of the exhaust increases, the net power increases due to the increase in available heat. Figure 12.9b shows the available exhaust heat and % conversion of waste heat into power. The conversion percentage of 1.05% was obtained at 75% engine load while using only one face of the AETEG unit for the installation of thermoelectric modules due to the availability of the limited number of modules. 4 modules connected in series were connected in parallel configuration to another set of 4 modules connected in series. The power out and conversion efficiency can be further improved by covering all the faces of AETEG. Moreover, the use of power conditioning circuit and matched loads would also enhance the peak output from the AETEG unit. With the improvement in the thermal operating range of the TEG modules, the amount of heat recovery can be further increased.

12.3.2 Simulation Results for Enhanced Heat Transfer

Figure 12.10 shows the temperature contours on the interface of the AETEG unit at 100% load condition. The increase in the number of baffles has tremendous impact on the interface temperature. The baffle-less geometry has an average temperature of 180 °C, whereas the geometry with 6 baffles has an average temperature of 238 °C. The increase of 58 °C can nearly double the power output of the AETEG unit. But

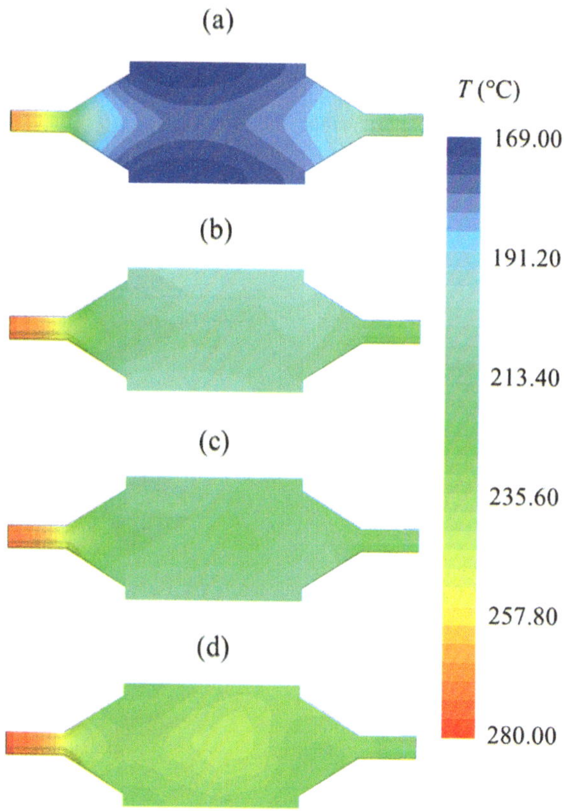

Fig. 12.10 Temperature contours for different internal geometries at 100% load condition **a** no baffles; **b** 3 baffles; **c** 4 baffles; and **d** 6 baffles

this increase in interface temperature is directly coupled with the increases in back pressure.

Table 12.4 shows the back pressure with varying internal geometry. The back pressure increased from 10.88 Pa for baffle-less geometry to 350 Pa for 6-baffle geometry. Increasing back pressure has adverse effects like an increase in pumping work and cylinder scavenging problems. The existing engines already have various exhaust components like diesel particulate filters, silencers, catalytic converters, etc., hence increasing the back pressure for AETEG unit beyond 1 kPa will have adverse effects on the engine performance (Mayer 2004).

Table 12.4 Back pressure for different internal geometries

Geometry	Back pressure (Pa)
No baffles	10.88
3 Baffles	32.81
4 Baffles	81.99
6 Baffles	349.85

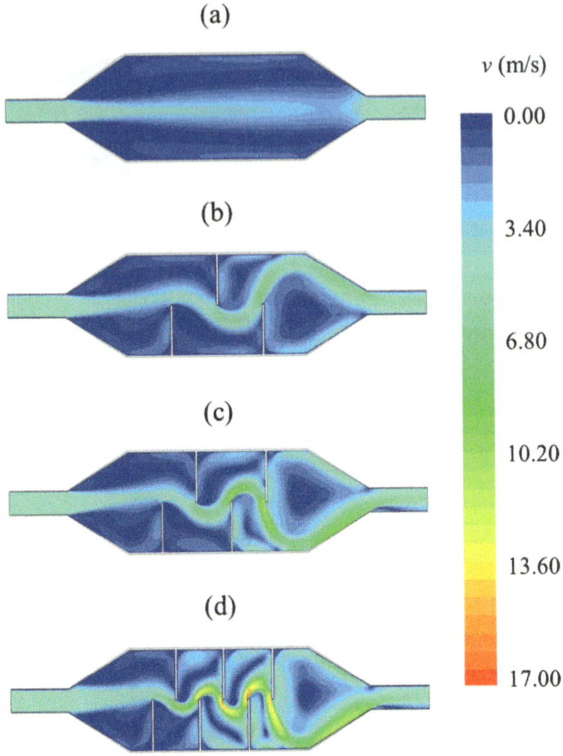

Fig. 12.11 Velocity magnitude for different internal geometries at 100% load condition. **a** No baffles; **b** 3 baffles; **c** 4 baffles; and **d** 6 baffles

Figure 12.11 shows the velocity profile on the central plane for the proposed AETEG units at 100% load condition. The inlet velocity calculated from the experiments was 4.85 m/s and served as an initial condition. The inclusion of baffles induces high turbulent flows, thus increasing the heat transfer coefficient.

12.3.3 Simulation Studies for Enhanced Acoustic Attenuation

Figure 12.12a shows the transmission loss (TL) for the different proposed geometries. The geometry with no internal baffles showed a very poor TL characteristic for all frequencies. This was mainly due to no reflection inside the geometry because of the absence of baffles or other reflecting geometries. The average transmission loss for this case was only 1.93 dBA. However, as the number of baffles increased, a high acoustic gain was observed. The average TL was 15.59, 27.55, and 53.81 dBA for 3, 4 and 6 baffles, respectively (see Fig. 12.12b). Another important finding was the frequency region for which high TL was observed. The geometries with

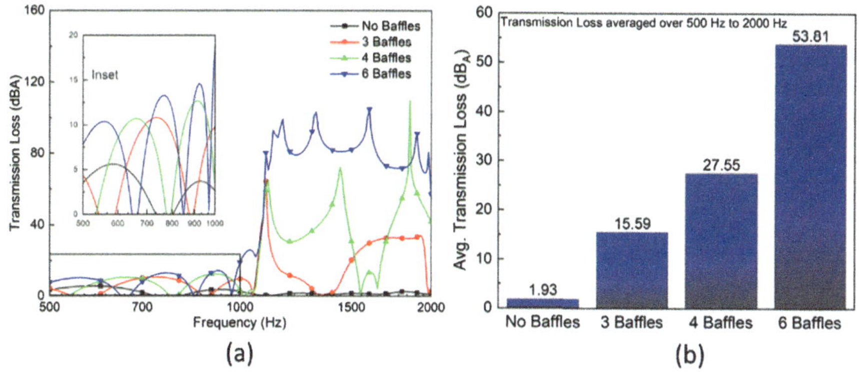

Fig. 12.12 **a** Transmission loss for different proposed geometries. Inset shows the transmission loss at frequencies less than 1000 Hz, and **b** averaged transmission loss for the frequency range of 500 to 2000 Hz

baffles showed very high TL for frequencies greater than 1 kHz, but the TL for lower frequencies was quite less.

12.4 Summary

The experimental investigation for the net power recovered from the engine exhaust gases using TEG modules was carried out with the aim to power auxiliary units of the automobile. Presently, the focus is to improve and optimize the thermal designs and the functionality of the systems for automotive applications. In our study, the maximum power and voltage achieved during the experiments were 18.83 W and 8.63 V. Therefore, due to low power and voltage output, an internal modification to the aluminum AETEG unit was proposed, and simulations were carried out implementing baffles to enhance the heat transfer. Increasing the number of baffles increases the heat transfer to the interface to a greater extent. A net increase of 58 °C in the mean interface temperature was observed with increasing the baffle number to 6. However, if the number of baffles is increased above 6, the backpressure in the exhaust line goes above 1 kPa, which is not recommended for present day automobile exhaust lines. The acoustic advantage of the AETEG geometry was also studied computationally, and it was found that the sound attenuation increases with the increasing number of baffles due to reoccurring reflections within the volume. The mean TL for the geometry with 6 baffles was over 50 dBA, while that for baffles less geometry was under 2 dBA.

In the present scenario, the conversion efficiency of TEG materials is low and can be inconvenient due to the additional power requirement for cooling of the TEG modules. This hinders the application of AETEG in the automotive industry. However, this adversity is due to the low figure of merit of thermoelectric materials and the ongoing aggressive research and development of high figure of merit thermoelectric materials can soon provide the solution to this intrinsic problem. Further, to reduce the excessive power usage for cooling the AETEG system, integration of the engine coolant loop with the AETEG coolant is another solution.

Acknowledgements The authors gratefully acknowledge the research funding and support provided by the Department of Science and Technology-Science and Engineering Research Board, Government of India vide Project No. ECR/2015/000135 titled "Study of Synergistic Use of Hydrogen and other Alternative Fuels in a Dual Fuel Engine for Emissions Reduction" at Indian Institute of Technology Mandi, India.

References

Angeline AA, Asirvatham LG, Hemanth DJ et al (2019) Performance prediction of hybrid thermoelectric generator with high accuracy using artificial neural networks. Sustain Energy Technol Assessments 33:53–60. https://doi.org/10.1016/j.seta.2019.02.008

Bass JC, Elsner NB, Leavitt FA (1994) Performance of the 1 kW thermoelectric generator for diesel engines. In: AIP conference proceedings, AIP, pp 295–298

Birol F (2016) Energy and air pollution—world energy outlook special report. IEA Publications, Paris

Crane D, Lagrandeur J, Jovovic V, et al (2013) TEG on-vehicle performance and model validation and what it means for further TEG development. J Electron Mater 42:1582–1591. https://doi.org/10.1007/s11664-012-2327-8

Demir ME, Dincer I (2017) Performance assessment of a thermoelectric generator applied to exhaust waste heat recovery. Appl Therm Eng 120:694–707. https://doi.org/10.1016/J.APPLTHERMALENG.2017.03.052

Deng YD, Liu X, Chen S, Tong NQ (2013) Thermal optimization of the heat exchanger in an automotive exhaust-based thermoelectric generator. J Electron Mater 42:1634–1640. https://doi.org/10.1007/s11664-012-2359-0

Deng Y, Liu C, Chu P (2016) Research on integration of automotive exhaust-based thermoelectric generator with front muffler. SAE Tech Pap 2016-01-0203. https://doi.org/10.4271/2016-01-0203

DiSalvo FJ (1999) Thermoelectric cooling and power generation. Science 285:703–706. https://doi.org/10.1126/science.285.5428.703

Espinosa N, Lazard M, Aixala L, Scherrer H (2010) Modeling a thermoelectric generator applied to diesel automotive heat recovery. J Electron Mater 39:1446–1455. https://doi.org/10.1007/s11664-010-1305-2

Gu J, Han D, Liu Y et al (2019) Performance analysis of a thermoelectric generator with closed and flooded passive cooling in small-scale space. Energy Sustain Dev 51:21–31. https://doi.org/10.1016/J.ESD.2019.05.002

Hale LE (1913) Thermo-electric battery for motor-vehicles

Hatami M, Ganji DD, Gorji-Bandpy M (2015) Experimental and numerical analysis of the optimized finned-tube heat exchanger for OM314 diesel exhaust exergy recovery. Energy Convers Manag 97:26–41. https://doi.org/10.1016/J.ENCONMAN.2015.03.032

He M, Wang E, Zhang Y et al (2020) Performance analysis of a multilayer thermoelectric generator for exhaust heat recovery of a heavy-duty diesel engine. Appl Energy 274:115298. https://doi.org/10.1016/j.apenergy.2020.115298

Henry A, Prasher R (2014) The prospect of high temperature solid state energy conversion to reduce the cost of concentrated solar power. Energy Environ Sci 7:1819–1828. https://doi.org/10.1039/c4ee00288a

Huang F, Zheng J, Baleynaud JM, Lu J (2017) Heat recovery potentials and technologies in industrial zones. J Energy Inst 90:951–961. https://doi.org/10.1016/J.JOEI.2016.07.012

Karri MA, Thacher EF, Helenbrook BT (2011) Exhaust energy conversion by thermoelectric generator: two case studies. Energy Convers Manag 52:1596–1611. https://doi.org/10.1016/j.enconman.2010.10.013

Karthick K, Suresh S, Joy GC, Dhanuskodi R (2019) Experimental investigation of solar reversible power generation in thermoelectric generator (TEG) using thermal energy storage. Energy Sustain Dev 48:107–114. https://doi.org/10.1016/J.ESD.2018.11.002

Kim S, Park S, Kim S, Rhi SH (2011) A thermoelectric generator using engine coolant for light-duty internal combustion Engine-Powered Vehicles. J Electron Mater 40:812–816. https://doi.org/10.1007/s11664-011-1580-6

Lee S, Bae C (2008) Design of a heat exchanger to reduce the exhaust temperature in a spark-ignition engine. Int J Therm Sci 47:468–478. https://doi.org/10.1016/j.ijthermalsci.2007.03.011

Liebl J, Neugebauer S, Eder A et al (2009) The thermoelectric generator from BMW is making use of waste heat. MTZ Worldw 70:4–11. https://doi.org/10.1007/bf03226939

Liu X, Deng YD, Zhang K et al (2014) Experiments and simulations on heat exchangers in thermoelectric generator for automotive application. Appl Therm Eng 71:364–370. https://doi.org/10.1016/J.APPLTHERMALENG.2014.07.022

Liu X, Deng YD, Li Z, Su CQ (2015) Performance analysis of a waste heat recovery thermoelectric generation system for automotive application. Energy Convers Manag 90:121–127. https://doi.org/10.1016/J.ENCONMAN.2014.11.015

Liu C, Pan X, Zheng X et al (2016) An experimental study of a novel prototype for two-stage thermoelectric generator from vehicle exhaust. J Energy Inst 89:271–281. https://doi.org/10.1016/J.JOEI.2015.01.019

Lu H, Wu T, Bai S et al (2013) Experiment on thermal uniformity and pressure drop of exhaust heat exchanger for automotive thermoelectric generator. Energy 54:372–377. https://doi.org/10.1016/J.ENERGY.2013.02.067

Luo D, Wang R, Yu W, Zhou W (2020) A numerical study on the performance of a converging thermoelectric generator system used for waste heat recovery. Appl Energy 270:115181. https://doi.org/10.1016/j.apenergy.2020.115181

Mayer A (2004) Number-based emission limits, VERT-DPF-verification procedure and experience with 8000 retrofits

Mulla R, Rabinal MHK (2019) Copper sulfides: earth-abundant and low-cost thermoelectric materials. Energy Technol 7:1800850. https://doi.org/10.1002/ente.201800850

Muralidhar N, Himabindu M, Ravikrishna RV (2018) Modeling of a hybrid electric heavy duty vehicle to assess energy recovery using a thermoelectric generator. Energy 148:1046–1059. https://doi.org/10.1016/J.ENERGY.2018.02.023

Nag S, Gupta A, Dhar A (2017) Sound attenuation in expansion chamber muffler using plane wave method and finite element analysis. Nonlinear Stud 24:69–78

Nag S, Dhar A, Gupta A (2018) Exhaust heat recovery using thermoelectric generators: a review. In: Srivastava D, Agarwal A, Datta A, Maurya R (eds) Advances in internal combustion engine research. Springer, Singapore, pp 193–206

Nag S, Sharma P, Gupta A, Dhar A (2019) Experimental study of engine performance and emissions for hydrogen diesel dual fuel engine with exhaust gas recirculation. Int J Hydrogen Energy 44:12163–12175. https://doi.org/10.1016/j.ijhydene.2019.03.120

Nour Eddine A, Chalet D, Faure X et al (2018) Optimization and characterization of a thermoelectric generator prototype for marine engine application. Energy 143:682–695. https://doi.org/10.1016/J.ENERGY.2017.11.018

Omer G, Yavuz AH, Ahiska R, Calisal KE (2020) Smart thermoelectric waste heat generator: design, simulation and cost analysis. Sustain Energy Technol Assessments 37:100623. https://doi.org/10.1016/j.seta.2019.100623

Orr B, Akbarzadeh A, Mochizuki M, Singh R (2016) A review of car waste heat recovery systems utilising thermoelectric generators and heat pipes. Appl Therm Eng 101:490–495. https://doi.org/10.1016/J.APPLTHERMALENG.2015.10.081

Pacheco N, Brito FP, Vieira R et al (2020) Compact automotive thermoelectric generator with embedded heat pipes for thermal control. Energy 197:117154. https://doi.org/10.1016/j.energy.2020.117154

Pang W, Yu H, Zhang Y, Yan H (2018) Electrical characteristics of a hybrid photovoltaic/thermoelectric generator system. Energy Technol 6:1248–1254. https://doi.org/10.1002/ente.201700801

Saqr KM, Mansour MK, Musa MN (2008) Thermal design of automobile exhaust based thermoelectric generators: objectives and challenges. Int J Automot Technol 9:155–160. https://doi.org/10.1007/s12239-008-0020-y

Serksnis AW (1976) Thermoelectric Generator for an automotive charging system. In: 11th Intersociety energy conversion engineering conference. Warrendale, PA United States, pp 1614–1661

Sharma S, Dwivedi VK, Pandit SN (2014) A review of thermoelectric devices for cooling applications. Int J Green Energy 11:899–909. https://doi.org/10.1080/15435075.2013.829778

Shu G, Ma X, Tian H et al (2018) Configuration optimization of the segmented modules in an exhaust-based thermoelectric generator for engine waste heat recovery. Energy 160:612–624. https://doi.org/10.1016/J.ENERGY.2018.06.175

Smil V (2017) Energy transitions: global and national perspectives, 2nd edn. Praeger, an imprint of ABC-CLIO, LLC

Soleimani Z, Zoras S, Ceranic B et al (2020) A review on recent developments of thermoelectric materials for room-temperature applications. Sustain Energy Technol Assessments 37:100604. https://doi.org/10.1016/j.seta.2019.100604

Stabler F (2009) Automotive thermal generator design issues. In: DOE Thermoelectric Applications Workshop

Thacher EF, Helenbrook BT, Karri MA, Richter CJ (2007) Testing of an automobile exhaust thermoelectric generator in a light truck. Proc Inst Mech Eng Part D J Automob Eng 221:95–107. https://doi.org/10.1243/09544070JAUTO51

The Automotive Research Association of India (2010) Document on test method, testing equipment and related procedures for testing type approval and confirmity of production (COP) of vehicle for emission as per CMV rules 115, 116 and 126. The Automotive Research Association of India, Pune, India

Tripathi G, Dhar A (2017) Exhaust heat recovery options for diesel locomotives. In: Agarwal AK, Dhar A, Gautam A, Pandey A (eds) Locomotives and rail road transportation. Springer, Singapore, pp 27–40

Tripathi G, Nag S, Dhar A, Patil DV (2018) Fuel injection equipment (FIE) design for the new-generation alternative fuel-powered diesel engines. In: Singh A, Agarwal R, Agarwal A et al (eds) Prospects of alternative transportation fuels. Energy, environment, and sustainability. Springer, Singapore, pp 387–405

Vining CB (2009) An inconvenient truth about thermoelectrics. Nat Mater 8:83–85. https://doi.org/10.1038/nmat2361

Wang Y, Li S, Xie X et al (2018) Performance evaluation of an automotive thermoelectric generator with inserted fins or dimpled-surface hot heat exchanger. Appl Energy 218:391–401. https://doi.org/10.1016/j.apenergy.2018.02.176

Zhao Y, Wang S, Ge M et al (2019) Performance investigation of an intermediate fluid thermoelectric generator for automobile exhaust waste heat recovery. Appl Energy 239:425–433. https://doi.org/10.1016/j.apenergy.2019.01.233

Chapter 13
Diesel Engine Waste Heat Recovery Schemes for Improved Fuel Economy and Reduced Emissions: Simulation Results

Subha Mondal and Sudipta De

Abstract Diesel engines are widely used for road and marine transports. Stationary diesel engines are also used for off-grid supply of electricity for households or to run the auxiliary equipment such as pump, compressor, etc. A significant portion of the thermal energy input to diesel engines is ultimately rejected as waste heat. Exhaust flue gas from marine diesel engines are at about 300 °C. On the other hand, jacket cooling water and scavenging air cooling water are available at below 100 °C. Available waste heat from road transport system may be utilized to produce cooling/heating effect for conditioning of the cabin environment. The waste heat may also be used in turbo chargers for a higher power density. In marine applications, the waste heat may drive a bottoming power cycle to supply the auxiliary power. Waste heat available form a stationary diesel power plant can even be used to run a cogeneration/polygeneration unit satisfying some of the localized energy needs. In the present chapter, simulation results of possible schemes and effects of waste heat recovery from diesel engines have been explored. Finally, generalized principles for the simulation of diesel engine waste heat recovery have been discussed.

Keywords Engine waste heat · Fuel economy · Cogeneration · Bottoming cycle · Simulation studies

Abbreviations

AFS	Annual fuel saving
AOH	Annual operating hours
AGHR	Annual greenhouse gas reduction
EF	Emission factor
EGHRU	Exhaust gas heat recovery unit

S. Mondal
Aliah University, Kolkata 700160, India

S. De (✉)
Jadavpur University, Kolkata 700032, India

A. K. Agarwal et al. (eds.), *Engine Modeling and Simulation*, Energy, Environment, and Sustainability, https://doi.org/10.1007/978-981-16-8618-4_13

EG	Exhaust gas
GWP	Global warming potential
HDDE	Heavy duty diesel engine
HT	High-temperature
JCW	Jacket cooling water
LO	Lubricating oil
LT	Low-temperature
ORC	Organic Rankine cycle
PT	Power turbine
PH	Pre-heater
R	Regenerator
SACW	Scavenging air cooling water
TEG	Thermo electric generator

13.1 Introduction

Diesel engine is an important device for modern transport systems. Diesel engines are widely used for road transport and marine transport. Diesel is also used as fuel in agricultural vehicles. Beside these, a diesel generator is an efficient option for off-grid small capacity secondary energy generation. In industrial sectors, diesel engines often supply the power for auxiliary equipment (such as pump, compressor etc.).

It should be noted that the share of liquid fuels in the electricity production sector is not significant. However, the transportation sector is the major end user of liquid fuels. International Energy Outlook (IEO), 2016 (IEO 2016) Projected about a 36 quadrillion Btu increment in the liquid fuel based energy input in the transportations between the years 2012 and 2040. Between these two specified years, out of all considered liquid fuels, the largest share may be for diesel (about 13 quadrillion Btu) (IEO 2016). It was also reported by IEO, 2016 (IEO 2016) that in the year 2040, diesel would generate close to 50 quadrillion Btu of primary energy for the transportation sector.

Emission from internal combustion (IC) engines is an issue of great concern. An IC engine produces NO_X, SO_X, particulate matter (PM), CO and CO_2 simultaneously. In the year 2010, the road transport sector (mainly heavy duty diesel engines) had been responsible for 14% of the total global greenhouse gas emission (Lion et al. 2017). On the other hand, in 2014 the marine transport sector had been the contributor of close to 3% of the annual global CO_2 emission (Balcombe et al. 2019). According to 2015 Paris agreement global temperature rise by the year 2100 should be kept well below 2 °C beyond the pre-industrial levels (Gao et al. 2017). To achieve this target global CO_2 emission should be reduced substantially. Thus, in 2016 the US government adopted a regulation with a target of 25% reduction in fuel consumption of heavy duty diesel engine (HDDE) by 2027, considering 2018 as the baseline year (EPA, NHTSA, DOT 2016). International Maritime Organization

(IMO) implemented Energy Efficiency Design Index (EEDI) and Ship Energy efficiency Management Plan (SEEMP) to control the emission from the marine ships (Ouyang et al. 2021).

In the current energy scenario, reducing automobile emissions has become a subject of interest of the researchers. Many of the recent studies indicated that using biodiesel would ensure improved emission qualities of diesel engines. Ettefaghi et al. (Ettefaghi et al. 2018) showed that using a mixture of diesel, bio-diesel, biodegradable nanoparticles and water as the engine fuel would reduce the brake specific fuel consumption, emissions of NO_X and unburned hydrocarbons, simultaneously. Dueso et al. (Dueso et al. 2018) experimentally showed that the sunflower-based biodiesel with and without bio-oil derived antioxidant additive would improve the emission measures of a diesel engine. Prakash et al. (Prakash et al. 2018) reported that the replacing of diesel by a mixture of bio-ethanol, diesel and castor oil would reduce the NO emission from 8.17 to 6.11 g/kWh.

Improving the energy-efficiency is also an effective way to decrease the fuel consumption and associated emissions from IC engines. Traditional IC engines can convert close to 40% of the thermal energy input through fuel into useful power. Thus, a considerable part of the primary energy input in IC engines is rejected with the exiting flue gas, jacket cooling water and scavenging air cooling water. This wasted energy may be partly converted into some useful energy utilities to enhance the overall energy utilization efficiency of the engine leading to greater fuel economy. It should be noted that the thermal energies carried away by different outgoing streams of an engine significantly differ in both quantity and quality. The nature of available waste heat also varies widely with the varying engine loads. Varying engine types also affect the temperature of the available waste heat. The temperature of the exhaust gas of a truck diesel engine is usually close to 500 °C. However, for a marine diesel engine, the exhaust gas temperature becomes close to 300 °C due to larger bore size and lower operating speed. Produced energy utilities by the waste heat recovery schemes integrated to the road / marine transport system are to be consumed by the vehicle itself. Thus, it is very challenging to implement a waste heat recovery schemes in an IC engine in the transport sector.

In the present chapter, possible diesel engine waste heat recovery schemes and their effects based on outputs of a few simulation studies will be explored.

13.2 Possible Waste Heats from Diesel Engines

Available waste heat from any source can be classified in three different categories (Forman et al. 2016). Heat available at or above 300 °C is considered as the high temperature waste heat and heat available at below 100 °C is termed as the low temperature waste heat. Heat available between these two temperature limits is medium temperature waste heat. Based on a few of the previously reported studies, temperatures of different waste heat streams exiting diesel engines are summarized in Table 13.1 (Ouyang et al. 2021; Liu et al. 2019; Wang et al. 2020; Boodaghi et al. 2021;

Table 13.1 Temperature of different waste heat stream exiting the diesel engine

Engine type	Researchers	Waste heat stream	Temperature range (°C)
Heavy duty diesel engine without turbo charger (for road transport)	Liu et al. (2019)	Exhaust flue gas	517.3
		Jacket cooling water	84
	Wang et al. (2020)	Exhaust flue gas	389.7
		Jacket cooling water	85
	Boodaghi et al. (2021)	Exhaust gas	530
Marine diesel engine with turbo charger	Song et al. (2015)	Exhaust flue gas	300
		Jacket cooling water	90
	Yang (2016)	Exhaust flue gas	290
		Jacket cooling water	90
		Scavenging air cooling water	76
	Ouyang et al. (2021)	Exhaust flue gas	285–326.3

Song et al. 2015; Yang 2016). It is observed from Table 13.1 that a diesel engine rejects high / medium temperature heat and low temperature heat simultaneously. Most of the heavy-duty diesel engines (HDDE) are four-stroke engines and usually produce exhaust gas at or above 350 °C. However, the exhaust gas temperature of HDDE with a turbocharger remains below 330 °C as shown in Table 13.1. For the marine application a two-stroke diesel engine is preferred due to its greater Power to weight ratio. The temperature of the exiting flue gas of a turbocharged marine diesel engine remains below 330 °C. Scavenging air cooling water exiting the marine diesel engine below 80 °C can be treated as a low temperature waste heat source. Temperatures of the jacket cooling water of most of the diesel engines vary between 80 and 90 °C. It is worth mentioning that, though temperatures of the jacket cooling water/scavenging air cooling water are not appreciably high, the heat associated with these streams may be useful due to their larger flow rates. It should be noted further that due to the presence of SO_2 in the exhaust flue gas of diesel engines, it cannot be cooled below the acid dew point temperature in the heat recovery device. The mass flow rates and temperatures of the diesel engine waste heat-carrying streams greatly depend on the rated output power and percentage of engine loads.

13.3 Generalized Principles of Diesel Engine Waste Heat Recovery Simulation

Before implementing the diesel engine waste heat recovery practically, possible advantages and drawbacks of the scheme are to be explored by simulating the waste heat recovery scheme. The simulation study should try to assess the thermodynamic and the economic feasibility simultaneously. Generalized principles of simulating a diesel engine waste heat utilization scheme are summarized as follows:

1. Different waste heat sources from a diesel engine are to be identified.
2. Possible utility outputs are to be selected based on the available end-users. For example, in the marine waste heat recovery system, all the produced utility outputs are to be consumed by the ship itself. Thus, the power, refrigeration effects, desalinated water may be some possible utility outputs for a marine diesel engine waste heat driven energy system, considering the ship itself as the end-user. Nature of the selected utility outputs also depends on the quality and quantity of available waste heat.
3. Possible technologies are to be identified to produce the selected utility outputs. Working fluids for the identified technologies should be less flammable, non-toxic and commercially available. They also should have lower global warming potential, zero ozone depletion potential.
4. Energy and exergy assessment are to be conducted to assess the technological feasibility. While conducting the energy and exergy analyses, losses across different components are to be taken into consideration by assigning either a specific efficiency or a loss coefficient to each of the components involved in the waste heat recovery. The efficiency and loss coefficient values may be assigned based on previously published literature or past experiences.
5. If the results of energy and exergy efficiencies are satisfactory, an economic analysis is to be conducted. During this economic analysis, bare module costs (BMC) of all equipment of the waste heat recovery scheme are to be estimated. To do this, equipment cost correlations proposed by Turton et al. (2009) may be used. Total establishment cost will be sum of the BMCs of individual equipment.
6. Based on annual fuel savings and reduced greenhouse gas emission, the payback period can also be calculated.
7. If the payback period estimated by the simulation is within a desirable limit, the experimental trial run may be recommended.

13.4 Technologies for the Diesel Engine Waste Heat Utilization

Increasing the power density by running a turbocharger is one of the conventional methods to recover a part of the exhaust gas waste heat in a diesel engine. Operating a thermodynamic cycle by the waste heat carried by the different exiting streams is another possible pathway to recover the diesel engine waste heat. Possible schemes for recovering the diesel engine waste heat can be summarized as follows:

13.4.1 Recovery of Waste Heat by Turbocompounding

In a turbocharged engine air is compressed before entering into the engine. Exhaust gas exiting the engine can be utilized to run a turbine supplying the power required to run the compressor. Both the turbine and the compressor are mounted on the same shaft. In a turbo charged engine, for a specified swept volume a larger mass of air enters into the combustion chamber. Increased fuel economy and a greater power density are two positive outcomes of the turbo charging.

In the turbocompound engine, the exhaust gas exiting the turbo charger is fed to a power turbine (PT) to produce an additional power output. The power turbine either may be mechanically coupled to the crankshaft of the engine or can be coupled directly to a generator producing electricity (Aghaali and Ångström 2015). Two possible simplified layouts of turbocompound diesel engines are shown in Fig. 13.1a, b respectively.

It is worth noting that incorporation of the turbocompounding in a diesel engine generates a back pressure in the exhaust manifold. This may affect the expelling process of burnt gases from the combustion chamber (Aghaali and Ångström 2015). The power consumed by the burnt gas expelling process from the combustion chamber in a turbocompound engine is somehow higher than that of the conventional engine.

A turbocompounding engine substantially improves the fuel economy, especially for the vehicle operating at a higher load over an extended period of time. The achievable overall efficiency of a modern marine ship with turbocompounding can be more than 50% (Hiereth and Prenninger 2007). In the year 1991, Scania had been the company that first implemented turbocompounding commercially in the

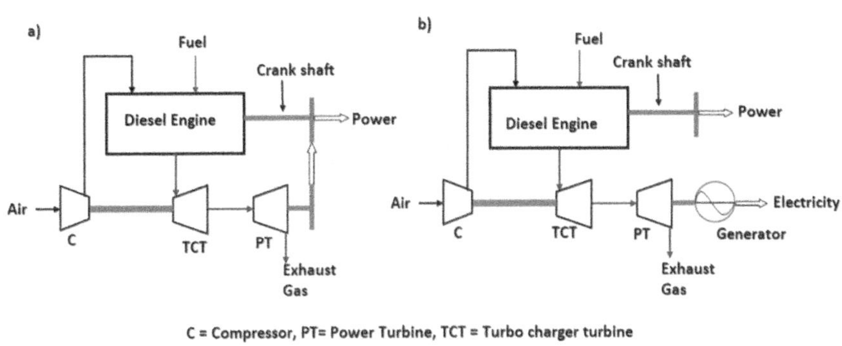

Fig. 13.1 **a** Turbo compounding with PT coupled to the crank shaft and **b** turbo compounding with PT coupled to the generator

6 cylinder 11 l DTC 1101 Diesel engine. This yielded 1–3% improvement in fuel economy at high load operating condition (Hiereth and Prenninger 2007). In the year 2009, the Bowman Power group introduced a turbocompounding system with a power turbine coupled to a permanent magnet type electric alternator (Ismail et al. 2012).

Some of the recently reported simulation study on turbocompounding of diesel engines and associated outputs are summarized as follows:

Zhao et al. (2016) simulated a model for the turbocompounded diesel engine to evaluate the effects of varying turbine efficiency, engine speed, exhaust gas temperature and pressure ratio on energy recovery, pumping loss and fuel saving. The simulated model was mostly based on the principles of thermodynamics. However, some empirical relations were also used in the modeling to estimate the engine friction losses and indicated thermal efficiency. According to the output of the simulation, turbine efficiency, exhaust gas temperature and generated back pressure greatly affected the fuel economy of the turbo compounded engine. On the other hand, varying engine speed did not have much influence on the fuel saving obtained through the turbo compounding.

Weerasinghe et al. (2010) conducted a simulation to make a comparison between turbocompounding and steam Rankine cycle. 2nd law efficiency and fuel savings were considered as the performance parameters of the simulation. The simulation was done using MATLAB based QSS tool box. According to the produced output, a steam Rankine cycle saves higher fuel compared to a turbo compounded engine. However, the use of a secondary fluid in the Rankine cycle would lead to a greater complexity.

Jye et al. (2013) evaluated the influence of the turbocompounding on a turbocharged diesel engine. To do so, they conducted simulation by the AVL BOOST module. The AVL BOOST is a code for one-dimensional simulation of the engine. During the simulation, conservation equations for mass, momentum and energy were solved along the mean path-line of the flow. According to the analyses made, mechanical turbocompounding with a bypass system would increase engine power by 1.2% and reduce the brake-specific fuel consumption (BSFC) by 1.9%. It was also concluded that the extra power produced by the turbo compounding was sufficient to satisfy the auxiliary power demand of the engine.

Finally, turbocompounding is a mature technology and would ensure an improved fuel economy for higher load applications over longer time period in the upcoming decades.

13.4.2 Waste Heat Recovery by Running a Thermodynamic Cycle

As discussed in Sect. 13.2, exhaust flue gas, jacket cooling water and scavenging air cooling water (For two-stroke engines only) are the streams carrying away most of the rejected heat from diesel engines. Many of the researchers suggested utilizing the waste heat carried by different streams to run a bottoming energy system.

The most preferred option with recovered waste heat is to produce some power output through a suitable power cycle. Thus, many researchers suggested the Organic Rankine cycle (ORC) and Transcritical CO_2 power cycle as the possible bottoming energy system to recover the wasted primary energy of the diesel engine. The vapour absorption refrigeration cycle and heat driven ejector refrigeration cycle are capable of producing cooling effects from the waste thermal energy of a diesel engine. The type of the bottoming energy system coupled to a diesel engine greatly depends on its end-users. For a stationary diesel engine, the output utilities of the waste heat recovery scheme can partially cater to some of the energy needs of the nearby locality. The generated power by the bottoming power cycle may be transmitted to the crankshaft of the HDDE for improving the fuel economy of a moving vehicle. Sometimes, this output power may be partly consumed by the compressor of a vapour compression refrigeration/heat pump system conditioning the cabin environment. An ejector refrigeration cycle or an absorption refrigeration cycle can also be operated using engine waste heat to condition the cabin environment of a HDDE. In marine application, a power cycle or combined cycle can be operated using the engine waste heat to satisfy the auxiliary energy needs of the marine ship.

Recently different layouts of diesel engine waste heat driven thermodynamic cycles were presented by various researches. The waste heat recovery potentials of these cycles were evaluated through some simulation studies. While conducting majority of the simulations, it was commonly assumed that:

- steady state operating conditions were applicable
- Frictional pressure losses were negligible
- Ambient operating condition was not varying
- All heat losses except the outflow of exhaust gas and coolant were negligible
- Isentropic efficiencies of turbo expanders would be between 0.7 and 0.9
- Isentropic efficiencies of pump and compressor would be between 0.7 and 0.85
- Mixing losses were negligible.

While simulating a thermodynamic cycle, properties of the working fluids at different state points are to be evaluated. This was usually done in most of the studies by using some standard software such as REFPROP (Lemmon et al. 2018), EES (Engineering equation solver) etc.

Some of the layouts of recently proposed thermodynamic cycles for the utilization of diesel engine waste heat and major outcomes of associated simulation studies can be summarized as follows:

ORCs for diesel engine waste heat utilization

A waste-heat recovery scheme to be introduced in a moving vehicle should be compact and light weight. Thus, the most simplified ORC layout is preferred for recovering the waste heat of the HDDE involved in road transport. This simplified layout of the ORC consists of a turbine; a condenser/cooler, a pump and a heat recovery unit (HRU), as presented in Fig. 13.2. In this layout, the exhaust gas exiting

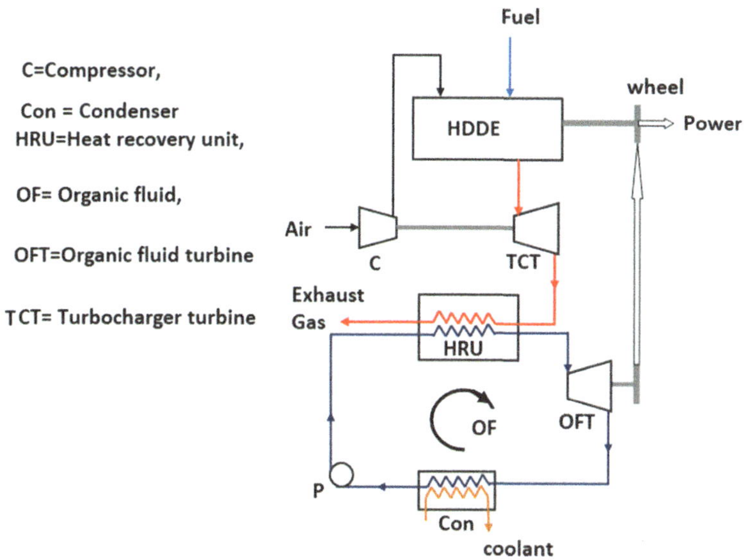

Fig. 13.2 Basic ORC as bottoming energy system of a HDDE

the turbocharger enters into the heat recovery unit. In the HRU, the organic fluid is heated up to the inlet temperature of the turbine by using the rejected heat of the exhaust gas.

The simplified layout presented in Fig. 13.2 may be modified to improve the overall performance. Yu et al. (2013) proposed a subcritical ORC layout to recover the waste heat from the exhaust gas and the jacket cooling water of a six cylinder diesel engine. The diesel engine was used to drive a generator set. The simplified layout of the heat recovery scheme proposed by Yu et al. (2013) is presented in Fig. 13.3. It is appearing from Fig. 13.3 that heat of the diesel engine exhaust gas was partly recovered in the thermal oil circuit. Heat rejected by the thermal oil circuit was utilized for heating the organic fluid to turbine inlet state. The heat of the jacket cooling water was used for preheating the working fluid entering into the oil working fluid heat exchanger. R245fa was considered as the working fluid of the ORC. Based on results of the simulated model, they reported that the use of the bottoming ORC in a diesel engine would ensure close to 6.1% improvement in thermal efficiency.

Yang (2016) proposed a transcritical ORC producing the secondary energy from the waste heat of a large marine ship. The proposed ORC recovered heat of the scavenging air cooling water (SACW), lubricating oil (LO), jacket cooling water (JCW) and exhaust gas (EG) simultaneously. The arrangement of the ORC proposed by Yang is depicted in Fig. 13.4 in a simplified manner. The layout consisted of three pre-heaters and an exhaust gas heat reccovery unit (EGHRU) for recovering the wasted thermal energy. The organic working fluid exiting the pump was divided into two streams. One fluid stream was heated by the lubricating oil in the pre-heater A. The heated stream coming out of the pre-heater was mixed with the 2nd stream.

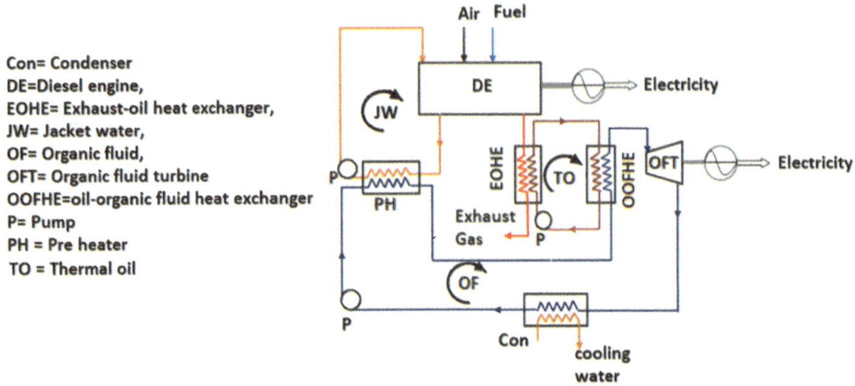

Fig. 13.3 Bottoming ORC of the DE used in a generator set (Yu et al. 2013)

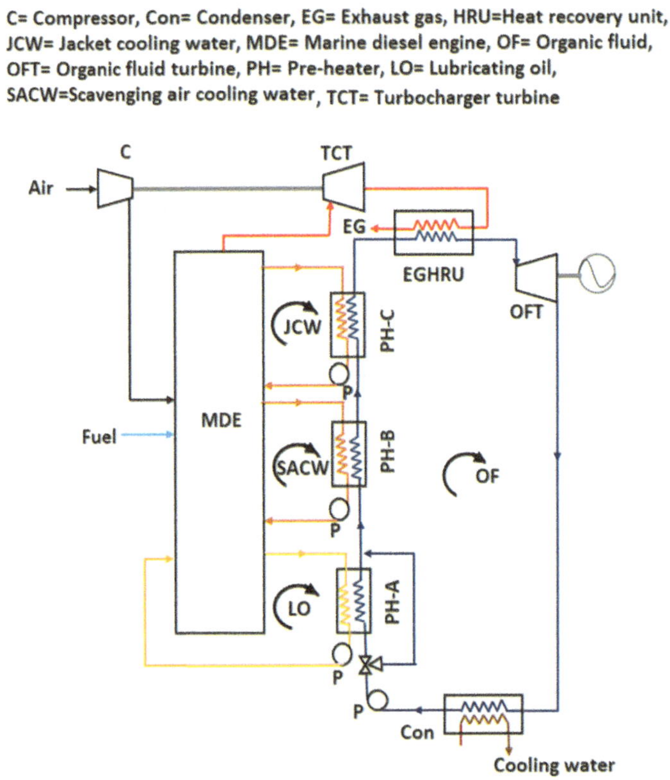

Fig. 13.4 Bottoming ORC for MDE waste heat recovery proposed by Yang (2016)

Finally, total mass of the organic fluid entered into the pre-heater B. In pre-heater B, the total mass of the organic fluid was heated to a few degrees by the low grade heat of the scavenging air cooling water. The organic working fluid was heated further by the jacket cooling water in the pre-heater C. The working fluid mass exiting the pre-heater C was heated to turbine inlet state by the engine exhaust gas in the exhaust gas heat recovery unit. Yang (2016) simulated the operation of the waste heat driven transcritical ORC using six different working fluids (namely R134a, R152a, R236fa, R290, R1234yf and R1234ze). Output power, thermal efficiency, levelized electricity cost and payback period had been considered as performance parameters in this simulation study. It was reported that the cycle yielded the lowest levelized energy cost while using R236fa as the working fluid. The associated payback period had been close to 3.02 years (Yang 2016).

Diesel engine waste heat recovery through CO_2 based power cycles

CO_2 is an inert, non-toxic and environment friendly working fluid. Thus, in many of the recent study, CO_2 had been considered as the working fluid in thermodynamic cycles (Song et al. 2012; Mondal and De 2015; Sahana et al. 2021). The compact design of the CO_2 turbo-machinery is also an added advantage of the CO_2-based power cycle (Hosseinpour et al. 2021). White et al. (White et al. 2021) pointed out that the CO_2-based power cycles would have a significant contribution to the future energy sector if the issues related to the designing of CO_2 turbo-machinery and heat exchangers are resolved. Compact structure of the CO2 based power cycle makes it suitable for the engine waste heat recovery in a vehicle.

A simplified arrangement of the supercritical CO_2 power cycle recovering the diesel engine waste heat simulated by Song et al. (2018) is presented in Fig. 13.5. It is appeared from Fig. 13.5 that the system consisted of two pre-heaters, one regenerator and one exhaust gas heat recovery unit to serve the purpose of waste heat recovery. In the low temperature (LT) pre-heater, the CO_2 mass exiting the compressor was heated by the jacket cooling water. Heat of the flue gas exiting the exhaust gas heat recovery unit was used to heat CO_2 stream further in the high temperature (HT) pre-heater. The CO_2 stream coming out of the HT pre-heater, after being heated by the exhaust CO_2 stream exiting the turbine, entered into the exhaust gas heat recovery unit (EGHRU). In the EGHRU, CO_2 was heated up to the turbine inlet state using the rejected heat of the hot flue gas. It was concluded that by incorporating the engine waste heat driven supercritical CO_2 power cycle, engine power output could be improved by 6.9% (Song et al. 2018).

Pan et al. (2020) modified the conventional recompression CO_2 power cycle to recover the waste heat carried by the exhaust gas of the diesel engine driving a container ship. The layout of the waste heat utilization scheme proposed by Pan et al. (2020) is presented in Fig. 13.6. The system consisted of two turbines, two compressors (i.e. the Main compressor and re-compressor), two regenerators (i.e. the LT regenerator and HT regenerator), an exhaust gas heat recovery unit (EGHRU) and a CO_2 cooler. Low temperature CO_2 mass exiting the LT regenerator was split into two streams. One stream was pressurized to the heat recovery pressure by the recompression compressor (RC). Another stream, after being cooled in the CO_2 cooler,

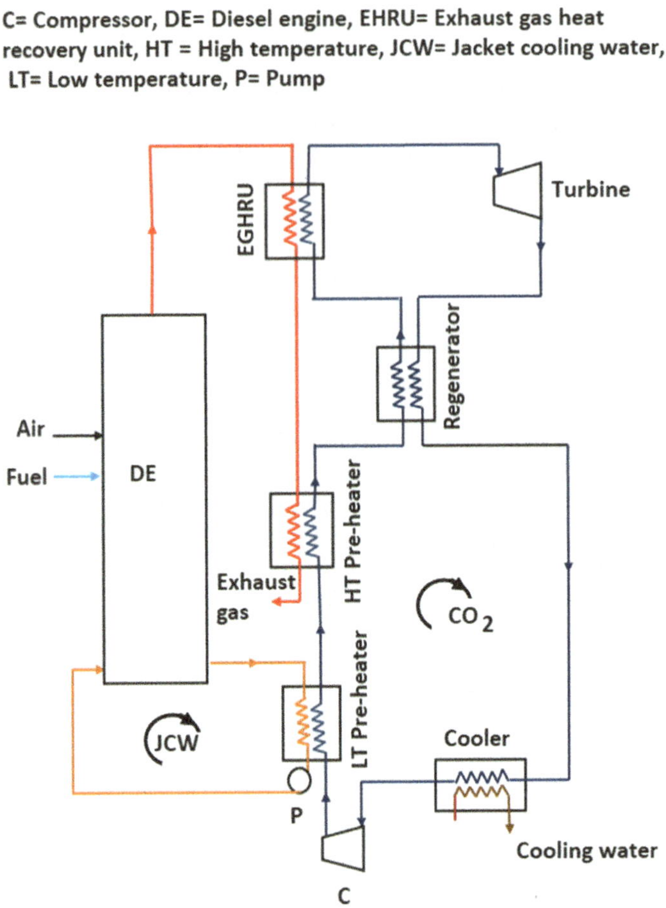

Fig. 13.5 Layout of the supercritical CO_2 power cycle for the DE waste heat utilization simulated by Song et al. (2018)

entered into the main compressor (MC). The main compressor also pressurized the CO_2 mass to the heat recovery pressure. The CO_2 stream exiting the MC was heated to some extent in the LT regenerator using the heat rejected by the low pressure CO_2 stream exiting the HT regenerator. After this, the two CO_2 streams (i.e. the main CO_2 stream and recompression CO_2 stream) were mixed and the total CO_2 mass was further heated up in the HT regenerator by the heat of the high temperature CO_2 stream exiting the turbine. The total mass of CO_2 exiting HT regenerator was heated by the hot engine exhaust gas in the EGHRU. The CO_2 mass exiting the EGHRU had been split into two streams to drive two separate turbines. A portion of the yielded turbine power was expensed to operate the MC and the RC. The remaining part was converted into electricity. Pan et al. (2020) showed through a simulation that by

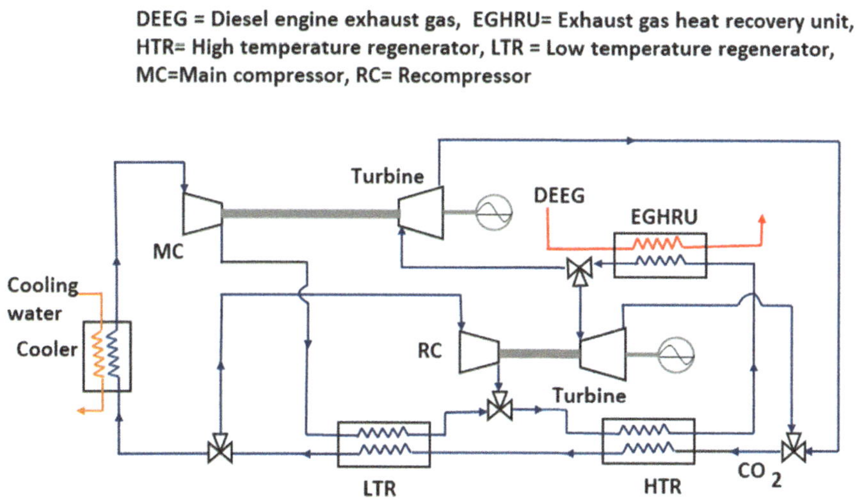

Fig. 13.6 Recompression CO_2 power cycle for the utilization of the diesel engine waste heat simulated by Pan et al. (2020)

implementing the proposed waste heat recovery scheme; about 1.01% reduction in auxiliary fuel consumption would be possible.

Maximizing the waste heat recovery in a diesel engine is a challenging task due to significant differences in mass flow rates and temperatures of different waste heat carrying streams of the diesel engine. Often a substantial part of the available low grade waste heat of a diesel engine remains unutilized during the waste heat recovery. Mondal et al. (2020) proposed the cascading between transcritical CO_2 power cycle and transcritical organic Rankine cycle to address this issue in a large marine diesel engine. R290, R600 and R1233zd (E) had been bottoming cycle working fluids of the cascading system due to their lower GWPs. The arrangement of the proposed cascading system is depicted in Fig. 13.7. The proposed system consisted of three pre-heaters (PH), one flue gas-CO_2 heat recovery unit (FG-CO_2HRU), one CO_2-organic fluid heat recovery unit (CO_2-OF HRU), two turbines and two coolers/condensers. CO_2 and organic fluid streams were pre-heated using the heat of the scavenging air cooling water in the pre-heaters A and C, respectively. Supercritical CO_2 exiting the pre-heater A was further heated by the heat of jacket cooling water and exhaust gas in the pre-heater B and exhaust gas CO_2 heat recovery unit (FG-CO_2 HRU). CO_2 was heated closer to the entering temperature of the flue gas to maximize thermal efficiency. Thus, from the energy conservation of the FG-CO_2 HRU, the CO_2 mass flow rate was estimated. The organic fluid exiting the pre-heater C was heated by the hot CO_2 stream coming out of the turbine in the CO_2-OF HRU. At a higher CO_2 turbine inlet pressure (i.e., >13 MPa), the heat of the jacket cooling water was sufficient to pre-heat the CO_2 stream from the pump exit state to the FG-CO_2 HRU inlet state. Hence, for a higher CO_2 turbine inlet pressure, the Pre-heater A was removed and the system consisted of only two pre-heaters. Usually, a CO_2 power cycle should be

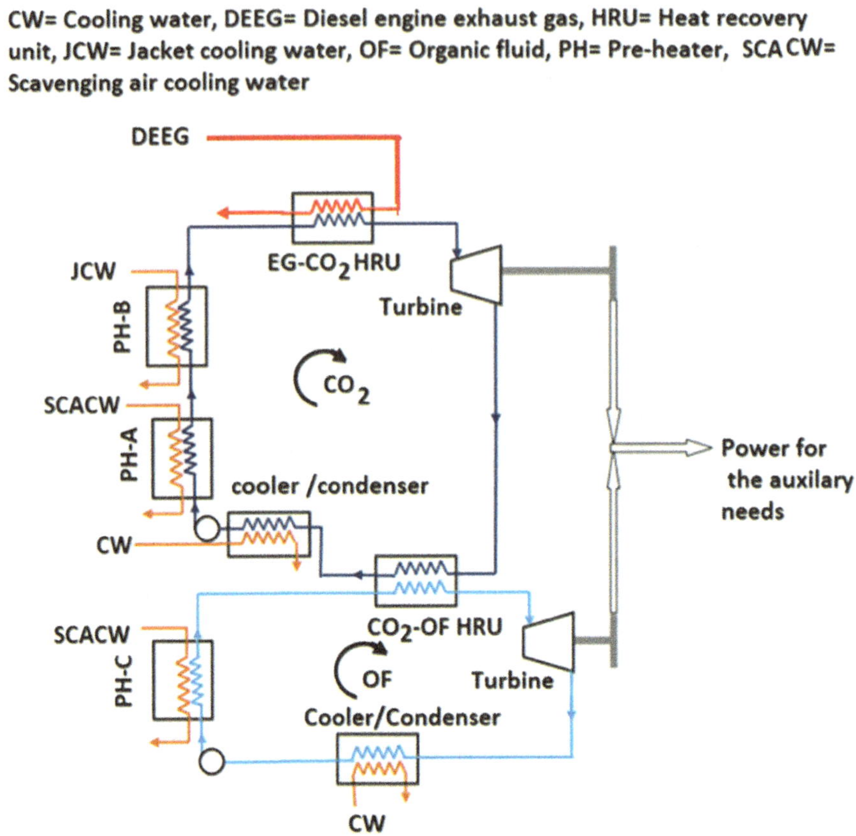

Fig. 13.7 CO_2 –organic fluid cascading power cycle for the utilization of the marine diesel engine waste heat simulated by Mondal et al. (2020)

operated with a higher turbine inlet pressure to ensure an acceptable power output. Controlling leakage of CO_2 at a very high turbine inlet pressure is a challenging task. A model was simulated to evaluate the thermo-economic performance of the cascading system. According to the outputs of the simulated model, the proposed cascading system yielded appreciably higher power output at a reasonably lower pressure at the inlet of the CO_2 turbine. The produced secondary energy of the waste heat driven cascading system was sufficient to satisfy the auxiliary power needs of the marine ship. According to the estimated output by the simulation, the marine waste heat driven CO_2-organic cascading system would ensure 8–9.5% fuel saving annually.

Diesel engine waste heat utilization through combined cycles
Many of the researchers suggested recovering the diesel engine waste heat by running a combined power and refrigeration cycle. Some of such cycles can be presented as follows:

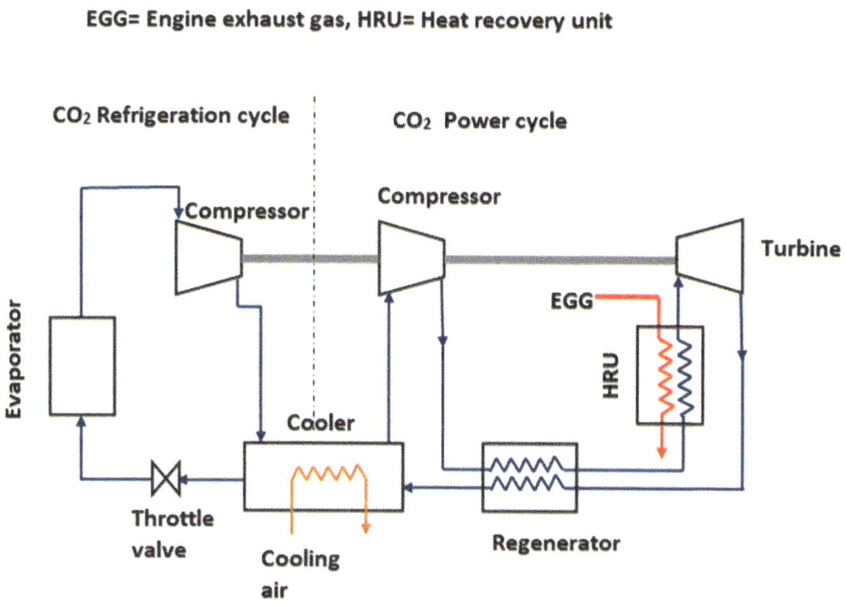

Fig. 13.8 Integration of CO_2 power and refrigeration cycle for preserving food in a moving vehicle proposed by Liang et al. (2020)

Liang et al. (2020) suggested the use of an engine waste heat driven CO_2 based combined power refrigeration cycle instead of a bulky absorption refrigeration cycle for preserving the food carried by a high way truck. The system was basically a combination of the supercritical CO_2 power cycle and transcritical CO_2 refrigeration cycle. Figure 13.8 is depicting the proposed scheme of Liang et al. (2020). It is appeared from Fig. 13.8 that engine waste heat was utilized to run a supercritical CO_2 power cycle. In the simulated model the power produced in the CO_2 turbine was used to run the compressors of both the power and refrigeration cycles.

Recently Yu et al. (2020) simulated the operation of a CO_2 based combined power refrigeration cycle recovering the diesel engine waste heat. The layout of a similar cycle is presented in Fig. 13.9. The system consisted of an exhaust gas heat recovery unit (EGHRU), one turbine, two compressors, two regenerators, one cooler, one evaporator, one subcooler and one throttle valve. In the proposed layout, the CO_2 mass exiting the CO_2 cooler was split-up into two streams. One stream was pressurized to heat recovery pressure by a high pressure (HP) compressor. The other stream entered into the throttling device after undergoing a subcooling process. The stream after undergoing a throttling process entered into the evaporator to produce some refrigeration effects. The CO_2 stream exiting the evaporator after being heated in the subcooler was pressurized to CO_2 cooler pressure in the low pressure (LP) compressor. The CO_2 stream exiting the HP compressor entered into the exhaust gas heat recovery unit (EGHRU) after being heated in two regenerators. In the EGHRU, the CO_2 mass was heated to the turbine inlet state by the exhaust gas of the diesel

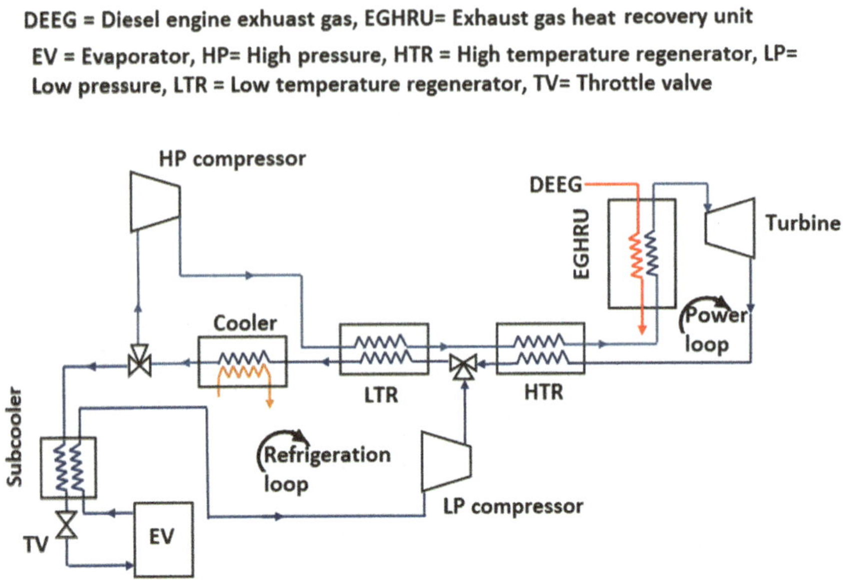

Fig. 13.9 Layout of the CO_2 based combined power refrigeration cycle for diesel engine waste heat recovery simulated by Yu et al. (2020)

engine. The exiting CO_2 stream from the turbine was cooled to some extent in the high temperature regenerator (HTR). Then this stream was mixed with the CO_2 stream exiting the LP compressor. After this mixing, the total mass of CO_2 was directed through the low temperature regenerator (LTR) as the high temperature fluid. For the proposed scheme of waste heat recovery, the average energy production cost was found to be 9.28 $/GJ.

Xia et al. (2016) integrated a supercritical CO_2 power cycle, an organic Rankine cycle (ORC) and a heat driven ejector refrigeration cycle to produce power and cooling effect simultaneously. Isobutane had been considered as the working fluid in both the ORC and ejector refrigeration cycle. The layout of the proposed combined cycle had been similar to that of the presented in Fig. 13.10. The diesel engine exhaust gas had been used in the two heat recovery units (i.e., The $EGCO_2$ HRU and EGOFHRU) to heat the supercritical CO_2 stream and subcritical organic fluid stream, respectively. The heated supercritical CO_2 was expanded through the turbine for producing a certain output power. The heated organic working fluid (OF) stream acted as the motive stream of the ejector to entrain the dry saturated organic fluid vapour exiting the evaporator of a refrigeration cycle. The total mass of the organic working fluid exiting the ejector was cooled in the organic fluid condenser. The mass of the organic fluid exiting the condenser was divided into two streams. One stream was pressurized by a pump up to the EGOFHRU pressure. The other stream after undergoing a throttling process entered into the refrigerator evaporator. In the ORC, the heat of the jacket cooling water was used to heat the OF stream to a two phase state

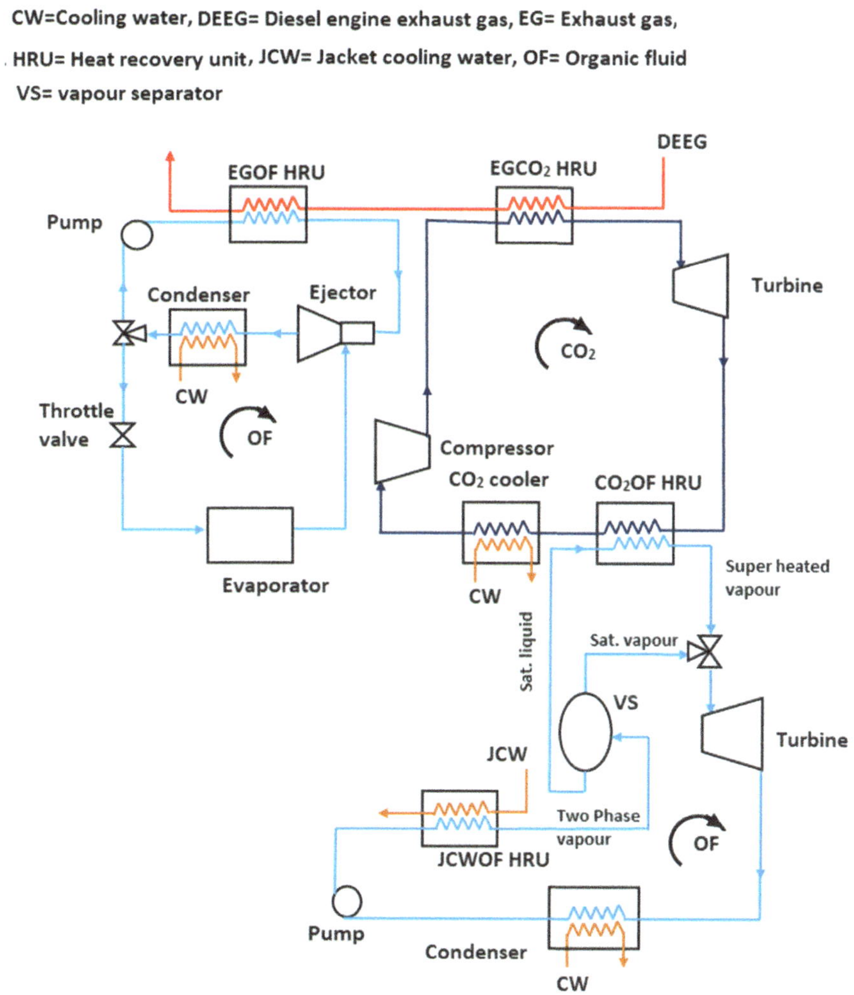

Fig. 13.10 Integration among Supercritical CO_2 power cycle, ORC and Ejector refrigeration cycle for MDE waste heat utilization (Xia et al. 2016)

from the pump exit condition. The two phase OF entered into a vapour separator. The saturated liquid leaving the vapour separator entered into the CO_2OF heat recovery unit (CO_2OF HRU). In the CO_2OF HRU, the OF was heated to a superheated vapour state by the exhaust CO_2 stream coming out of the turbine. The saturated OF vapour from the vapour separator was mixed with the superheated OF vapour exiting the CO_2OF HRU. After the mixing, the total OF mass was expanded through the OF turbine to produce some power output. For the optimized combined cycle simulated by Xia et al. (2016) the average cost rate per unit exergy had been found to be 65.53 \$/MW h.

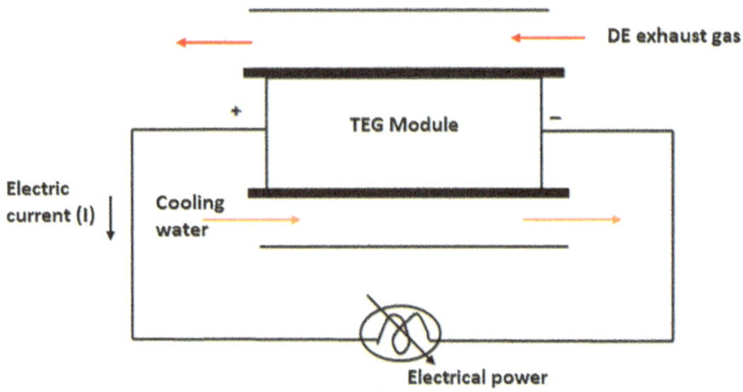

Fig. 13.11 Thermo electric generator (TEG) to recover the diesel engine (DE) waste heat

13.4.3 Thermoelectric Generator to Recover the Waste Heat of Diesel Engine

A thermoelectric generator (TEG) is capable of converting heat directly into electricity. It is basically the outcome of the Seebeck effect observed in some specific pair of materials. The Selection of a thermoelectric material greatly depends on the operating temperature. For less than 150 °C, usually, Bi_2Te_3 is used as the material of TEG. Lead telluride (PbTe) is suitable for the high-temperature heat recovery as its melting temperature is about 1190 K. Silicon-Germanium alloys can also be used as the material of TEG exposed to a higher temperature (Jouhara et al. 2021).

Though the thermoelectric conversion efficiency varies in the range of 3–5% (LeBlanc 2014), it may be preferred in engine waste heat recovery application due to its compactness and emission free operations. However, the unavailability of suitable thermoelectric materials at a reasonable cost is a bottle neck in the path of commercialization of the TEG technology. A possible layout to recover the diesel engine waste heat using a TEG module is represented in Fig. 13.11.

Some of the recently reported simulation studies on TEG technology recovering the diesel engine waste heat and associated outputs are also summarized in the following section:

Kumar et al. (2013) developed a numerical model simulating the process of secondary energy production through thermo-electric conversion of the engine waste heat. The simulated model could used to estimate the overall heat transfer rate, electrical power output and associated pressure drop for specified exhaust inlet condition and volume of the TEG. To ensure the accuracy of the calculation, the TEG was discretized along the path of the exhaust gas flow by applying the finite volume method. The output of the model was also validated against the experimental results.

Hsiao et al. (2010) developed a one dimensional resistance model to simulate the operation of a TEG embedded either on the exhaust pipe or on the radiator. The simulation results were verified against the experimental outputs. The study was conducted for the HZ-2 and TGM-127 thermoelectric modules. According to the output of the simulation, the maximum output power density could be obtained with the HZ-2 thermoelectric module. For this HZ-2 module, 51.13 MW/cm^2 power density was achievable for close to 290 °C temperatures difference between the hot and cold junctions. It was also concluded that a better output could be obtained by embedding the TEG on the exhaust pipe compared to the output obtained with the TEG embedded on the radiator.

Fernández-Yáñez et al. (2018) reported that during the recovery of waste heat from the engine operating at a low power point, the performance of the TEG would be superior compared to that of an electric turbo generator. However, at higher power mode operation of the engine, an electric turbo generator would outperform the TEG.

He et al. (2020) applied the finite volume method to simulate the operation of a multi-layer thermo electric generator (TEG) recovering the diesel engine waste heat. In this simulation study, TEP1-1264-3.4 developed by Thermonamic Co., Ltd. was used as the thermoelectric module. It was revealed during the simulation study that the output of the TEG improved with increasing the engine load. However, the effect of varying engine speed on the TEG output was insignificant. According to this study, the thermoelectric conversion efficiency would depend on the engine operating condition and its value would be between 1.41 and 4.12%.

Eddine et al. (2019) proposed a dynamic model to evaluate the performance of a TEG recovering the diesel engine waste heat. The produced electricity of the TEG was utilized to charge a 12 V battery. Bismuth Telluride was considered as material of the TEG during the simulation. It was observed that the TEG was capable of delivering maximum 42 W of electricity.

13.5 Emission Reduction Through Diesel Engine Waste Heat Utilization

Partial conversion of the waste heat of a diesel engine into power enhances the overall efficiency. This will reduce the engine size for same power and associated fuel consumption. Reduced fuel consumption reduces emission too. However, available waste heat from any engine depends on its rated capacity. An engine of reduced capacity releases lesser waste heat that affects the output of the waste heat recovery scheme. Thus, to estimate the fuel saving and associated emission reduction, an iterative method suggested by Mondal et al. (2020) is to be implemented. The iterative method applied to estimate the fuel saving and associated emission reduction due to the waste heat recovery in a large marine ship is presented in Fig. 13.12. According to the model presented in Fig. 13.12, if the output power of the waste heat recovery scheme is more than the auxiliary power needed for the ship, the additional power is fed to the propeller shaft. While implementing the model, the engine should be assumed to be operated at a specified mean speed or load.

It is worthy to mention that most of the reported studies on diesel engine waste heat utilization did not present much data on emission reduction. However, some of

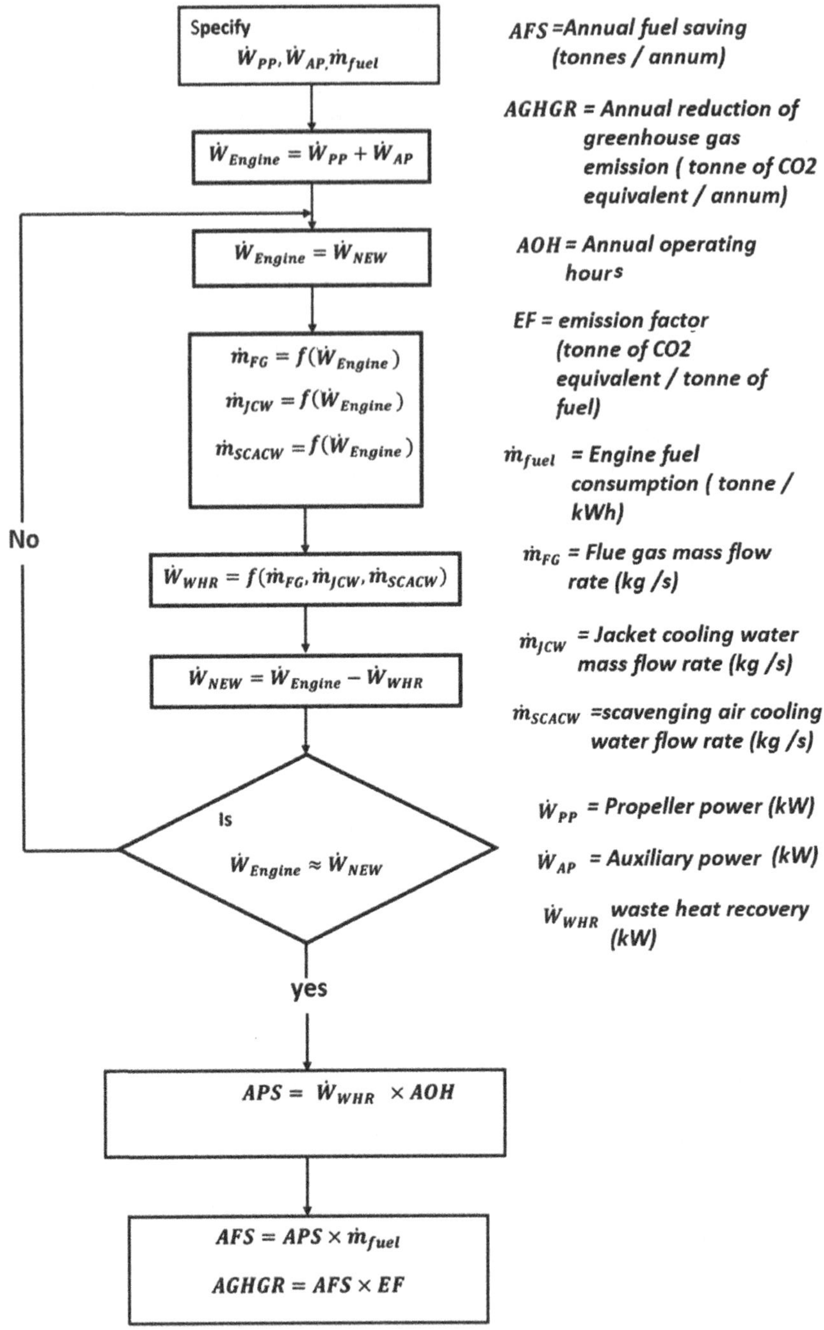

Fig. 13.12 Steps of calculating annual reduction of greenhouse gas emission due to the marine diesel engine waste heat recovery

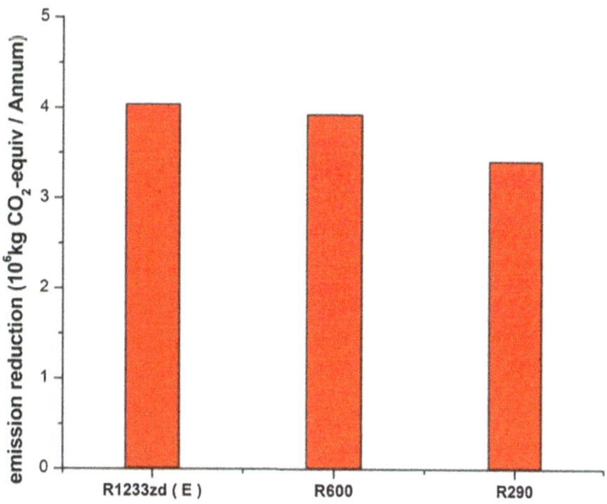

Fig. 13.13 Annual reduction in greenhouse gas emission for various bottoming cycle working fluids of the cascading waste heat recovery system of Mondal et al. (2020)

the studies presented results related to annual fuel savings. Thus, using the reported values of annual fuel savings (or annual power production), annual reduction of greenhouse gas emission can be estimated by applying the iterative method presented in Fig. 13.12. For example, Mondal et al. (2020) reported that their proposed waste heat recovery scheme would produce output power in the range of 5.90–6.97 MW. They assumed that 0.167 kg/kWh fuel was consumed by the marine diesel engine. If the above iterative method is applied to the power output data presented by Mondal et al. (2020), the annual reduction of green house gas emission will fall in the range of 3420–4040 tonnes of CO_2-equiv/annum as presented in Fig. 13.13. In this calculation, the annual operating hours of the marine ship is 7200 h. The emission factor of diesel is taken as 0.482 kg of CO_2-equiv/kg of fuel (Dones et al. 2004).

Yang (2016) reported that a marine diesel engine waste heat driven transcritical organic Rankine cycle with R152a would avoid 7810 tonnes of CO_2 emission annually. Annually avoidable CO_2 emission for various working fluids of marine diesel engine waste heat driven transcritical ORCS based on the study of Yang (2016) are summarized in Fig. 13.14.

13.6 Conclusion

Diesel engines are widely used prime movers for generating the secondary energies (i.e., both mechanical power and electricity). However, the emission from a diesel engine is an issue of great concern. As the efficiency of the diesel engine is close to 40%, it produces a substantial amount of waste heat. The waste heat carried by different outgoing streams of a diesel engine may be utilized to produce some of the energy utilities. In this chapter possible technologies of recovering the diesel engine waste heat are summarized based on simulation study conducted by various

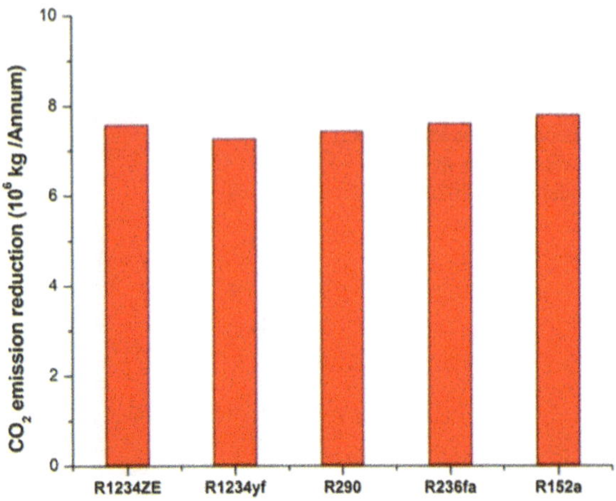

Fig. 13.14 Annual CO_2 emission reduction using marine diesel engine waste heat driven transcritical ORC with different organic fluids (Yang 2016)

researchers. Turbo compounding, thermodynamic power cycles, combined power refrigeration cycle, thermoelectric generator are some of such technologies. It is evident from the outputs of models simulated by various researchers that implementing waste heat recovery in the diesel engine would be helpful in reducing the engine fuel consumption and associated emission. Finally, the diesel engine waste heat recovery is having a great potential of becoming a future sustainable energy option in the current phase of energy transition. However, prototype development and experimental run based on previously conducted simulation study are necessary to commercially implement the diesel engine waste heat recovery.

References

Aghaali H, Ångström H-E (2015) A review of turbocompounding as a waste heat recovery system for internal combustion engines. Renew Sustain Energy Rev 49:813–824

Balcombe P, Brierley J, Lewis C, Skatvedt L, Speirs J, Hawkes A, Staffell I (2019) How to decarbonise international shipping: options for fuels, technologies and policies. Energy Convers Manage 182:72–88

Boodaghi H, Etghani MM, Sedighi K (2021) Performance analysis of a dual-loop bottoming organic Rankine cycle (ORC) for waste heat recovery of a heavy-duty diesel engine, Part I: thermodynamic analysis. Energy Convers Manage 241:113830

Dones R, Heck T, Hirschberg S (2004) Greenhouse gas emissions from energy systems, comparison and overview. In: Encyclopedia of energy, pp 77–95. https://doi.org/10.1016/B0-12-176480-X/00397-1

Dueso C, Muñoz M, Moreno F, Arroyo J, Gil-Lalaguna N, Bautista AG, Sánchez JL (2018) Performance and emissions of a diesel engine using sunflower biodiesel with a renewable antioxidant additive from bio-oil. Fuel 234:276–285

Eddine N, Sara H, Chalet D, Farue X, Aixala L, Cormerias M (2019) Modeling and simulation of a thermoelectric generator using bismuth telluride for waste heat recovery in automotive diesel engines. J ElecMater 48:2036–2045

EPA, NHTSA, DOT (2016) Greenhouse gas emissions and fuel efficiency standards for medium- and heavy-duty engines and vehicles—Phase 2

Ettefaghi E, Ghobadian B, Rashidi A, Najafi G, Khoshtaghaza MH, Rashtchi M, Sadeghian S (2018) A novel bio-nano emulsion fuel based on biodegradable nanoparticles to improve diesel engines performance and reduce exhaust emissions. Renew Energy 125:64–72

Fernández-Yáñez P, Armas O, Kiwan R, Stefanopoulou AG, Boehman AL (2018) A thermoelectric generator in exhaust systems of spark-ignition and compression-ignition engines. A comparison with an electric turbo-generator. Appl Energy 229:80–87

Forman C, Muritala IK, Pardemann R, Meyer B (2016) Estimating the global waste heat potential. Renew Sustain Energy Rev 57:1568–1579

Gao Y, Gao X, Zhang X (2017) The 2 °C global temperature target and the evolution of the long-term goal of addressing climate change—from the United Nations framework convention on climate change to the Paris agreement. Engineering 3:272–278

He M, Wang E, Zhang Y, Zhang W, Zhang F, Zhao C (2020) Performance analysis of a multilayer thermoelectric generator for exhaust heat recovery of a heavy-duty diesel engine. Appl Energy 274:115298

Hiereth H, Prenninger P (2007) Charging the internal combustion engine. Springer, New York

Hosseinpour J, Howard J, Chen J, Engeda A (2021) Challenges for developing and marketing a Brayton-cycle based power genset gas-turbine using supercritical CO_2 and a compressor design for simple recuperated cycle. J Energy Resour Technol 144:032101-1-9

Hsiao YY, Chang WC, Chen SL (2010) A mathematic model of thermoelectric module with applications on waste heat recovery from automobile engine. Energy 35:1447–1454

IEO (2016) With Projections to 2040. U.S. Energy Information Administration, Washington, DC 20585, https://www.eia.gov/outlooks/ieo/pdf/0484(2016).pdf. Last accessed on 2021/05/23

Ismail Y, Durrieu D, Menegazzi P, Chesse P, Chalet D (2012) Potential of exhaust heat recovery by turbocompounding. In: International power trains, fuels and lubricants meeting, SAE Technical Paper 2012-01-1603. https://doi.org/10.4271/2012-01-1603

Jouhara H, Żabnieńska-Góra A, Khordehgah N, Doraghi Q, Ahmad L, Norman L, Axcell B, Wrobel L, Dai S (2021) Thermoelectric generator (TEG) technologies and applications. Int J Thermofluids 9:100063

Jye AETS, Pesiridis A, Rajoo S (2013) Effects of mechanical turbo compounding on a turbocharged diesel engine. In: 17th Asia Pacific automotive engineering conference, APAC 2013, Bangkok. https://doi.org/10.4271/2013-01-0103

Kumar S, Heister SD, Xu X, Salvador JR, Meisner GP (2013) Thermoelectric generators for automotive waste heat recovery systems part I: numerical modeling and baseline model analysis. J ElecMater 42:665–674. https://doi.org/10.1007/s11664-013-2471-9

LeBlanc S (2014) Thermoelectric generators: Linking material properties and systems engineering for waste heat recovery applications. Sustain Mater Technol 1–2:26–35

Lemmon EW, Bell IH, Huber ML, McLinden MO (2018) NIST standard reference database 23: reference fluid thermodynamic and transport properties-REFPROP, Version 10.0, National Institute of Standards and Technology, Standard Reference Data Program, Gaithersburg

Liang Y, Sun Z, Dong M, Lu J, Yu Z (2020) Investigation of a refrigeration system based on combined supercritical CO_2 power and transcritical CO_2 refrigeration cycles by waste heat recovery of engine. Int J Refrig 118:470–482

Lion S, Michos CN, Vlaskos I, Rouaud C, Taccani C (2017) A review of waste heat recovery and Organic Rankine Cycles (ORC) in on-off highway vehicle Heavy Duty Diesel Engine applications. Renew Sustain Energy Rev 79:691–708

Liu P, Shu G, Tian H (2019) How to approach optimal practical Organic Rankine cycle (OP-ORC) by configuration modification for diesel engine waste heat recovery. Energy 174:543–552

Mondal S, De S (2015) CO_2 based power cycle with multi-stage compression and intercooling for low temperature waste heat recovery. Energy 90:1132–1143

Mondal S, Datta S, De S (2020) Auxiliary power through marine waste heat recovery using a CO_2-organic cascading cycle. Clean Technol Environ Policy 22:893–906

Ouyang T, Wang Z, Zhao Z, Lu J, Zhang M (2021) An advanced marine engine waste heat utilization scheme: electricity-cooling cogeneration system integrated with heat storage device. Energy Convers Manage 235:113955

Pan R, Yuan C, Sun Y, Yan X, Lu M, Bucknall R (2020) Thermo-economic analysis and multi-objective optimization of S-CO_2 Brayton cycle waste heat recovery system for an ocean-going 9000 TEU container ship. Energy Convers Manage 221:113077

Prakash T, Geo VE, Martin LJ, Nagalingam B (2018) Effect of ternary blends of bio-ethanol, diesel and castor oil on performance, emission and combustion in a CI engine. Renew Energy 122:301–309

Sahana C, De S, Mondal S (2021) Integration of CO_2 power and refrigeration cycles with a desalination unit to recover geothermal heat in an oilfield. Appl Therm Eng 189:116744

Song Y, Wang J, Dai Y, Zhou E (2012) Thermodynamic analysis of a transcritical CO_2 power cycle driven by solar energy with liquified natural gas as its heat sink. Appl Energy 92:194–203

Song J, Song Y, Gu C (2015) Thermodynamic analysis and performance optimization of an Organic Rankine Cycle (ORC) waste heat recovery system for marine diesel engines. Energy 82:976–985

Song J, Li X, Ren X, Gu C (2018) Performance improvement of a preheating supercritical CO_2 (S-CO_2) cycle based system for engine waste heat recovery. Energy Convers Manage 161:225–233

Turton R, Bailie RC, Whiting WB, Shaeiwitz JA (2009) Analysis, synthesis and design of chemical processes. Prentice Hall, New Jersey

Wang R, Shu G, Wang X, Tian H, Li X, Wang M, Cai J (2020) Dynamic performance and control strategy of CO_2-mixture transcritical power cycle for heavy-duty diesel engine waste-heat recovery. Energy Convers Manage 205:112389

Weerasinghe WMSR, Stobart RK, Hounsham SM (2010) Thermal efficiency improvement in high output diesel engines a comparison of a Rankine cycle with turbo-compounding. Appl Therm Eng 30:2253–2256

White MT, Bianchi G, Chai L, Tassou SA, Sayma AI (2021) Review of supercritical CO_2 technologies and systems for power generation. Appl Therm Eng 185:116447

Xia J, Wang J, Lou J, Zhao P, Dai Y (2016) Thermo-economic analysis and optimization of a combined cooling and power (CCP) system for engine waste heat recovery. Energy Convers Manage 128:303–316

Yang M (2016) Optimizations of the waste heat recovery system for a large marine diesel engine based on transcritical Rankine cycle. Energy 113:1109–1124

Yu G, Shu G, Tian H, Wei H, Liu L (2013) Simulation and thermodynamic analysis of a bottoming Organic Rankine Cycle (ORC) of diesel engine (DE). Energy 51:281–290

Yu A, Su W, Lin X, Zhou N, Zhao L (2020) Thermodynamic analysis on the combination of supercritical carbon dioxide power cycle and transcritical carbon dioxide refrigeration cycle for the waste heat recovery of shipboard. Energy Convers Manage 221:113214

Zhao R, Zhug W, Zhang Y, Yin Y, Zhao Y, Chen Z (2016) Parametric study of a turbocompound diesel engine based on an analytical model. Energy 115:435–445

The manufacturer's authorised representative in the EU is Springer Nature Customer Service Centre GmbH, Europaplatz 3, 69115 Heidelberg, Germany. If you have any concerns regarding our products, please contact ProductSafety@springernature.com

Printed and bound by CPI Group (UK) Ltd, Croydon, CR0 4YY
24/09/2026
02222176-0001